Linear Algebra

Textbooks in Mathematics

Series editors:

Al Boggess, Kenneth H. Rosen

Linear Algebra

James R. Kirkwood, Bessie H. Kirkwood

Real Analysis

With Proof Strategies

Daniel W. Cunningham

Train Your Brain

Challenging Yet Elementary Mathematics

Bogumil Kaminski, Pawel Pralat

Contemporary Abstract Algebra, Tenth Edition

Joseph A. Gallian

Geometry and Its Applications

Walter J. Meyer

Linear Algebra

What You Need to Know

Hugo J. Woerdeman

Introduction to Real Analysis, 3rd Edition

Manfred Stoll

Discovering Dynamical Systems Through Experiment and Inquiry

Thomas LoFaro, Jeff Ford

Functional Linear Algebra

Hannah Robbins

Introduction to Financial Mathematics

With Computer Applications

Donald R. Chambers, Qin Lu

Linear Algebra

An Inquiry-based Approach

Jeff Suzuki

https://www.routledge.com/Textbooks-in-Mathematics/book-series/CANDHTEXBOOMTH

Linear Algebra
What You Need to Know

Hugo J. Woerdeman

CRC Press
Taylor & Francis Group
Boca Raton London New York

CRC Press is an imprint of the
Taylor & Francis Group, an **informa** business

A CHAPMAN & HALL BOOK

First edition published 2021
by CRC Press
6000 Broken Sound Parkway NW, Suite 300, Boca Raton, FL 33487-2742

and by CRC Press
2 Park Square, Milton Park, Abingdon, Oxon, OX14 4RN

ISBN: 9780367697389 (hbk)
ISBN: 9780367684730 (pbk)

Typeset in Computer Modern font
by KnowledgeWorks Global Ltd

To my very supportive family:
Dara, Sloane, Sam, Ruth, and Myra.

Contents

Preface

Preface to the Instructor

This book is intended for a first linear algebra course. There are many great books that cover this topic very well, and I have used many of them in my classes. I wrote this book due to the desire to have a concise source of all the material I can cover in this course. In addition, I always hope to get my students used to complex numbers towards the end of the course, and thus I include complex vectors in the last chapter. If I can achieve that my students understand this material in a one term course, I am very happy. I have found that teaching a class partially by using slides (the more theoretical parts and/or longer examples) and lecturing on the board (emphasizing the highlights and going through examples) works very well. Making the slides available beforehand is especially appreciated. In this way I can cover all the material (skipping some of the proofs, though).

I also made an effort to include some information regarding **applications**. These include:

- Balancing chemical equations (see Exercise 1.6.20).

- Linear equations appearing in analyzing electrical circuits (see Exercise 1.6.21).

- Minimal rank completion and the Netflix challenge (see Exercise 2.6.21).

- Matrices and graph theory (see Exercises 2.6.22 and 3.7.28).

- The Leontief input-output economic model (see Exercise 3.7.27).

- Determinants and volumes (see Section 4.4).

- Interpolation and the discrete cosine transform (see Exercise 5.5.22).

- Finite impulse response (FIR) filtering (see Exercise 6.5.34).

- Error detecting codes (see Exercise 6.5.35).

- Systems of differential equations (see Section 7.4).

- Markov chains and Google's PageRank (see Exercise 7.5.25).

- Linear recurrence relations (see Exercises 7.5.26, 7.5.27, and 7.5.28).

- Curve fitting/regression line (see Section 8.5).

- Quadratic forms (see Exercise 8.8.43).

- Image compression (see Exercise 8.8.56).

- Principal component analysis (see Exercise 8.8.57).

- Robust principal component analysis (see Exercise 8.8.58).

There are of course many more applications, and one way to get students to explore them is to have them write a paper or make a video about their favorite application (one of those listed above or one they find on their own*).

Regarding **computational software** there are some pointers in the 'Preface for the students' followed up by specific suggestions in the exercises of the first few chapters. Several suggestions of different software are made. While MATLAB® is still my personal favorite choice, it seems that Sage (especially among algebraists) and the Python libraries are increasingly popular (and quite easy to try out online).

In my view, the **core material** in this book is the following and takes about **28 academic hours**† of lectures:

- **Chapter 1**: Row reduction, the vector space $\mathbb{R}^n$, linear combinations and span, the equation $A\mathbf{x} = \mathbf{b}$. (4 hours)

- **Chapter 2**: Subspaces in $\mathbb{R}^n$, column, row and null spaces, linear independence, basis, dimension, coordinate systems. (5 hours)

- **Chapter 3, Sections 3.1–3.4 + part of Section 3.6**: Matrix multiplication, transpose, inverses, elementary matrices, introduce triangular matrices. (4 hours)

---

*For instance, how Linear Algebra is used in streamlining kidney donor/recipient matches (see Hamouda E, El-Metwally S, Tarek M (2018) Ant Lion Optimization algorithm for kidney exchanges. PLoS ONE 13(5): e0196707. doi.org/10.1371/journal.pone.0196707), or in community detection (see Newman, M. E. J. (2006) Finding community structure in networks using the eigenvectors of matrices, Phys. Rev. E 74(3): 036104. link.aps.org/doi/10.1103/PhysRevE.74.036104), or in risk management (see Georgescu Dan I., Higham Nicholas J. and Peters Gareth W. (2018) Explicit solutions to correlation matrix completion problems, with an application to risk management and insurance, R. Soc. open sci. 5172348 doi.org/10.1098/rsos.172348).

†Academic hour = 50 minutes.

- **Chapter 4, Section 4.1**: Definition of determinants and how to compute them. (1 hour)

- **Chapter 5**: Definition of a vector space, examples of vector spaces over $\mathbb{R}$, repeat of linear independence, basis, dimension, coordinate systems, but now for general vector spaces over $\mathbb{R}$. (4 hours)

- **Chapter 6, Sections 6.1–6.4**: Linear maps, range and kernel, matrix representations, change of basis. (4 hours)

- **Chapter 7, Sections 7.1–7.3**: Eigenvalues, diagonalizability, complex eigenvalues. (3 hours)

- **Chapter 8, Sections 8.1–8.3**: Dot product, norm, orthogonality, Gram–Schmidt. (3 hours)

The remaining sections are more optional.

- **Chapter 3, Sections 3.5–3.6**: Block matrices and LU factorization. (2 hours)

- **Chapter 4, Sections 4.2**: In this section the properties of the determinants are proven. Depending on the audience this section can be covered (1 hour), or only briefly mentioned.

- **Chapter 4, Sections 4.3–4.4**: Cramer's rule is not very important, but my experience is that students like it. Mentioning the connection with volumes is useful but also not essential. (1 hour)

- **Chapter 7, Section 7.4**: The solution of systems of differential equations using eigenvalues and eigenvectors can be discussed fairly quickly. (less than 1 hour)

- **Chapter 8, Sections 8.4**: Interpreting Gram–Schmidt as QR factorization and unitary matrices should ideally be covered in the course. (1 hour)

- **Chapter 8, Sections 8.5–8.7**: These sections on least squares (with curve fitting), symmetric matrices and the singular value decomposition can be covered or left out at the end of the course. Section 8.5 can be skipped, as Sections 8.6 and 8.7 do not rely on it. (1 hour each)

I hope that you find this to be a useful text. If you have a chance to give me feedback on it, I would be very appreciative.

Preface to the Student

With the world becoming increasingly digital, Linear Algebra is gaining importance. When we send texts, share video, do internet searches, there are Linear Algebra algorithms in the background that make it work. But there are so many more instances where Linear Algebra is used and where it could be used, ranging from traditional applications such as analyzing physical systems, electrical networks, filtering signals, modeling economies to very recent applications such as guessing consumer preferences, facial recognition, extracting information from surveillance video, etc. In fact, any time we organize data in sequences or arrays, there is an opportunity for Linear Algebra to be used. Some applications are introduced in exercises (see the 'Preface to the Instructor for a list), and a search such as 'Image Processing and Linear Algebra' will provide you with many more resources on them. Aside from its wide applicability, Linear Algebra also has the appeal of a strong mathematical theory. It is my hope that this text will help you master this theory, so that one day you will be able to apply it in your area of interest.

There are many computational software programs that can help you do the Linear Algebra calculations. These include[‡] great commercial products such as MATLAB®, Maple™ or Mathematica®, and great open source products, such as Sage (sagemath.org), R (www.r-project.org), and the Python libraries NumPy (numpy.org), SciPy (scipy.org) and SymPy (sympy.org). It is definitely worthwhile to get used to one of these products, so that the threshold to use them in the future will be low. They are easy to search for, and some products have online calculators. Here are some to try (without having to download anything or sign in):

- www.wolframalpha.com/examples/mathematics/linear-algebra/. You will see a box 'Multiply matrices'; if you click on the $=$ sign with the content $\{\{2, -1\}, \{1, 3\}\}.\{\{1, 2\}, \{3, 4\}\}$, you get the product of these two matrices. In addition, if you scroll down you see other capabilities. For instance, you can type 'Are (2, -1) and (4, 2) linearly independent?' and hit 'return' or click on the $=$ sign.

- sagecell.sagemath.org. To enter a matrix A and multiply it by itself, you can enter for instance

  A = matrix([[0,1,2], [3,4,5], [6,7,9]])
  A*A

  and hit 'Evaluate'. To compute its inverse, you add

  A.inverse()

[‡]The Wikipedia pages 'List of computer algebra systems' and 'List of numerical-analysis software' have more complete lists.

and hit 'Evaluate'. For more information, see doc.sagemath.org/html/en/ prep/Quickstarts/Linear-Algebra.html.

- live.sympy.org. To enter matrices A and B and multiply them, do

    A=Matrix(2, 3, [1, 2, 3, 4, 5, 6])

    B=Matrix(3, 3, [1, 2, 3, 4, 5, 6, 7, 8, 9])

    A*B

For more, see docs.sympy.org/latest/modules/matrices/matrices.html.

For the purpose of this course, you can use these programs to check answers (in the first few chapters we give some specific suggestions to get you started) and use it in those exercises where computation by hand would be too involved. In addition, you can generate your own exercises in case you would like more practice. At the level of this course you will not see major differences between these programs other than logistical ones, such as formatting conventions. Once you get to more specialized use, one program may perform better than the other depending on the specifics. Regardless, though, if you have experience with one of the programs it will be easier to learn another.

Finally, let me mention that Linear Algebra is an active area of research. The International Linear Algebra Society (ILAS; ilasic.org) and the Society for Industrial and Applied Mathematics (SIAM; siam.org) are societies of researchers that organize Linear Algebra conferences, publish Linear Algebra scientific journals, publish newsletters and that maintain active websites that also advertise research opportunities for students. Both societies offer free membership for students, so please check them out.

Acknowledgments

In the many years that I have taught I have used many textbooks for linear algebra courses, both at the undergraduate and graduate level. All these different textbooks have influenced me. In addition, discussions with students and colleagues, sitting in on lectures, reading papers, etc., have all shaped my linear algebra courses. I wish I had a way to thank you all specifically for how you helped me, but I am afraid that is simply impossible. So I hope that a general thank you to all of you who have influenced me, will do: THANK YOU!

I am also very thankful to the reviewers of my book proposal who provided me with very useful feedback and to those at CRC Press who helped me bring this manuscript to publication. In particular, I would like to thank Professors Mark Hunacek, ChuanRen Liu, Dennis Merino, Kenneth H. Rosen, and Fuzhen Zhang for their detailed feedback, and Senior Editor Robert Ross for all his help in shaping a final product. Finally, I would like to express my appreciation to my family for all their love and support, and especially my wife Dara for also providing insightful specific feedback.

Notation

Here are some often-used notations:

- $\mathbb{N} = \{1, 2, 3, \ldots\}$
- $\mathbb{R}$ = the field of real numbers
- $\mathbb{C}$ = the field of complex numbers
- $\mathbf{a}_j$ = the jth column of the matrix A
- $A = (a_{ij})_{i=1, j=1}^{m, \ n}$ = is the $m \times n$ matrix with entries a_{ij}
- $\mathrm{col}_j(A)$ = the jth column of the matrix A (alternative notation)
- $\mathrm{row}_i(A)$ = the ith row of the matrix A
- Col A = the column space of a matrix A
- Nul A = the null space of a matrix A
- Row A = the row space of a matrix A
- $\det(A)$ = the determinant of the matrix A
- $\mathrm{tr}(A)$ = the trace of a matrix A (= the sum of its diagonal entries)
- $\mathrm{adj}(A)$ = the adjugate of the matrix A
- $\mathrm{rank}(A)$ = the rank of a matrix A
- $\mathbf{x} = (x_i)_{i=1}^{n}$ = the column vector $\mathbf{x} \in \mathbb{R}^n$ with entries $x_1, \ldots, x_n$.
- $E^{(1)}$, $E^{(2)}$, $E^{(3)}$ are elementary matrices of type I, II, and III, respectively.
- $\mathrm{rref}(A)$ = the reduced row echelon form of A.
- $\mathbb{R}^n$ = the vector space of column vectors with n real entries
- $\mathbb{R}[t]$ = the vector space of polynomials in t with coefficients in $\mathbb{R}$
- $\mathbb{R}_n[t]$ = the vector space of polynomials of degree $\leq n$ in t with coefficients in $\mathbb{R}$

- $\mathbb{R}^{m \times n}$ = the vector space of $m \times n$ matrices with entries in $\mathbb{R}$
- $\mathbb{C}^n$ = the vector space of column vectors with n complex entries
- $\mathbb{C}^{m \times n}$ = the vector space of $m \times n$ matrices with entries in $\mathbb{C}$
- $\mathbf{0}$ = the zero vector
- $\dim V$ = the dimension of the vector space V
- $\mathrm{Span}\,\{\mathbf{v}_1, \ldots, \mathbf{v}_n\}$ = the span of the vectors $\mathbf{v}_1, \ldots, \mathbf{v}_n$
- $\{\mathbf{e}_1, \ldots, \mathbf{e}_n\}$ = the standard basis in $\mathbb{R}^n$ (or $\mathbb{C}^n$)
- $[\mathbf{v}]_{\mathcal{B}}$ = the vector of coordinates of $\mathbf{v}$ relative to the basis $\mathcal{B}$
- $\mathrm{Ker}\,T$ = the kernel (or nullspace) of a linear map (or matrix) T
- $\mathrm{Ran}\,T$ = the range of a linear map (or matrix) T
- id = the identity map
- $[T]_{\mathcal{C} \leftarrow \mathcal{B}}$ = matrix representation of T with respect to the bases $\mathcal{B}$ and $\mathcal{C}$
- I_n = the $n \times n$ identity matrix
- $\dotplus$ = direct sum
- $p_A(t)$ = the characteristic polynomial of the matrix A
- $\deg p(t)$ = the degree of the polynomial $p(t)$
- $\equiv$ emphasizes that the equality holds for all t (for example, $\mathbf{0}(t) \equiv 0$).
- A^T = the transpose of the matrix A
- A^* = the conjugate transpose of the matrix A
- $\langle \cdot, \cdot \rangle$ = the dot product
- $\|\mathbf{x}\|$ = the Euclidean norm of the vector $\mathbf{x}$
- $\mathrm{Re}\,z$ = real part of z
- $\mathrm{Im}\,z$ = imaginary part of z
- $\bar{z}$ = complex conjugate of z
- $|z|$ = absolute value (modulus) of z
- $e^{it} = \cos t + i \sin t$, which is Euler's notation
- $\sigma_j(A)$ = the jth singular value of the matrix A, where $\sigma_1(A) = \|A\|$ is the largest singular value
- $\mathbb{F}$ = a generic field

List of Figures

1

Matrices and Vectors

CONTENTS

In this chapter we will introduce matrices and vectors in the context of systems of linear equations. Let us start with a system of two linear equations with two unknowns, such as

$$\begin{cases} 2x + 3y &= 4, \\ 4x - y &= -6. \end{cases}$$

To solve this system, one would take a combination of the two equations and eliminate one of the variables. We can take the second equation and subtract two times the first, giving us

$$(4x - y) - 2(2x + 3y) = -6 - 2 \cdot 4 \quad \text{which simplifies to} \quad -7y = -14,$$

yielding the equivalent system

$$\begin{cases} 2x + 3y &= 4, \\ -7y &= -14. \end{cases}$$

Now we can solve for y (giving $y = 2$) and substitute this in the first equation and obtain

$$\begin{cases} 2x &= 4 - 6, \\ y &= 2, \end{cases}$$

and we find $x = -1$ and $y = 2$.

We will develop a systematic way to do this for any number of equations with any number of unknowns (often denoted as $x_1, x_2, x_3, \ldots$) through the use of matrices and vectors. We will start by introducing matrices.

1.1   Matrices and Linear Systems

A real **matrix** is a rectangular array of real numbers. Here is an example:

$$
A = \begin{bmatrix} 2 & -2 & -1 & 7 & 1 & 1 \\ 1 & -1 & -1 & 6 & -5 & 4 \\ 0 & 0 & 1 & -1 & 3 & -3 \\ 1 & -1 & 0 & 3 & 1 & 0 \end{bmatrix} \tag{1.1}
$$

The **size** of this matrix is 4×6. It has 4 rows and 6 columns. The 4th **row** of this matrix is

$$
\begin{bmatrix} 1 & -1 & 0 & 3 & 1 & 0 \end{bmatrix},
$$

and its 2nd **column** is

$$
\begin{bmatrix} -2 \\ -1 \\ 0 \\ -1 \end{bmatrix}.
$$

The numbers in the matrix are its **entries**. We indicate the location of an entry by a pair of integers (k, l), where k denotes the row the entry is in, and l denotes the column the entry is in. For instance, the $(2, 4)$th **entry** of the above matrix is the number 6. We denote the (i, j)th entry of a matrix A by a_{ij}. Thus $a_{24} = 6$. The set of all $m \times n$ matrices with real entries will be denoted as $\mathbb{R}^{m \times n}$. Thus the matrix A above belongs to $\mathbb{R}^{4 \times 6}$; we write $A \in \mathbb{R}^{4 \times 6}$.

As we will see, a matrix can represent many different things. However, in this section we will focus on how matrices can be used to represent systems of linear equations. But even in the context of linear systems of equations, the same matrix can represent different situations. Here are a few different interpretations of matrix (1.1):

- **Interpretation 1**: It could stand for the system of equations

$$
\begin{cases} 2x_1 - 2x_2 - x_3 + 7x_4 + x_5 & = & 1, \\ x_1 - x_2 - x_3 + 6x_4 - 5x_5 & = & 4, \\ x_3 - x_4 + 3x_5 & = & -3, \\ x_1 - x_2 + \quad\quad 3x_4 + x_5 & = & 0, \end{cases}
$$

If so, we write

$$
\left[\begin{array}{cccccc|c} 2 & -2 & -1 & 7 & 1 & 1 \\ 1 & -1 & -1 & 6 & -5 & 4 \\ 0 & 0 & 1 & -1 & 3 & -3 \\ 1 & -1 & 0 & 3 & 1 & 0 \end{array} \right],
$$

and we refer to this matrix as an **augmented matrix**. The part on the left of the vertical line is referred to as the **coefficient matrix**, and the part on the right of the vertical line is considered the augmented part. Thus an augmented matrix represents a system of equations with unknowns, which are typically named $x_1, x_2, \ldots$ (or $y_1, y_2, \ldots$, if we have already used the x's). There are as many unknowns as there columns in the coefficient matrix.

- **Interpretation 2**: Matrix (1.1) could stand for the two systems of equations

$$\begin{cases} 2x_1 - 2x_2 - x_3 + 7x_4 &=& 1, \\ x_1 - x_2 - x_3 + 6x_4 &=& -5, \\ x_3 - x_4 &=& 3, \\ x_1 - x_2 + 3x_4 &=& 1, \end{cases} \quad \& \quad \begin{cases} 2y_1 - 2y_2 - y_3 + 7y_4 &=& 1, \\ y_1 - y_2 - y_3 + 6y_4 &=& 4, \\ y_3 - y_4 &=& -3, \\ y_1 - y_2 + 3y_4 &=& 0. \end{cases}$$

If so, we write:

$$\left[\begin{array}{cccc|cc} 2 & -2 & -1 & 7 & 1 & 1 \\ 1 & -1 & -1 & 6 & -5 & 4 \\ 0 & 0 & 1 & -1 & 3 & -3 \\ 1 & -1 & 0 & 3 & 1 & 0 \end{array} \right].$$

Here it is critical that the systems have the same coefficient matrix.

- **Interpretation 3**: Matrix (1.1) could stand for the system of equations

$$\begin{cases} 2x_1 - 2x_2 - x_3 + 7x_4 + x_5 + x_6 &=& 0, \\ x_1 - x_2 - x_3 + 6x_4 - 5x_5 + 4x_6 &=& 0, \\ x_3 - x_4 + 3x_5 - 3x_6 &=& 0, \\ x_1 - x_2 + 3x_4 + x_5 &=& 0. \end{cases}$$

If so, we should write:

$$\left[\begin{array}{cccccc|c} 2 & -2 & -1 & 7 & 1 & 1 & 0 \\ 1 & -1 & -1 & 6 & -5 & 4 & 0 \\ 0 & 0 & 1 & -1 & 3 & -3 & 0 \\ 1 & -1 & 0 & 3 & 1 & 0 & 0 \end{array} \right].$$

We don't always write the zeros, though! A system of linear equations where the right hand sides are all zeros is called a **homogeneous** system of linear equations.

- **Interpretation 4**: Matrix (1.1) could stand for the system of equations

$$\begin{cases} 2x_1 - 2x_2 - x_3 + 7x_4 + x_5 + x_6 &=& a, \\ x_1 - x_2 - x_3 + 6x_4 - 5x_5 + 4x_6 &=& b, \\ x_3 - x_4 + 3x_5 - 3x_6 &=& c, \\ x_1 - x_2 + 3x_4 + x_5 &=& d. \end{cases}$$

where a, b, c, d are arbitrary real numbers. If so, we should write:

$$\begin{bmatrix} 2 & -2 & -1 & 7 & 1 & 1 & | & a \\ 1 & -1 & -1 & 6 & -5 & 4 & | & b \\ 0 & 0 & 1 & -1 & 3 & -3 & | & c \\ 1 & -1 & 0 & 3 & 1 & 0 & | & d \end{bmatrix}.$$

Similar as before, we may not always write the arbitrary a, b, c, d; rather, we keep them in the back of our mind that they are there.

A question you may have is: why do we allow the same object to represent different situations? The answer to this question is: we will develop techniques to manipulate matrices which will provide useful insights in these and other interpretations of the matrices. In fact, a typical scenario is the following:

- We start with a problem.

- We represent the essential features of the problem as a matrix.

- We manipulate the matrix by techniques we will develop.

- Finally, we use the results of the manipulations to draw conclusions that are relevant to the problem we started with.

This type of process will appear many times in Linear Algebra!

1.2 Row Reduction: Three Elementary Row Operations

The following three systems of linear equations (1.2)–(1.4) have the same set of solutions. Which system gives a clearer vision of what the set of solutions may be?

$$\begin{cases} 2x_1 - 2x_2 - x_3 + 7x_4 + x_5 & = & 1, \\ x_1 - x_2 - x_3 + 6x_4 - 5x_5 & = & 4, \\ x_3 - x_4 + 3x_5 & = & -3, \\ x_1 - x_2 + 3x_4 + x_5 & = & 0, \end{cases} \tag{1.2}$$

or

$$\begin{cases} x_1 - x_2 + \quad\quad 3x_4 + x_5 &=& 0, \\ -x_3 + 3x_4 - 6x_5 &=& 4, \\ 2x_4 - 3x_5 &=& 1, \\ 2x_5 &=& -2, \end{cases} \tag{1.3}$$

or

$$\begin{cases} x_1 - x_2 &=& 4, \\ x_3 &=& -1, \\ x_4 &=& -1, \\ x_5 &=& -1. \end{cases} \tag{1.4}$$

We hope you find (1.3) clearer than (1.2), and that (1.4) is the most clear. In fact from (1.4) you can immediately read off that $x_3 = -1$, $x_4 = -1$, $x_5 = -1$, and that x_2 can be any real number as long as we then set $x_1 = 4 + x_2$. It is now easy to check that this solution also works for (1.2) and (1.3). In the future we will write this set of solutions as

$$\begin{bmatrix} x_1 \\ x_2 \\ x_3 \\ x_4 \\ x_5 \end{bmatrix} = \begin{bmatrix} 4 \\ 0 \\ -1 \\ -1 \\ -1 \end{bmatrix} + x_2 \begin{bmatrix} 1 \\ 1 \\ 0 \\ 0 \\ 0 \end{bmatrix}, \quad \text{with } x_2 \in \mathbb{R} \text{ free.} \tag{1.5}$$

To be a free variable in a system of linear equations means that we can choose its value to be any real number. The other variables (in this example x_1, x_3, x_4 and x_5) are called basic variables. Once values for the free variables have been chosen, the basic variables are fixed.

In this process we are making several conventions. The unknowns are numbered (typically, $x_1, x_2, \ldots$), and the coefficients of x_j appear in the jth column of the coefficient matrix. Also, in deciding which variables are free we will have a convention. In the above system x_1 and x_2 depend on one another (choosing one will fix the other). Our convention will be that if we have a choice, we view the variables with the higher index as the free variable(s). Thus, due to our convention, in the above example x_2 is a free variable and x_1 a basic variable. In the process that we will explain next this convention is built in, so it is important to stick to it.

The process of converting system (1.2) to (1.3) and subsequently to (1.4) is referred to as **row reduction** (or **Gaussian elimination**). The following three elementary row operations to its augmented matrix will be allowed:

Operation 1: Switch two rows.
Operation 2: Replace row(i) by row(i) + α row(k), $i \neq k$, where $\alpha \in \mathbb{R}$.
Operation 3: Multiply a row by a non-zero number $\beta \neq 0$.

Notice that each operation can be undone by applying the same operation. When we undo the action of Operation 2, we should apply Operation 2 again with α replaced by $-\alpha$. When we undo the action of Operation 3, we should apply Operation 3 again with β replaced by $\frac{1}{\beta}$. Consequently, the set of solutions of a system of linear equations does not change when performing these operations. Indeed, a solution to the original system will also be a solution to the converted system, and vice versa.

Let us perform these manipulations to the matrix (1.1):

$$\begin{bmatrix} 2 & -2 & -1 & 7 & 1 & 1 \\ 1 & -1 & -1 & 6 & -5 & 4 \\ 0 & 0 & 1 & -1 & 3 & -3 \\ 1 & -1 & 0 & 3 & 1 & 0 \end{bmatrix}.$$

Switching rows 1 and 4, we get

$$\begin{bmatrix} \boxed{1} & -1 & 0 & 3 & 1 & 0 \\ 1 & -1 & -1 & 6 & -5 & 4 \\ 0 & 0 & 1 & -1 & 3 & -3 \\ 2 & -2 & -1 & 7 & 1 & 1 \end{bmatrix}. \tag{1.6}$$

We put a box around the (1,1) entry, as our next objective is to make all entries below it equal to 0. In terms of the system of equations, this means that we are eliminating the variable x_1 from all equations except the top one. To create a 0 in the (2,1) position, we replace row(2) by row(2) $-$ row(1). In other words, we apply Operation 2 with $i = 2$, $k = 1$, and $\alpha = -1$, giving the following result.

$$\begin{bmatrix} \boxed{1} & -1 & 0 & 3 & 1 & 0 \\ 0 & 0 & -1 & 3 & -6 & 4 \\ 0 & 0 & 1 & -1 & 3 & -3 \\ 2 & -2 & -1 & 7 & 1 & 1 \end{bmatrix}. \tag{1.7}$$

Next, we would like to make a 0 in the (4,1) entry (thus eliminating the variable x_1 from the fourth equation). To do this, we apply again Operation 2 (with $i = 4$, $k = 1$ and $\alpha = -2$) and replace row(4) by row(4) -2 row(1):

$$\begin{bmatrix} \boxed{1} & -1 & 0 & 3 & 1 & 0 \\ 0 & 0 & \boxed{-1} & 3 & -6 & 4 \\ 0 & 0 & 1 & -1 & 3 & -3 \\ 0 & 0 & -1 & 1 & -1 & 1 \end{bmatrix}. \tag{1.8}$$

Notice that we next boxed entry (2,3). The reason is that at this point we have achieved our goal in column 1, and as it happens, the corresponding rows in column 2 already have zeros. Thus we are going to focus on column 3, and use the (2,3) entry to make zeros below it. We do this by applying Operation 2 (now with $i = 3$, $k = 2$ and $\alpha = 1$), first replacing row(3) by row(3) + row(2), which gives:

$$
\begin{bmatrix}
\boxed{1} & -1 & 0 & 3 & 1 & 0 \\
0 & 0 & \boxed{-1} & 3 & -6 & 4 \\
0 & 0 & 0 & 2 & -3 & 1 \\
0 & 0 & -1 & 1 & -1 & 1
\end{bmatrix}.
\tag{1.9}
$$

Next, we replace row(4) by row(4) − row(2):

$$
\begin{bmatrix}
\boxed{1} & -1 & 0 & 3 & 1 & 0 \\
0 & 0 & \boxed{-1} & 3 & -6 & 4 \\
0 & 0 & 0 & \boxed{2} & -3 & 1 \\
0 & 0 & 0 & -2 & 5 & -3
\end{bmatrix}.
\tag{1.10}
$$

We boxed the $(3, 4)$ entry and make a zero below it by replacing row(4) by row(4) + row(3):

$$
\begin{bmatrix}
\boxed{1} & -1 & 0 & 3 & 1 & 0 \\
0 & 0 & \boxed{-1} & 3 & -6 & 4 \\
0 & 0 & 0 & \boxed{2} & -3 & 1 \\
0 & 0 & 0 & 0 & \boxed{2} & -2
\end{bmatrix}.
\tag{1.11}
$$

We boxed entry $(4, 5)$. We are going to call the boxed entries **pivots**. Going from left to right, each pivot is the first nonzero entry in its row. Moreover, each pivot is strictly to the right of the pivot in the row above it. The matrix (1.11) is in **row echelon form**, defined below.

> **Definition 1.2.1.** We say that a matrix is in **row echelon form** if
>
> 1. all nonzero rows (rows with at least one nonzero element) are above any rows of all zeroes, and
>
> 2. the leading coefficient (the first nonzero number from the left, also called the **pivot**) of a nonzero row is always strictly to the right of the leading coefficient of the row above it.

The columns that contain a pivot, will be called a **pivot column**. In the above example, columns 1, 3, 4, and 5 are the pivot columns. The variables corresponding to the pivot columns are the **basic variables**. In the system above x_1, x_3, x_4 and x_5 are the basic variables. The remaining variables, which are the ones that correspond to non-pivot columns, are the **free variables**. In the example above, x_2 is the free variable.

Notice that the matrix (1.11) corresponds to the linear system (1.3). Next, we would like to further reduce it to arrive at system (1.4). To do this, we first make all pivots equal to 1 by applying Operation 3. We multiply row 2 by −1, row 3 by $\frac{1}{2}$ and row 4 by $\frac{1}{2}$, to obtain:

$$\begin{bmatrix} \boxed{1} & -1 & 0 & 3 & 1 & 0 \\ 0 & 0 & \boxed{1} & -3 & 6 & -4 \\ 0 & 0 & 0 & \boxed{1} & -\frac{3}{2} & \frac{1}{2} \\ 0 & 0 & 0 & 0 & \boxed{1} & -1 \end{bmatrix}. \tag{1.12}$$

Our final objective is make zeros above every pivot, starting with the most right pivot. To do this, we apply Operation 2 and (i) replace row(3) by row(3)+ $\frac{3}{2}$row(4), (ii) replace row(2) by row(2) − 6 row(4), (iii) replace row(1) by row(1) − row(4):

$$\begin{bmatrix} \boxed{1} & -1 & 0 & 3 & 0 & 1 \\ 0 & 0 & \boxed{1} & -3 & 0 & 2 \\ 0 & 0 & 0 & \boxed{1} & 0 & -1 \\ 0 & 0 & 0 & 0 & \boxed{1} & -1 \end{bmatrix}. \tag{1.13}$$

Next, we (i) replace row(2) by row(2)+3 row(3), (ii) replace row(1) by row(1)− 3 row(3):

$$\begin{bmatrix} \boxed{1} & -1 & 0 & 0 & 0 & 4 \\ 0 & 0 & \boxed{1} & 0 & 0 & -1 \\ 0 & 0 & 0 & \boxed{1} & 0 & -1 \\ 0 & 0 & 0 & 0 & \boxed{1} & -1 \end{bmatrix}. \tag{1.14}$$

Now the matrix is in **reduced row echelon form**, defined below. Notice that (1.14) corresponds to system (1.4).

Definition 1.2.2. We say that a matrix is in **reduced row echelon form** if

1. it is in row echelon form.

2. every leading coefficient (=pivot) is 1 and is the only nonzero entry in its column.

Let us do another example.

Example 1.2.3. Put the matrix A below in reduced row echelon form.

$$A = \begin{bmatrix} 2 & -4 & 3 & 0 \\ 1 & -1 & -1 & 2 \\ 3 & -5 & 2 & 2 \\ 1 & -1 & 0 & 3 \end{bmatrix} \rightarrow \begin{bmatrix} \boxed{2} & -4 & 3 & 0 \\ 0 & 1 & -\frac{5}{2} & 2 \\ 0 & 1 & -\frac{5}{2} & 2 \\ 0 & 1 & -\frac{3}{2} & 3 \end{bmatrix} \rightarrow$$

$$\begin{bmatrix} \boxed{2} & -4 & 3 & 0 \\ 0 & \boxed{1} & -\frac{5}{2} & 2 \\ 0 & 0 & 0 & 0 \\ 0 & 0 & 1 & 1 \end{bmatrix} \rightarrow \begin{bmatrix} \boxed{2} & -4 & 3 & 0 \\ 0 & \boxed{1} & -\frac{5}{2} & 2 \\ 0 & 0 & \boxed{1} & 1 \\ 0 & 0 & 0 & 0 \end{bmatrix}.$$

Notice that we have achieved row echelon form. To bring it to reduced row echelon form, we continue

$$
\rightarrow
\begin{bmatrix}
\boxed{1} & -2 & \frac{3}{2} & 0 \\
0 & \boxed{1} & -\frac{5}{2} & 2 \\
0 & 0 & \boxed{1} & 1 \\
0 & 0 & 0 & 0
\end{bmatrix}
\rightarrow
\begin{bmatrix}
\boxed{1} & -2 & 0 & -\frac{3}{2} \\
0 & \boxed{1} & 0 & \frac{9}{2} \\
0 & 0 & \boxed{1} & 1 \\
0 & 0 & 0 & 0
\end{bmatrix}
\rightarrow
\begin{bmatrix}
\boxed{1} & 0 & 0 & \frac{15}{2} \\
0 & \boxed{1} & 0 & \frac{9}{2} \\
0 & 0 & \boxed{1} & 1 \\
0 & 0 & 0 & 0
\end{bmatrix}.
$$

If we interpret the matrix A as the augmented matrix of a linear system, then we obtain the solution $x_1 = \frac{15}{2}, x_2 = \frac{9}{2}, x_3 = 1$. The first row of A would correspond to the equation $2x_1 - 4x_2 + 3x_3 = 0$, and indeed $x_1 = \frac{15}{2}$, $x_2 = \frac{9}{2}$, $x_3 = 1$ satisfies this equation. It works for the other rows as well.

$\square$

In the Gaussian elimination process one can view the effect of an operation on the matrix as a whole, or one can view it as an effect on each column. For instance, if we switch rows 1 and 2 in the matrix, it means that we switch rows 1 and 2 in each column. This leads to the following observation, which will turn out to be useful.

Proposition 1.2.4. *Let A be an $m \times n$ matrix, and let B be obtained from A by removing columns $j_1, \ldots, j_r$. Perform elementary row operations on A and obtain the matrix C, and perform the same elementary row operations on B to get the matrix D. Then D can be obtained from C by removing columns $j_1, \ldots, j_r$.*

Proof. Performing an elementary row operation on a matrix can be viewed as performing the same elementary row operation on each column individually. Thus removing a column before an operation or afterwards does not affect the other columns. $\square$

Let us end this section by providing the Gaussian elimination process in pseudo code. First, we need to explain the notation we will use:

- by $(a_{rj})_{r=i}^{n} \neq \mathbf{0}$ we mean that at least one of $a_{ij}, a_{i+1,j}, \ldots, a_{nj}$ is nonzero.

- by $x \leftarrow y$ we mean that the outcome y gets stored in the variable x.

- by $E^{(1)}_{r \leftrightarrow i} A$ we mean that we switch rows r and i in A.

- by $E^{(2)}_{(s,i,c)} A$ we mean that row(s) in A gets replaced by row$(s) + c$ row(i).

- by $E^{(3)}_{(i,\beta)} A$ we mean that row(i) in A gets multiplied by β.

- the quantity rank keeps track of how many pivots have been identified.

- in the array pivot we store the pivot locations.

- rref(A) is shorthand for the 'reduced row echelon form'* of A.

Algorithm 1 Gaussian elimination: part 1

1: **procedure** GAUSS1(A)   ▷ Finds row echelon form of $m \times n$ matrix A
2:     $i \leftarrow 1$, $j \leftarrow 1$, rank $\leftarrow 0$, pivot $\leftarrow$ empty array
3:     **while** $j \leq n$ **do**   ▷ We have the answer if $j = n + 1$
4:         **if** $(a_{rj})_{r=i}^{n} \neq \mathbf{0}$ **then**
5:             choose $i \leq r \leq n$ with $a_{rj} \neq 0$
6:             $A \leftarrow E_{r \leftrightarrow i}^{(1)} A$
7:             rank $\leftarrow$ rank $+ 1$
8:             pivot(rank) $\leftarrow (i, j)$
9:             **for** $s = i + 1, \dots, n$ **do** $A \leftarrow E_{(s,i,-a_{sj}/a_{ij})}^{(2)} A$
10:            $i \leftarrow i + 1$, $j \leftarrow j + 1$
11:        **else**
12:            $j \leftarrow j + 1$
13:    **return** A, pivot, rank   ▷ A is in echelon form, pivot positions, rank

---

Algorithm 2 Gaussian elimination: part 2

1: **procedure** GAUSS2(A, pivot, rank)   ▷ Continues to find rref(A)
2:     $r \leftarrow$ rank
3:     **while** $r > 0$ **do**   ▷ We have the answer if $r = 0$
4:         $(i, j) \leftarrow$ pivot(r)
5:         $A \leftarrow E_{(i,1/a_{ij})}^{(3)} A$
6:         **for** $s = 1, \dots, i - 1$ **do** $A \leftarrow E_{(s,i,-a_{sj})}^{(2)} A$
7:         $r \leftarrow r - 1$
8:     **return** A ▷ A is in its reduced echelon form

In Algorithm 1 above we have in line 5 a choice when we choose the next pivot. In the method of **partial pivoting** we choose among a_{rj}, $i \leq r \leq n$, the entry with the largest absolute value. The advantage of this choice is that it reduces the effect of round-off errors.

---

*It is also the command in MATLAB® and Maple™ and other software to find the reduced row echelon form.

1.3 Vectors in $\mathbb{R}^n$, Linear Combinations and Span

A real vector is a convenient way to group together a set of real numbers in a particular order. There are many contexts where this concept is useful. A vector can represent a point in space, a digital sound signal, measurements (from a science experiment, value of a stock at different times, temperatures at different locations, etc.), a text message, a code, sampled values of a function, etc. Mathematically, it is a matrix with just one column, and thus defined as follows.

A **vector** in $\mathbb{R}^n$ has the form

$$\mathbf{u} = \begin{bmatrix} u_1 \\ u_2 \\ \vdots \\ u_n \end{bmatrix} \in \mathbb{R}^n, \text{ with } u_1, \ldots, u_n \in \mathbb{R}.$$

Sometimes it is convenient to write $\mathbf{u} = (u_i)_{i=1}^n$ for the vector above. To be more precise, the vector $\mathbf{u}$ is in fact a column vector. We occasionally use row vectors (in $\mathbb{R}^{1 \times n}$), which are of the form

$$\begin{bmatrix} u_1 & u_2 & \cdots & u_n \end{bmatrix}, \text{ with } u_1, \ldots, u_n \in \mathbb{R}.$$

In this section, however, we focus on column vectors and refer to them as just 'vectors'. The theory for row and column vectors is the same, so it suffices to just present it for column vectors.

We denote a vector by a boldface lower case letter, and for its entries we use the corresponding letter in italics with indices corresponding to their location. For instance, $\mathbf{x} \in \mathbb{R}^n$ has entries $x_1, \ldots, x_n$. We define two operations on vectors, **addition** and **scalar multiplication**, as follows:

$$\text{(i) } \mathbf{u} + \mathbf{v} = \begin{bmatrix} u_1 \\ u_2 \\ \vdots \\ u_n \end{bmatrix} + \begin{bmatrix} v_1 \\ v_2 \\ \vdots \\ v_n \end{bmatrix} := \begin{bmatrix} u_1 + v_1 \\ u_2 + v_2 \\ \vdots \\ u_n + v_n \end{bmatrix} \in \mathbb{R}^n.$$

$$\text{(ii) } \alpha \mathbf{u} = \alpha \begin{bmatrix} u_1 \\ u_2 \\ \vdots \\ u_n \end{bmatrix} := \begin{bmatrix} \alpha u_1 \\ \alpha u_2 \\ \vdots \\ \alpha u_n \end{bmatrix} \in \mathbb{R}^n \text{ for } \alpha \in \mathbb{R}.$$

The number α is referred to as a **scalar**. For example,

$$\begin{bmatrix} 5 \\ -1 \\ 1 \end{bmatrix} + \begin{bmatrix} 2 \\ 6 \\ -10 \end{bmatrix} = \begin{bmatrix} 7 \\ 5 \\ -9 \end{bmatrix} \in \mathbb{R}^3, \quad 3 \begin{bmatrix} 2 \\ 1 \\ 0 \\ 1 \end{bmatrix} = \begin{bmatrix} 6 \\ 3 \\ 0 \\ 3 \end{bmatrix} \in \mathbb{R}^4, \quad 2 \begin{bmatrix} 4 \\ 1 \end{bmatrix} - \begin{bmatrix} 5 \\ 4 \end{bmatrix} = \begin{bmatrix} 3 \\ -2 \end{bmatrix} \in \mathbb{R}^2.$$

These operations satisfy the following rules:

1. **Commutativity of addition**: for all $\mathbf{u}, \mathbf{v} \in \mathbb{R}^n$ we have that $\mathbf{u} + \mathbf{v} = \mathbf{v} + \mathbf{u}$.

2. **Associativity of addition**: for all $\mathbf{u}, \mathbf{v}, \mathbf{w} \in \mathbb{R}^n$ we have that $(\mathbf{u} + \mathbf{v}) + \mathbf{w} = \mathbf{u} + (\mathbf{v} + \mathbf{w})$.

3. **Existence of a neutral element for addition**: there exists a $\mathbf{0} \in \mathbb{R}^n$ so that $\mathbf{u} + \mathbf{0} = \mathbf{u} = \mathbf{0} + \mathbf{u}$ for all $\mathbf{u} \in \mathbb{R}^n$.

4. **Existence of an additive inverse**: for every $\mathbf{u} \in \mathbb{R}^n$ there exists a $-\mathbf{u} \in \mathbb{R}^n$ so that $\mathbf{u} + (-\mathbf{u}) = \mathbf{0} = (-\mathbf{u}) + \mathbf{u}$.

5. **First distributive law**: for all $c \in \mathbb{R}$ and $\mathbf{u}, \mathbf{v} \in \mathbb{R}^n$ we have that $c(\mathbf{u} + \mathbf{v}) = c\mathbf{u} + c\mathbf{v}$.

6. **Second distributive law**: for all $c, d \in \mathbb{R}$ and $\mathbf{u} \in \mathbb{R}^n$ we have that $(c + d)\mathbf{u} = c\mathbf{u} + d\mathbf{u}$.

7. **Associativity for scalar multiplication**: for all $c, d \in \mathbb{R}$ and $\mathbf{u} \in \mathbb{R}^n$ we have that $c(d\mathbf{u}) = (cd)\mathbf{u}$.

8. **Unit multiplication rule**: for every $\mathbf{u} \in \mathbb{R}^n$ we have that $1\mathbf{u} = \mathbf{u}$.

Here

$$\mathbf{0} = \begin{bmatrix} 0 \\ 0 \\ \vdots \\ 0 \end{bmatrix}$$

is the **zero vector**. You will see these rules again (with $\mathbb{R}^n$ and $\mathbb{R}$ replaced by more general sets V and $\mathbb{F}$) when we give the definition of a vector space. For now we can just say that since these rules hold, $\mathbb{R}^n$ is a vector space over $\mathbb{R}$.

The following notion typically takes some getting used to. We say that a vector $\mathbf{y} \in \mathbb{R}^n$ is a **linear combination** of vectors $\mathbf{u}_1, \ldots, \mathbf{u}_p \in \mathbb{R}^n$ if there exist scalars $x_1, \ldots, x_p \in \mathbb{R}$ so that

$$\mathbf{y} = x_1 \mathbf{u}_1 + \cdots + x_p \mathbf{u}_p.$$

For instance, $\begin{bmatrix} -7 \\ 5 \end{bmatrix}$ is a linear combination of $\begin{bmatrix} 4 \\ 1 \end{bmatrix}$, $\begin{bmatrix} 5 \\ -1 \end{bmatrix}$ because

$$2\begin{bmatrix} 4 \\ 1 \end{bmatrix} - 3\begin{bmatrix} 5 \\ -1 \end{bmatrix} = \begin{bmatrix} -7 \\ 5 \end{bmatrix}.$$

Here are some typical problems.

Example 1.3.1. Is $\begin{bmatrix} -1 \\ 13 \\ -21 \end{bmatrix}$ a linear combination of $\begin{bmatrix} 5 \\ -1 \\ 1 \end{bmatrix}$, $\begin{bmatrix} 2 \\ 6 \\ -10 \end{bmatrix}$?

We need find out whether there exist scalars x_1, x_2 so that

$$x_1\begin{bmatrix} 5 \\ -1 \\ 1 \end{bmatrix} + x_2\begin{bmatrix} 2 \\ 6 \\ -10 \end{bmatrix} = \begin{bmatrix} -1 \\ 13 \\ -21 \end{bmatrix}.$$

In other words, we would like to solve the system

$$\begin{cases} 5x_1 + 2x_2 = -1, \\ -x_1 + 6x_2 = 13, \\ x_1 - 10x_2 = -21. \end{cases} \tag{1.15}$$

Setting up the augmented matrix and doing row reduction, we get

$$\begin{bmatrix} 5 & 2 & | & -1 \\ -1 & 6 & | & 13 \\ 1 & -10 & | & -21 \end{bmatrix} \rightarrow \begin{bmatrix} \boxed{1} & -10 & | & -21 \\ 0 & -4 & | & -8 \\ 0 & 52 & | & 104 \end{bmatrix} \rightarrow \begin{bmatrix} \boxed{1} & 0 & | & -1 \\ 0 & \boxed{1} & | & 2 \\ 0 & 0 & | & 0 \end{bmatrix}.$$

This has a solution $x_1 = -1$ and $x_2 = 2$. Thus, the answer is **yes**, because

$$-\begin{bmatrix} 5 \\ -1 \\ 1 \end{bmatrix} + 2\begin{bmatrix} 2 \\ 6 \\ -10 \end{bmatrix} = \begin{bmatrix} -1 \\ 13 \\ -21 \end{bmatrix}.$$

$\square$

Example 1.3.2. Is $\begin{bmatrix} 2 \\ 9 \\ -5 \end{bmatrix}$ in a linear combination of $\begin{bmatrix} 1 \\ 1 \\ 1 \end{bmatrix}$ and $\begin{bmatrix} 3 \\ 4 \\ 5 \end{bmatrix}$?

We need find out whether there exist scalars x_1, x_2 so that

$$x_1\begin{bmatrix} 1 \\ 1 \\ 1 \end{bmatrix} + x_2\begin{bmatrix} 3 \\ 4 \\ 5 \end{bmatrix} = \begin{bmatrix} 2 \\ 9 \\ -5 \end{bmatrix}.$$

In other words, we would like to solve the system

$$\begin{cases} x_1 + 3x_2 = 2, \\ x_1 + 4x_2 = 9, \\ x_1 + 5x_2 = -5. \end{cases} \tag{1.16}$$

Setting up the augmented matrix and doing row reduction, we get

$$\left[\begin{array}{cc|c} \boxed{1} & 3 & 2 \\ 1 & 4 & 9 \\ 1 & 5 & -5 \end{array}\right] \rightarrow \left[\begin{array}{cc|c} \boxed{1} & 3 & 2 \\ 0 & \boxed{1} & 7 \\ 0 & 2 & -7 \end{array}\right] \rightarrow \left[\begin{array}{cc|c} \boxed{1} & 3 & 2 \\ 0 & \boxed{1} & 7 \\ 0 & 0 & \boxed{-21} \end{array}\right].$$

This corresponds to the system

$$\begin{cases} x_1 + 3x_2 &=& 2, \\ x_2 &=& 7, \\ 0 &=& -21, \end{cases}$$

which does not have a solution. Thus the answer is **no**. ☐

Definition 1.3.3. Given vectors $\mathbf{u}_1, \ldots, \mathbf{u}_p \in \mathbb{R}^n$ we define

$$\text{Span}\{\mathbf{u}_1, \ldots, \mathbf{u}_p\} := \{x_1\mathbf{u}_1 + \cdots + x_p\mathbf{u}_p \; : \; x_1, \ldots, x_p \in \mathbb{R}\} \subseteq \mathbb{R}^n.$$

Thus, $\text{Span}\{\mathbf{u}_1, \ldots, \mathbf{u}_p\}$ consists of all linear combinations of the vectors $\mathbf{u}_1, \ldots, \mathbf{u}_p$.

In this terminology, the examples above can be summarized as

$$\begin{bmatrix} -1 \\ 13 \\ -21 \end{bmatrix} \in \text{Span}\left\{\begin{bmatrix} 5 \\ -1 \\ 1 \end{bmatrix}, \begin{bmatrix} 2 \\ 6 \\ -10 \end{bmatrix}\right\}, \qquad \begin{bmatrix} 2 \\ 9 \\ -5 \end{bmatrix} \notin \text{Span}\left\{\begin{bmatrix} 1 \\ 1 \\ 1 \end{bmatrix}, \begin{bmatrix} 3 \\ 4 \\ 5 \end{bmatrix}\right\}.$$

Here are some typical problems.

Example 1.3.4. Is $\text{Span}\left\{\begin{bmatrix} 2 \\ 5 \end{bmatrix}, \begin{bmatrix} 4 \\ 10 \end{bmatrix}, \begin{bmatrix} 1 \\ -1 \end{bmatrix}\right\} = \mathbb{R}^2$?

Thus, we need to check whether an arbitrary vector $\begin{bmatrix} a \\ b \end{bmatrix}$ belongs to $\text{Span}\left\{\begin{bmatrix} 2 \\ 5 \end{bmatrix}, \begin{bmatrix} 4 \\ 10 \end{bmatrix}, \begin{bmatrix} 1 \\ -1 \end{bmatrix}\right\}$. In other words, does the system with the augmented matrix

$$\left[\begin{array}{ccc|c} 2 & 4 & 1 & a \\ 5 & 10 & -1 & b \end{array}\right]$$

have a solution regardless of the values of a and b? Row reducing we get

$$\left[\begin{array}{ccc|c} \boxed{2} & 4 & 1 & a \\ 0 & 0 & \boxed{-\dfrac{7}{2}} & b - \dfrac{5}{2}a \end{array}\right].$$

Consequently, we arrive at the system

$$\begin{cases} 2x_1 + 4x_2 + x_3 &= a, \\ -\frac{7}{2}x_3 &= b - \frac{5}{2}a. \end{cases}$$

We can now see that this system has a solution, regardless of the values of a and b. Indeed, from the second equation we can solve for $x_3 (= -\frac{2}{7}(b - \frac{5}{2}a))$ and then we can choose x_2 freely and subsequently solve for $x_1 (= \frac{1}{2}(a - 4x_2 - x_3))$. Thus the answer is **yes**:

$$\text{Span}\left\{ \begin{bmatrix} 2 \\ 5 \end{bmatrix}, \begin{bmatrix} 4 \\ 10 \end{bmatrix} \begin{bmatrix} 1 \\ -1 \end{bmatrix} \right\} = \mathbb{R}^2.$$

When we did the row reduction, there was not a real need to keep track of the augmented part: all we cared about was 'can we solve the system regardless of the value on the right hand side of the equations?' The answer was yes in this case, because in the coefficient matrix every row had a pivot. This allowed us to solve for the unknowns x_1, x_2, x_3 starting with the one of the highest index first (x_3), then address x_2 next and finally x_1.

□

Example 1.3.5. Is $\text{Span}\left\{ \begin{bmatrix} 1 \\ 1 \\ 2 \end{bmatrix}, \begin{bmatrix} 2 \\ 3 \\ 10 \end{bmatrix} \begin{bmatrix} 3 \\ 4 \\ 12 \end{bmatrix} \right\} = \mathbb{R}^3$?

We need to check whether an arbitrary vector $\begin{bmatrix} a \\ b \\ c \end{bmatrix}$ belongs to $\text{Span}\left\{ \begin{bmatrix} 1 \\ 1 \\ 2 \end{bmatrix}, \right.$ $\begin{bmatrix} 2 \\ 3 \\ 10 \end{bmatrix} \begin{bmatrix} 3 \\ 4 \\ 12 \end{bmatrix} \left. \right\}$. In other words, does the system with the augmented matrix

$$\begin{bmatrix} 1 & 2 & 3 & a \\ 1 & 3 & 4 & b \\ 2 & 10 & 12 & c \end{bmatrix}$$

have a solution regardless of the values of a, b and c? Row reducing we get

$$\begin{bmatrix} \boxed{1} & 2 & 3 & a \\ 0 & \boxed{1} & 1 & b - a \\ 0 & 6 & 6 & c - 2a \end{bmatrix} \rightarrow \begin{bmatrix} \boxed{1} & 2 & 3 & a \\ 0 & \boxed{1} & 1 & b - a \\ 0 & 0 & 0 & c - 6b + 4a \end{bmatrix}. \qquad (1.17)$$

We arrive at the system

$$\begin{cases} x_1 + 2x_2 + 3x_3 &= a, \\ x_2 + x_3 &= b - a, \\ 0 &= c - 6b + 4a. \end{cases}$$

We can now see that if $c - 6b + 4a \neq 0$, then there is no solution. Consequently, the answer here is **no**:

$$\text{Span}\left\{ \begin{bmatrix} 1 \\ 1 \\ 2 \end{bmatrix}, \begin{bmatrix} 2 \\ 3 \\ 10 \end{bmatrix}, \begin{bmatrix} 3 \\ 4 \\ 12 \end{bmatrix} \right\} \neq \mathbb{R}^3.$$

For instance, if we choose $a = b = 0$ and $c = 1$, we find that the system is not solvable and thus

$$\begin{bmatrix} 0 \\ 0 \\ 1 \end{bmatrix} \notin \text{Span}\left\{ \begin{bmatrix} 1 \\ 1 \\ 2 \end{bmatrix}, \begin{bmatrix} 2 \\ 3 \\ 10 \end{bmatrix}, \begin{bmatrix} 3 \\ 4 \\ 12 \end{bmatrix} \right\}.$$

There are plenty of other vectors that do not belong to this span. The reason why the answer turned out to be no is that in (1.17) the coefficient matrix does not have a pivot in every row. Because of this we can choose a right hand side that gives us an equation with 0 on the left hand side and a nonzero on the right hand side (here $0 = c - 6b + 4a(\neq 0)$), which is an impossibility.
□

Let us summarize the type of problems presented above.

Problem 1. Check whether $\mathbf{y} \in \text{Span}\{\mathbf{u}_1, \ldots, \mathbf{u}_p\}$.
Solution. Row reduce the augmented matrix $\begin{bmatrix} \mathbf{u}_1 & \cdots & \mathbf{u}_p & | & \mathbf{y} \end{bmatrix}$. If in the echelon form all the pivots are in the coefficient matrix, then the answer is yes. If in the echelon form there is a pivot in the augmented part, then the answer is no.

Problem 2. Check whether $\text{Span}\{\mathbf{u}_1, \ldots, \mathbf{u}_p\} = \mathbb{R}^n$.
Solution. Row reduce the coefficient matrix $\begin{bmatrix} \mathbf{u}_1 & \cdots & \mathbf{u}_p \end{bmatrix}$. If in the echelon form of the coefficient matrix every row has a pivot, then the answer is yes. If in the echelon form of the coefficient matrix there is row without a pivot, then the answer is no.

Note that in Problem 2 we only need to worry about the coefficient matrix. There is no need to include the augmented part in the calculations (remember Interpretation 4 in Section 1.1!). Another useful observation is that if in Problem 2 we have that $p < n$ (thus the coefficient matrix has fewer columns than rows), then there can never be a pivot in every row and thus the answer is automatically no. As an example, without doing any computations, we can conclude that

$$\text{Span}\left\{ \begin{bmatrix} 1 \\ 5 \\ 1 \\ 2 \end{bmatrix}, \begin{bmatrix} 0 \\ 1 \\ 3 \\ -7 \end{bmatrix}, \begin{bmatrix} 7 \\ -3 \\ 4 \\ 17 \end{bmatrix} \right\} \neq \mathbb{R}^4,$$

because the row reduced coefficient matrix will never have a pivot in row 4.

Finally, let us use the above to characterize the pivot columns of a matrix A.

Proposition 1.3.6. *Let* $A = \begin{bmatrix} \mathbf{a}_1 & \cdots & \mathbf{a}_n \end{bmatrix}$ *be an* $m \times n$ *matrix. Then the* k*th column of* A *is a pivot column if and only if* $\mathbf{a}_k \notin \mathrm{Span}\{\mathbf{a}_1, \ldots, \mathbf{a}_{k-1}\}$.

Proof. Suppose that $\mathbf{a}_k \notin \mathrm{Span}\{\mathbf{a}_1, \ldots, \mathbf{a}_{k-1}\}$. This means that when we row reduce the matrix $\begin{bmatrix} \mathbf{a}_1 & \cdots & \mathbf{a}_{k-1} & | & \mathbf{a}_k \end{bmatrix}$ there will be a pivot in the last column. But when we use the same row operations on A, it gives that the kth column of A is a pivot column.

Conversely, suppose that the kth column of A is a pivot column. By Proposition 1.2.4 this means that row reducing the matrix $\begin{bmatrix} \mathbf{a}_1 & \cdots & \mathbf{a}_{k-1} & | & \mathbf{a}_k \end{bmatrix}$ will yield a pivot in the last column. But then $\mathbf{a}_k \notin \mathrm{Span}\{\mathbf{a}_1, \ldots, \mathbf{a}_{k-1}\}$. $\square$

1.4 Matrix Vector Product and the Equation $Ax = b$

We define the product of a matrix and a vector as follows. Given are $A = \begin{bmatrix} \mathbf{a}_1 & \cdots & \mathbf{a}_n \end{bmatrix}$ and $\mathbf{x} = \begin{bmatrix} x_1 \\ \vdots \\ x_n \end{bmatrix}$. Thus $\mathbf{a}_j$ denotes the jth column of A. We define

$$A\mathbf{x} := x_1 \mathbf{a}_1 + \cdots + x_n \mathbf{a}_n. \tag{1.18}$$

Alternatively, if we write

$$A = \begin{bmatrix} a_{11} & \cdots & a_{1n} \\ \vdots & & \vdots \\ a_{m1} & \cdots & a_{mn} \end{bmatrix},$$

then

$$A\mathbf{x} := x_1 \begin{bmatrix} a_{11} \\ \vdots \\ a_{m1} \end{bmatrix} + x_2 \begin{bmatrix} a_{12} \\ \vdots \\ a_{m2} \end{bmatrix} + \cdots + x_n \begin{bmatrix} a_{1n} \\ \vdots \\ a_{mn} \end{bmatrix} = \begin{bmatrix} \sum_{j=1}^{n} a_{1j} x_j \\ \vdots \\ \sum_{j=1}^{n} a_{mj} x_j \end{bmatrix}.$$

For example,

$$
\begin{bmatrix} 1 & 9 \\ -2 & -5 \\ -1 & 4 \end{bmatrix} \begin{bmatrix} 2 \\ -3 \end{bmatrix} = \begin{bmatrix} -25 \\ 11 \\ -14 \end{bmatrix}, \quad \begin{bmatrix} 2 & -2 & -1 & 7 & 1 & 1 \\ 1 & -1 & -1 & 6 & -5 & 4 \\ 0 & 0 & 1 & -1 & 3 & -3 \\ 1 & -1 & 0 & 3 & 1 & 0 \end{bmatrix} \begin{bmatrix} 1 \\ -2 \\ 0 \\ 3 \\ 0 \\ 0 \end{bmatrix} = \begin{bmatrix} 27 \\ 21 \\ -3 \\ 12 \end{bmatrix}.
$$

It is clear from (1.18) that $A\mathbf{x}$ is a linear combination of the columns $\mathbf{a}_1, \ldots, \mathbf{a}_n$ of A. In other words, $A\mathbf{x} \in \text{Span}\{\mathbf{a}_1, \ldots, \mathbf{a}_n\}$. An important equation in Linear Algebra is the equation $A\mathbf{x} = \mathbf{b}$, where we are looking for a solution vector $\mathbf{x}$. We observe that this equation has a solution if and only if $\mathbf{b}$ is a linear combination of the columns $\mathbf{a}_1, \ldots, \mathbf{a}_n$ of A. The equation $A\mathbf{x} = \mathbf{0}$, where the right hand side is the zero vector, is called a **homogeneous** linear equation. A homogeneous equation always has at least one solution (namely, $\mathbf{x} = \mathbf{0}$).

It is easy to check that the matrix vector product satisfies the following rules.

Proposition 1.4.1. *Let A be an $m \times n$ matrix, $\mathbf{u}, \mathbf{v} \in \mathbb{R}^n$, and $c \in \mathbb{R}$. Then*

(i) $A(\mathbf{u} + \mathbf{v}) = A\mathbf{u} + A\mathbf{v}$.

(ii) $A(c\mathbf{u}) = c(A\mathbf{u})$.

Proof. We have

$$A(\mathbf{u} + \mathbf{v}) = (u_1 + v_1)\mathbf{a}_1 + \cdots + (u_n + v_n)\mathbf{a}_n.$$

Using the second distributive law and subsequently the commutativity of addition, we can rewrite this as

$$u_1\mathbf{a}_1 + v_1\mathbf{a}_1 + \cdots + u_n\mathbf{a}_n + v_n\mathbf{a}_n = u_1\mathbf{a}_1 + \cdots + u_n\mathbf{a}_n + v_1\mathbf{a}_1 + \cdots + v_n\mathbf{a}_n.$$

The latter is equal to $A\mathbf{u} + A\mathbf{v}$, proving (i). It should be noted that we also used the associative law since we added several vectors together without specifying in which order to add them. Due to the associative law, any order in which you add several vectors gives the same result.

We will leave the proof of (ii) as an exercise. □

Let us put this new concept to good use in what we have done before.

Example 1.4.2. Consider the system

$$\begin{cases} 2x_1 - 2x_2 - x_3 + 7x_4 &= 1, \\ x_1 - x_2 - x_3 + 6x_4 &= 0, \\ x_3 - x_4 &= 1, \\ x_1 - x_2 + 3x_4 &= 1. \end{cases} \quad (1.19)$$

If we let

$$A = \begin{bmatrix} 2 & -2 & -1 & 7 \\ 1 & -1 & -1 & 6 \\ 0 & 0 & 1 & -1 \\ 1 & -1 & 0 & 3 \end{bmatrix}, \mathbf{x} = \begin{bmatrix} x_1 \\ x_2 \\ x_3 \\ x_4 \end{bmatrix}, \mathbf{b} = \begin{bmatrix} 1 \\ 0 \\ 1 \\ 1 \end{bmatrix},$$

Then (1.19) is the equation $A\mathbf{x} = \mathbf{b}$. How does one find all solutions $\mathbf{x}$?

Put the corresponding augmented matrix in reduced row echelon form:

$$[A \mid \mathbf{b}] = \begin{bmatrix} 2 & -2 & -1 & 7 & \mid & 1 \\ 1 & -1 & -1 & 6 & \mid & 0 \\ 0 & 0 & 1 & -1 & \mid & 1 \\ 1 & -1 & 0 & 3 & \mid & 1 \end{bmatrix} \to \cdots \to \begin{bmatrix} \boxed{1} & -1 & 0 & 0 & \mid & 1 \\ 0 & 0 & \boxed{1} & 0 & \mid & 1 \\ 0 & 0 & 0 & \boxed{1} & \mid & 0 \\ 0 & 0 & 0 & 0 & \mid & 0 \end{bmatrix}.$$

This gives that x_2 is a free variable, and we find

$$\begin{cases} x_1 &= 1 + x_2 \\ x_2 &= x_2 \quad \text{(free variable)} \\ x_3 &= 1 \\ x_4 &= 0 \end{cases} \quad \text{or} \quad \begin{bmatrix} x_1 \\ x_2 \\ x_3 \\ x_4 \end{bmatrix} = \begin{bmatrix} 1 \\ 0 \\ 1 \\ 0 \end{bmatrix} + x_2 \begin{bmatrix} 1 \\ 1 \\ 0 \\ 0 \end{bmatrix}.$$

Thus the set of solutions is given by

$$\left\{ \mathbf{x} = \begin{bmatrix} 1 \\ 0 \\ 1 \\ 0 \end{bmatrix} + x_2 \begin{bmatrix} 1 \\ 1 \\ 0 \\ 0 \end{bmatrix} : x_2 \in \mathbb{R} \right\}.$$

$\square$

If we let $\mathbf{p} = \begin{bmatrix} 1 \\ 0 \\ 1 \\ 0 \end{bmatrix}$ and $\mathbf{v} = x_2 \begin{bmatrix} 1 \\ 1 \\ 0 \\ 0 \end{bmatrix}$, then one easily checks that $A\mathbf{p} = \mathbf{b}$ and $A\mathbf{v} = \mathbf{0}$. We refer to $\mathbf{p}$ as a **particular** solution of the equation $A\mathbf{x} = \mathbf{b}$, and to $\mathbf{v}$ as a solution to the homogeneous equation $A\mathbf{x} = \mathbf{0}$. We see that any solution to $A\mathbf{x} = \mathbf{b}$ is of the form $\mathbf{x} = \mathbf{p} + \mathbf{v}$. This is a general principle as we now state in the next proposition.

Proposition 1.4.3. *Let A be an $m \times n$ matrix and $\mathbf{b} \in \mathbb{R}^m$. Consider the equation $A\mathbf{x} = \mathbf{b}$, where $\mathbf{x} \in \mathbb{R}^n$ is unknown.*

(i) If $\mathbf{p}$ is a particular solution of $A\mathbf{x} = \mathbf{b}$ and $\mathbf{v}$ is a solution of the homogeneous equation $A\mathbf{x} = \mathbf{0}$, then $\mathbf{p} + \mathbf{v}$ is a solution of $A\mathbf{x} = \mathbf{b}$.

(ii) If $\mathbf{p}_1$ and $\mathbf{p}_2$ are particular solutions of $A\mathbf{x} = \mathbf{b}$, then $\mathbf{p}_1 - \mathbf{p}_2$ is a solution to the homogeneous equation $A\mathbf{x} = \mathbf{0}$.

Proof. (i) If $A\mathbf{p} = \mathbf{b}$ and $A\mathbf{v} = \mathbf{0}$, then

$$A(\mathbf{p} + \mathbf{v}) = A\mathbf{p} + A\mathbf{v} = \mathbf{b} + \mathbf{0} = \mathbf{b}.$$

(ii) If $A\mathbf{p}_1 = \mathbf{b}$ and $A\mathbf{p}_2 = \mathbf{b}$, then $A(\mathbf{p}_1 - \mathbf{p}_2) = A\mathbf{p}_1 - A\mathbf{p}_2 = \mathbf{b} - \mathbf{b} = \mathbf{0}$. $\square$

The matrix vector equation $A\mathbf{x} = \mathbf{b}$ (or, equivalently, the system of linear equations) is called **consistent** if a solution exists. If no solution exists, the system is called **inconsistent**. For a consistent system there are two options: there is exactly one solution, or there are infinitely many solutions. The result is the following.

Theorem 1.4.4. *Let A be an $m \times n$ matrix and $\mathbf{b} \in \mathbb{R}^m$. Consider the equation $A\mathbf{x} = \mathbf{b}$, where $\mathbf{x} \in \mathbb{R}^n$ is unknown. There are three possibilities:*

- *The system is inconsistent. This happens when $\mathbf{b}$ is not a linear combination of the columns of A.*

- *The system is consistent and there is a unique solution. This happens when $\mathbf{b}$ is a linear combination of the columns of A, and $A\mathbf{x} = \mathbf{0}$ has only $\mathbf{0}$ as its solution.*

- *The system is consistent and there are infinitely many solutions. This happens when $\mathbf{b}$ is a linear combination of the columns of A, and $A\mathbf{x} = \mathbf{0}$ has a nonzero solution.*

Proof. It is clear that consistency of the system corresponds exactly to $\mathbf{b}$ being a linear combination of the columns of A. If there is a solution to $A\mathbf{x} = \mathbf{b}$, then there are two options: (i) $A\mathbf{x} = \mathbf{0}$ has only the solution $\mathbf{x} = \mathbf{0}$. (ii) $A\mathbf{x} = \mathbf{0}$ has a solution $\mathbf{x}_0 \neq \mathbf{0}$.

Assume (i), and let $\mathbf{p}_1$ and $\mathbf{p}_2$ be solutions to $A\mathbf{x} = \mathbf{b}$. Then by Proposition 1.4.3(ii) we have that $\mathbf{p}_1 - \mathbf{p}_2$ is a solution to the homogeneous equation $A\mathbf{x} = \mathbf{0}$. Since the only solution to the homogeneous equation is $\mathbf{0}$, we obtain $\mathbf{p}_1 - \mathbf{p}_2 = \mathbf{0}$. Thus $\mathbf{p}_1 = \mathbf{p}_2$, yielding uniqueness of the solution to $A\mathbf{x} = \mathbf{b}$.

Next, assume (ii). Then for a particular solution $\mathbf{p}$ to $A\mathbf{x} = \mathbf{b}$, we get that for any $c \in \mathbb{R}$, the vector $\mathbf{p} + c\,\mathbf{x}_0$ is also a solution to $A\mathbf{x} = \mathbf{b}$. Thus $A\mathbf{x} = \mathbf{b}$ has infinitely many solutions. $\qquad\qquad\qquad\qquad\qquad\qquad\qquad\qquad\qquad\qquad\qquad$ □

Let us also summarize what we have seen previously regarding the span using the new notation.

Theorem 1.4.5. *Let* $A = \begin{bmatrix} \mathbf{a}_1 & \cdots & \mathbf{a}_n \end{bmatrix}$ *be a* $m \times n$ *matrix. The following are equivalent.*

1. For each $\mathbf{b} \in \mathbb{R}^m$ *the equation* $A\mathbf{x} = \mathbf{b}$ *has a solution.*

2. Each $\mathbf{b} \in \mathbb{R}^m$ *is a linear combination of the columns of* A.

3. $\mathbb{R}^m = \mathrm{Span}\{\mathbf{a}_1, \ldots, \mathbf{a}_n\}$. *(i.e., the columns of* A *span* $\mathbb{R}^m$.)

4. The echelon form of the coefficient matrix A *has a pivot in every row.*

Finally, let us introduce the identity matrix, which plays a special role in multiplication. The $n \times n$ **identity matrix** is defined as

$$I_n = \begin{bmatrix} 1 & 0 & \cdots & 0 \\ 0 & 1 & \cdots & 0 \\ \vdots & \vdots & \ddots & \vdots \\ 0 & 0 & \cdots & 1 \end{bmatrix}.$$

The entries in positions (i,i), $i = 1, \ldots, n$, are all equal to 1, while the other entries are all 0. We refer to the entries (i,i), $i = 1, \ldots, n$, as the **main diagonal** of the matrix. The identity matrix is the unique matrix with the property that $I_n\mathbf{x} = \mathbf{x}$ for all $\mathbf{x} \in \mathbb{R}^n$. Note that indeed,

$$I_n\mathbf{x} = x_1 \begin{bmatrix} 1 \\ 0 \\ \vdots \\ 0 \end{bmatrix} + x_2 \begin{bmatrix} 0 \\ 1 \\ \vdots \\ 0 \end{bmatrix} + \cdots + x_n \begin{bmatrix} 0 \\ 0 \\ \vdots \\ 1 \end{bmatrix} = \begin{bmatrix} x_1 \\ x_2 \\ \vdots \\ x_n \end{bmatrix} = \mathbf{x}.$$

We will use the columns of the identity matrix often, so it is worthwhile to introduce the notation

$$\mathbf{e}_1 = \begin{bmatrix} 1 \\ 0 \\ \vdots \\ 0 \end{bmatrix}, \mathbf{e}_2 = \begin{bmatrix} 0 \\ 1 \\ \vdots \\ 0 \end{bmatrix}, \ldots, \mathbf{e}_n = \begin{bmatrix} 0 \\ 0 \\ \vdots \\ 1 \end{bmatrix}.$$

1.5   How to Check Your Work

In Linear Algebra you are going to do a lot of matrix manipulations, and you
are likely to make some calculation errors. One fortunate thing about Linear
Algebra is that, in general, it is not a lot of work to check your work. Let us
start by describing how this can be done with row reduction.

Suppose we start with

$$A = \begin{bmatrix} 4 & 4 & 4 & 4 & 4 & 10 \\ 13 & 7 & -5 & -2 & 13 & 10 \\ 13 & 6 & -8 & 1 & 24 & 15 \\ 3 & 2 & 0 & 1 & 4 & 5 \\ 7 & 6 & 4 & 5 & 8 & 15 \end{bmatrix} = \begin{bmatrix} \mathbf{a}_1 & \cdots & \mathbf{a}_6 \end{bmatrix}$$

and we work hard to compute its reduced row echelon form

$$B = \begin{bmatrix} \boxed{1} & 0 & -2 & 0 & 4 & 0 \\ 0 & \boxed{1} & 3 & 0 & -5 & 0 \\ 0 & 0 & 0 & \boxed{1} & 2 & 0 \\ 0 & 0 & 0 & 0 & 0 & \boxed{1} \\ 0 & 0 & 0 & 0 & 0 & 0 \end{bmatrix} = \begin{bmatrix} \mathbf{b}_1 & \cdots & \mathbf{b}_6 \end{bmatrix}.$$

How can we get some confidence that we did it correctly? Well, it is easy to
see that the columns of B satisfy

$$\mathbf{b}_3 = -2\mathbf{b}_1 + 3\mathbf{b}_2 \quad \text{and} \quad \mathbf{b}_5 = 4\mathbf{b}_1 - 5\mathbf{b}_2 + 2\mathbf{b}_4.$$

The row operations do not change these linear relations between the columns,
so we must also have that

$$\mathbf{a}_3 = -2\mathbf{a}_1 + 3\mathbf{a}_2 \quad \text{and} \quad \mathbf{a}_5 = 4\mathbf{a}_1 - 5\mathbf{a}_2 + 2\mathbf{a}_4.$$

And, indeed

$$\begin{bmatrix} 4 \\ -5 \\ -8 \\ 0 \\ 4 \end{bmatrix} = -2\begin{bmatrix} 4 \\ 13 \\ 13 \\ 3 \\ 7 \end{bmatrix} + 3\begin{bmatrix} 4 \\ 7 \\ 6 \\ 2 \\ 6 \end{bmatrix}, \quad \begin{bmatrix} 4 \\ 13 \\ 24 \\ 4 \\ 8 \end{bmatrix} = 4\begin{bmatrix} 4 \\ 13 \\ 13 \\ 3 \\ 7 \end{bmatrix} - 5\begin{bmatrix} 4 \\ 7 \\ 6 \\ 2 \\ 6 \end{bmatrix} + 2\begin{bmatrix} 4 \\ -2 \\ 1 \\ 1 \\ 5 \end{bmatrix}.$$

Next, suppose you are given

$$A = \begin{bmatrix} 8 & 9 & -6 & -1 \\ 4 & 7 & 2 & -3 \\ -3 & 0 & 9 & -3 \\ 3 & 6 & 3 & -3 \end{bmatrix} \quad \text{and} \quad \mathbf{b} = \begin{bmatrix} 28 \\ 24 \\ 3 \\ 21 \end{bmatrix}.$$

And you solve the equation $A\mathbf{x} = \mathbf{b}$ by finding the reduced row echelon form of the augmented matrix

$$[A \mid \mathbf{b}] = \begin{bmatrix} 8 & 9 & -6 & -1 & 28 \\ 4 & 7 & 2 & -3 & 24 \\ -3 & 0 & 9 & -3 & 3 \\ 3 & 6 & 3 & -3 & 21 \end{bmatrix} \rightarrow \begin{bmatrix} \boxed{1} & 0 & -3 & 1 & -1 \\ 0 & \boxed{1} & 2 & -1 & 4 \\ 0 & 0 & 0 & 0 & 0 \\ 0 & 0 & 0 & 0 & 0 \end{bmatrix}.$$

And you find the solutions

$$\mathbf{x} = \begin{bmatrix} -1 \\ 4 \\ 0 \\ 0 \end{bmatrix} + x_3 \begin{bmatrix} 3 \\ -2 \\ 1 \\ 0 \end{bmatrix} + x_4 \begin{bmatrix} -1 \\ 1 \\ 0 \\ 1 \end{bmatrix}, \quad \text{where } x_3, x_4 \in \mathbb{R} \text{ are free.}$$

How can you check your solution? Well, we need to have that

$$A \begin{bmatrix} -1 \\ 4 \\ 0 \\ 0 \end{bmatrix} = \mathbf{b}, \quad A \begin{bmatrix} 3 \\ -2 \\ 1 \\ 0 \end{bmatrix} = \mathbf{0}, \quad A \begin{bmatrix} -1 \\ 1 \\ 0 \\ 1 \end{bmatrix} = \mathbf{0}.$$

And, indeed

$$\begin{bmatrix} 8 & 9 & -6 & -1 \\ 4 & 7 & 2 & -3 \\ -3 & 0 & 9 & -3 \\ 3 & 6 & 3 & -3 \end{bmatrix} \begin{bmatrix} -1 \\ 4 \\ 0 \\ 0 \end{bmatrix} = \begin{bmatrix} 28 \\ 24 \\ 3 \\ 21 \end{bmatrix},$$

$$\begin{bmatrix} 8 & 9 & -6 & -1 \\ 4 & 7 & 2 & -3 \\ -3 & 0 & 9 & -3 \\ 3 & 6 & 3 & -3 \end{bmatrix} \begin{bmatrix} 3 \\ -2 \\ 1 \\ 0 \end{bmatrix} = \mathbf{0}, \quad \begin{bmatrix} 8 & 9 & -6 & -1 \\ 4 & 7 & 2 & -3 \\ -3 & 0 & 9 & -3 \\ 3 & 6 & 3 & -3 \end{bmatrix} \begin{bmatrix} -1 \\ 1 \\ 0 \\ 1 \end{bmatrix} = \mathbf{0}.$$

In future problems there will also be ways to check your answers. It will be useful to figure out these ways and use them.

1.6   Exercises

Exercise 1.6.1. Are the following matrices in row echelon form? In reduced row echelon form?

(a) $\begin{bmatrix} 1 & 0 & 3 \\ 0 & 1 & 4 \\ 0 & 0 & 0 \end{bmatrix}.^{\dagger}$

(b) $\begin{bmatrix} -2 & -4 & 0 & 4 & 2 \\ 0 & 5 & 6 & 3 & 14 \\ 0 & 0 & 2 & -6 & 6 \end{bmatrix}.$

(c) $\begin{bmatrix} 1 & 0 & 0 & 5 & 2 \\ 0 & 1 & 0 & -16 & 14 \\ 1 & 0 & 1 & 0 & 6 \end{bmatrix}.$

(d) $\begin{bmatrix} 1 & 0 & 0 & 10 & 2 \\ 0 & 0 & 0 & 0 & 0 \\ 0 & 1 & 0 & -3 & 5 \\ 0 & 0 & 1 & 2 & 1 \end{bmatrix}.$

(e) $\begin{bmatrix} 0 & 1 & 0 & -4 & 2 \\ 0 & 0 & 1 & 0 & 0 \\ 0 & 0 & 0 & 0 & 5 \\ 0 & 0 & 0 & 0 & 0 \end{bmatrix}.$

Exercise 1.6.2. Put the following matrices in reduced row echelon form.

(a) $\begin{bmatrix} 1 & 1 & 3 \\ 2 & 1 & 4 \\ 2 & 2 & 6 \end{bmatrix}.$

(b) $\begin{bmatrix} -2 & -4 & 0 & -6 & 2 \\ 1 & 5 & 6 & -6 & 14 \\ -1 & -1 & 2 & -6 & 6 \end{bmatrix}.$

(c) Explain how you can check your computations by hand‡.

Exercise 1.6.3. Determine whether the following systems of linear equations are consistent, and if so find the set of all solutions.

(a) $\begin{cases} 2x_1 + x_2 & = 1, \\ 2x_1 + 2x_2 + x_3 = 0. \end{cases}$

(b) $\begin{cases} x_1 - x_2 & = & 5, \\ x_2 - x_3 & = & 7, \\ x_1 & - x_3 & = & 13. \end{cases}$

---

†To check your answer using computer software, go for instance to live.sympy.org and enter 'A=Matrix(3, 3, [1, 0, 3, 0, 1, 4, 0, 0, 0])' and next enter 'A.is_echelon'.

‡To check your answer using computer software, go for instance to www.wolframalpha.com and enter 'row reduce $\{\{2, -4, 0, -6, 2\}, \{1, 5, 6, -6, 14\}, \{-1, -1, 2, -6, 6\}\}$'.

(c) $\begin{cases} 2x_1 - 2x_2 - x_3 &= 2, \\ x_1 - x_2 - x_3 &= -1. \end{cases}$

(d) $\begin{cases} x_1 + x_2 + x_3 &= 0, \\ 3x_1 + 2x_2 + x_3 &= 0, \\ 4x_1 + 3x_2 + 2x_3 &= 0. \end{cases}$

(e) $\begin{cases} 2x_1 - 2x_2 - x_3 + 11x_4 &= 1, \\ x_1 - x_2 - x_3 + 6x_4 &= 4, \\ x_3 - x_4 &= -7, \\ x_1 - x_2 + 5x_4 &= -3. \end{cases}$

Exercise 1.6.4. Consider the system

$$\begin{cases} x_1 + 2x_2 &= 5, \\ ax_1 + bx_2 &= c. \end{cases}$$

(a) Draw the line $x_1 + 2x_2 = 5$ in $\mathbb{R}^2$.

(b) For which values of a, b, and c, does $ax_1 + bx_2 = c$ give the same line as $x_1 + 2x_2 = 5$?

(c) For which values of a, b, and c, does $ax_1 + bx_2 = c$ give a line parallel to $x_1 + 2x_2 = 5$, but not the same line?

(d) For which values of a, b, and c, does $ax_1 + bx_2 = c$ give a line that intersects $x_1 + 2x_2 = 5$ at a point?

(e) For which values of a, b, and c, does the system have no solution, infinitely many solutions, or a unique solution. How does this compare to the answers under (b), (c) and (d)?

Exercise 1.6.5. Find all solutions to $\begin{bmatrix} 2 & -1 & 2 \\ -1 & 0 & 1 \\ 1 & 1 & 4 \end{bmatrix} \begin{bmatrix} x_1 \\ x_2 \\ x_3 \end{bmatrix} = \begin{bmatrix} 4 \\ -1 \\ 6 \end{bmatrix}$.

Exercise 1.6.6. Find all solutions to the system whose augmented matrix is given by

$$\begin{bmatrix} 2 & 4 & 0 & | & 6 \\ 1 & 2 & 4 & | & 7 \\ -1 & -1 & 4 & | & 5 \end{bmatrix}.$$

Exercise 1.6.7. Prove the following rules for $\mathbf{u}, \mathbf{v} \in \mathbb{R}^n$, and $c, d \in \mathbb{R}$.

(a) $\mathbf{u} + \mathbf{v} = \mathbf{v} + \mathbf{u}$.

(b) $c(\mathbf{u} + \mathbf{v}) = c\mathbf{u} + c\mathbf{v}$.

(c) $c\mathbf{u} + d\mathbf{u} = (c + d)\mathbf{u}$.

Exercise 1.6.8. For the following write, if possible, **b** as a linear combination of $\mathbf{a}_1, \mathbf{a}_2, \mathbf{a}_3$.

(a) $\mathbf{a}_1 = \begin{bmatrix} 1 \\ 1 \\ 2 \\ 1 \end{bmatrix}, \mathbf{a}_2 = \begin{bmatrix} 0 \\ 3 \\ -1 \\ -3 \end{bmatrix}, \mathbf{a}_3 = \begin{bmatrix} -1 \\ 2 \\ -3 \\ 3 \end{bmatrix}, \mathbf{b} = \begin{bmatrix} 0 \\ 0 \\ 0 \\ 1 \end{bmatrix}$.

(b) $\mathbf{a}_1 = \begin{bmatrix} 1 \\ 1 \\ 2 \\ 1 \end{bmatrix}, \mathbf{a}_2 = \begin{bmatrix} 0 \\ 3 \\ 3 \\ -3 \end{bmatrix}, \mathbf{a}_3 = \begin{bmatrix} 0 \\ 2 \\ 2 \\ 3 \end{bmatrix}, \mathbf{b} = \begin{bmatrix} 0 \\ 0 \\ 0 \\ 1 \end{bmatrix}$.

(c) $\mathbf{a}_1 = \begin{bmatrix} 1 \\ -2 \\ 0 \end{bmatrix}, \mathbf{a}_2 = \begin{bmatrix} -3 \\ 0 \\ 9 \end{bmatrix}, \mathbf{a}_3 = \begin{bmatrix} -1 \\ -4 \\ 3 \end{bmatrix}, \mathbf{b} = \begin{bmatrix} 0 \\ -6 \\ 10 \end{bmatrix}$.

Exercise 1.6.9. For the following, determine for which value(s) of h is **b** a linear combination of $\mathbf{a}_1$ and $\mathbf{a}_2$?

(a) $\mathbf{a}_1 = \begin{bmatrix} 2 \\ 1 \\ -1 \end{bmatrix}, \mathbf{a}_2 = \begin{bmatrix} 4 \\ -3 \\ -1 \end{bmatrix}, \mathbf{b} = \begin{bmatrix} 1 \\ h \\ 0 \end{bmatrix}$.

(b) $\mathbf{a}_1 = \begin{bmatrix} 2 \\ 1 \\ 0 \end{bmatrix}, \mathbf{a}_2 = \begin{bmatrix} -4 \\ 3 \\ 2 \end{bmatrix}, \mathbf{b} = \begin{bmatrix} 2 \\ 6 \\ h \end{bmatrix}$.

Exercise 1.6.10. For which h, k and m are there two free variables in the general solution to the equation whose augmented matrix is given by

$$\begin{bmatrix} 1 & 4 & 1 & 5 \\ 2 & k & 2 & 10 \\ h & 8 & 2 & m \end{bmatrix}.$$

Exercise 1.6.11. For which h and k does the system described by the following augmented matrix have 0,1 and ∞ number of solutions?

$$A = \begin{bmatrix} 2 & 4 & 0 & 2 \\ 1 & -2 & h & -1 \\ 0 & -8 & 4 & k \end{bmatrix}.$$

Exercise 1.6.12. Compute the following products.

(a) $\begin{bmatrix} 2 & 2 \\ 2 & -10 \end{bmatrix} \begin{bmatrix} 6 \\ 1 \end{bmatrix}$.

(b) $\begin{bmatrix} 2 & 1 & -1 \\ -2 & 3 & -2 \\ 7 & 5 & -1 \end{bmatrix} \begin{bmatrix} 1 \\ 2 \\ -1 \end{bmatrix}$.§

(c) $\begin{bmatrix} 1 & 1 & 0 & -1 \\ 2 & 1 & 1 & 5 \end{bmatrix} \begin{bmatrix} 1 \\ 0 \\ 2 \\ 1 \end{bmatrix}$.

Exercise 1.6.13. Prove that

$$A(c\mathbf{u}) = cA\mathbf{u},$$

where $A \in \mathbb{R}^{m \times n}$, $\mathbf{u} \in \mathbb{R}^n$, and $c \in \mathbb{R}$.

Exercise 1.6.14. Prove that

$$A(c\mathbf{u} + d\mathbf{v}) = cA\mathbf{u} + dA\mathbf{v},$$

where $A \in \mathbb{R}^{m \times n}$, $\mathbf{u}, \mathbf{v} \in \mathbb{R}^n$, and $c, d \in \mathbb{R}$.

Exercise 1.6.15. Let

$$\begin{bmatrix} -1 & 2 & -3 \\ 2 & -5 & 2 \\ 1 & -4 & -5 \end{bmatrix}.$$

Do the columns of A span $\mathbb{R}^3$?

Exercise 1.6.16. Do the following vectors span $\mathbb{R}^3$?

$$\begin{bmatrix} 1 \\ 2 \\ 0 \end{bmatrix}, \begin{bmatrix} -1 \\ -1 \\ 1 \end{bmatrix}, \begin{bmatrix} 0 \\ 1 \\ 1 \end{bmatrix}, \begin{bmatrix} -1 \\ 2 \\ 1 \end{bmatrix}.$$

Exercise 1.6.17. Let

$$\mathbf{a}_1 = \begin{bmatrix} 2 \\ h \end{bmatrix}, \mathbf{a}_2 = \begin{bmatrix} -4 \\ 10 \end{bmatrix}.$$

For which value(s) of h do $\mathbf{a}_1, \mathbf{a}_2$ span $\mathbb{R}^2$?

Exercise 1.6.18.

For the following sets of vectors in $\mathbb{R}^4$, determine whether they span $\mathbb{R}^4$.

(a) $\begin{bmatrix} 1 \\ 0 \\ 1 \\ 0 \end{bmatrix}, \begin{bmatrix} 1 \\ 1 \\ 0 \\ 0 \end{bmatrix}, \begin{bmatrix} 1 \\ 0 \\ 1 \\ 2 \end{bmatrix}, \begin{bmatrix} 1 \\ 1 \\ 0 \\ 0 \end{bmatrix}, \begin{bmatrix} 1 \\ 3 \\ -2 \\ 0 \end{bmatrix}.$

---

§You can check your answer using for instance sagecell.sagemath.org and enter 'A = matrix([[2,1,-1], [-2,3,-2], [7,5,-1]])' and 'v=vector([1,2,-1])' and 'A*v', and hit Evaluate.

(b) $\begin{bmatrix} 5 \\ -1 \\ 1 \\ 2 \end{bmatrix}, \begin{bmatrix} 1 \\ 7 \\ 3 \\ -4 \end{bmatrix}, \begin{bmatrix} 10 \\ 0 \\ 2 \\ 2 \end{bmatrix}, \begin{bmatrix} 1 \\ 0 \\ 1 \\ 0 \end{bmatrix}.$

(c) $\begin{bmatrix} 4 \\ -1 \\ 1 \\ 2 \end{bmatrix}, \begin{bmatrix} 1 \\ 8 \\ -3 \\ -5 \end{bmatrix}, \begin{bmatrix} 9 \\ -3 \\ 0 \\ 2 \end{bmatrix}.$

Exercise 1.6.19. True or False? Justify each answer.

(i) A homogeneous system of 4 linear equations with 5 unknowns always has infinitely many solutions.

(ii) If $\hat{\mathbf{x}}, \tilde{\mathbf{x}}$ are two solutions to the equation $A\mathbf{x} = \mathbf{b}$, then $\hat{\mathbf{x}} - \tilde{\mathbf{x}}$ is a solution to the homogeneous equation $A\mathbf{x} = \mathbf{0}$.

(iii) A system with fewer linear equations than variables has always a solution.

(iv) If both vectors $\mathbf{x}_1$ and $\mathbf{x}_2$ are solutions to the equation $A\mathbf{x} = \mathbf{b}$, then $\frac{1}{2}(\mathbf{x}_1 + \mathbf{x}_2)$ is also a solution to $A\mathbf{x} = \mathbf{b}$.

(v) The system $A\mathbf{x} = \mathbf{0}$ always has a solution.

(vi) A system with more linear equations than variables does not have a solution.

(vii) A consistent system of 4 equations with 3 unknowns can have infinitely many solutions.

(viii) If $A = \begin{bmatrix} \mathbf{a}_1 & \cdots & \mathbf{a}_4 \end{bmatrix}$, has reduced row echelon form

$$\begin{bmatrix} 1 & 1 & 0 & 0 \\ 0 & 0 & 1 & 1 \\ 0 & 0 & 0 & 0 \\ 0 & 0 & 0 & 0 \end{bmatrix},$$

then $\mathbf{a}_1 = \mathbf{a}_2$ and $\mathbf{a}_3 = \mathbf{a}_4$.

(ix) If the columns of the $m \times n$ matrix A span $\mathbb{R}^m$, then the equation $A\mathbf{x} = \mathbf{b}$ with $\mathbf{b} \in \mathbb{R}^m$ always has a solution.

(x) A system of 3 linear equations with 4 unknowns always has infinitely many solutions.

Exercise 1.6.20. In chemical equations the number of atoms of the reactants and the products need to be balanced. Balancing chemical equations comes down to solving a homogeneous system of linear equations. Consider the reaction

$$C_3H_8 + O_2 \to H_2O + CO_2.$$

Let x_1, x_2, x_3 and x_4 be the coefficients (= the number of molecules) of C_3H_8, O_2, H_2O, and CO_2, respectively. The number of carbon (C), hydrogen (H), and oxygen (O) atoms does not change in the chemical reaction. This leads to the system

$$\begin{cases} 3x_1 &= & x_4, \\ 8x_1 &= & 2x_3, \\ 2x_2 &= & x_3 + 2x_4. \end{cases}$$

Thus we find the equation $A\mathbf{x} = \mathbf{0}$, where $\mathbf{x} = (x_i)_{i=1}^4$ and

$$A = \begin{bmatrix} 3 & 0 & 0 & -1 \\ 8 & 0 & -2 & 0 \\ 0 & 2 & -1 & -2 \end{bmatrix}.$$

Solving the system we find solutions

$$\begin{bmatrix} x_1 \\ x_2 \\ x_3 \\ x_4 \end{bmatrix} = x_4 \begin{bmatrix} 1/3 \\ 5/3 \\ 4/3 \\ 1 \end{bmatrix}.$$

Choosing $x_4 = 3$ (to avoid fractions) we obtain $x_1 = 1$, $x_2 = 5$, $x_3 = 4$, and thus the balanced equation is

$$C_3H_8 + 5O_2 \to 4H_2O + 3CO_2.$$

Balance the following equations.

(a) $KMnO_4 + HCl \to KCl + MnCl_2 + H_2O + Cl_2$.

(b) $C_6H_5COOH + O_2 \to CO_2 + H_2O$.

Exercise 1.6.21. Systems of linear equations appear in analyzing electrical circuits. We illustrate this using the electrical circuit in Figure 1.1, which has resisters and voltage sources. We will set up equations for the currents (indicated by I_j) that flow through the system. We need the following physics laws.

- Kirchoff's first law: at each node the incoming and outgoing current is the same. For instance, at node C we have incoming current $I_1 + (I_2 - I_1)$ and outgoing current I_2, which are indeed equal.

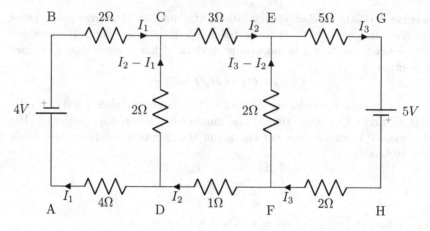

Figure 1.1: An electrical circuit with three loops.

- Kirchhoff's second law: in any closed loop network, the sum of all the voltage drops around a loop equals zero.

- Ohm's law: the current (I in Ampere) through an electrical conductor is directly proportional to the voltage drop (V in Volt). This proportionality is indicated by the resistance R in Ohm. In other words, $V = IR$.

If we apply this to the loop ABCD in Figure 1.1, we obtain that

$$-4 + 2I_1 + 2(-I_2 + I_1) + 4I_1 = 0.$$

Indeed, going from A to B, the voltage increases by 4, and thus the voltage drop is -4. Going from B to C the voltage drop is $2I_1$ as current I_1 flows through the resister with resistance 2 Ohm. From C to D we have that the current $-(I_2 - I_1)$ flows through a resister of 2 Ohm. Finally from D to A we have the current I_1 flows through a resister of 4 Ohm. The loop CEFD gives

$$3I_2 + 2(I_2 - I_3) + I_2 + 2(I_2 - I_1) = 0,$$

and finally the loop $EGHF$ gives

$$5I_3 - 5 + 2I_3 + 2(I_3 - I_2) = 0.$$

Thus we arrive at the system

$$\begin{bmatrix} 8 & -2 & 0 \\ -2 & 8 & -2 \\ 0 & -2 & 9 \end{bmatrix} \begin{bmatrix} I_1 \\ I_2 \\ I_3 \end{bmatrix} = \begin{bmatrix} 4 \\ 0 \\ 5 \end{bmatrix}.$$

Using our favorite electronic computational software, we find $I_1 = \frac{73}{127}, I_2 =$

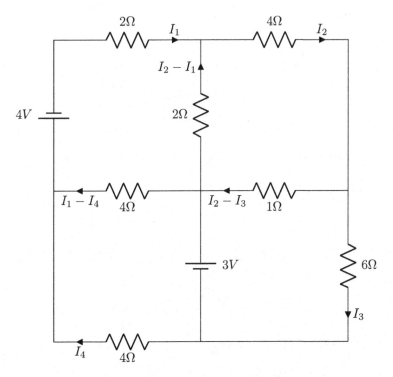

Figure 1.2: An electrical circuit with four loops.

$\frac{38}{127}$, and $I_3 = \frac{79}{127}$. With A as the reference node, we find the voltages (relative to the voltage at A; thus $V_A = 0$),

$$V_B = 4, V_C = \frac{362}{127}, V_E = \frac{248}{127}, V_G = \frac{-147}{127}, V_H = \frac{488}{127}, V_F = \frac{330}{127}, V_D = \frac{292}{127}.$$

(a) Set up the equations for the circuit in Figure 1.2.

(b) For Figure 1.2 compute the currents I_j, $j = 1, \ldots, 4$, and the voltages at the nodes.

2

Subspaces in $\mathbb{R}^n$, Basis and Dimension

CONTENTS

2.1   Subspaces in $\mathbb{R}^n$

We introduced addition and scalar multiplication as operations on vectors. Nonempty subsets of $\mathbb{R}^n$ that are closed under these operations are called subspaces. Here is the definition.

Definition 2.1.1. A set $S \subseteq \mathbb{R}^n$ is called a **subspace** of $\mathbb{R}^n$ if

(i) $S \neq \emptyset$,

(ii) $\mathbf{u}, \mathbf{v} \in S$ and $c, d \in \mathbb{R}$ imply $c\mathbf{u} + d\mathbf{v} \in S$.

It is easy to see that any subspace must contain the zero vector. Indeed, since S is not empty it contains an element $\mathbf{u}$. But then by (ii), we must have that $\mathbf{0} = 0\mathbf{u} + 0\mathbf{u} \in S$. Also, one can check condition (ii) in two steps, by checking closure for addition and closure for scalar multiplication separately. We summarize these observations in the following proposition.

Proposition 2.1.2. *A subset S of $\mathbb{R}^n$ is a subspace of $\mathbb{R}^n$ if and only if*

1. $\mathbf{0} \in S$,

2. *if $\mathbf{u}, \mathbf{v} \in S$, then $\mathbf{u} + \mathbf{v} \in S$ (closure under addition),*

> *3. if $c \in \mathbb{R}$ and $\mathbf{u} \in S$, then $c\mathbf{u} \in S$ (closure under scalar multiplication).*

Let us look at a few examples.

Example 2.1.3. Let $S = \left\{ \begin{bmatrix} x_1 \\ x_2 \end{bmatrix} : x_1 = 2x_2 \right\}$. Is S a subspace of $\mathbb{R}^2$?

Clearly $\begin{bmatrix} 0 \\ 0 \end{bmatrix} \in S$, since $0 = 2 \cdot 0$. Next, suppose that $\mathbf{x} = \begin{bmatrix} x_1 \\ x_2 \end{bmatrix} \in S$ and

$\mathbf{y} = \begin{bmatrix} y_1 \\ y_2 \end{bmatrix} \in S$. Then $x_1 = 2x_2$ and $y_1 = 2y_2$. If we let $c, d \in \mathbb{R}$, then

$$c\mathbf{x} + d\mathbf{y} = \begin{bmatrix} cx_1 + dy_1 \\ cx_2 + dy_2 \end{bmatrix}.$$

Since $2(cx_2 + dy_2) = c(2x_2) + d(2y_2) = cx_1 + dy_1$, we get that $c\mathbf{x} + d\mathbf{y} \in S$. Thus S satisfies (i) and (ii) in the definition, giving that S is a subspace of $\mathbb{R}^2$. □

Example 2.1.4. Let $S = \left\{ \begin{bmatrix} x_1 \\ x_2 \end{bmatrix} : x_1 = x_2^2 \right\}$. Is S a subspace of $\mathbb{R}^2$?

Clearly $\begin{bmatrix} 0 \\ 0 \end{bmatrix} \in S$, since $0 = 0^2$. Next, suppose that $\mathbf{x} = \begin{bmatrix} x_1 \\ x_2 \end{bmatrix} \in S$ and $\mathbf{y} =$

$\begin{bmatrix} y_1 \\ y_2 \end{bmatrix} \in S$. Then $x_1 = x_2^2$ and $y_1 = y_2^2$. If we let $c, d \in \mathbb{R}$, then

$$c\mathbf{x} + d\mathbf{y} = \begin{bmatrix} cx_1 + dy_1 \\ cx_2 + dy_2 \end{bmatrix}.$$

Now $(cx_2 + dy_2)^2 = c^2x_2^2 + 2cdx_2y_2 + d^2y_2^2 = c^2x_1 + 2cdx_2y_2 + d^2y_1$, but does this necessarily equal $cx_1 + dy_1$? The answer is no. So let us look for vectors $\mathbf{x}, \mathbf{y} \in S$ and $c, d \in \mathbb{R}$ where rule (ii) fails. We can take for instance $\mathbf{x} = \begin{bmatrix} 1 \\ 1 \end{bmatrix} = \mathbf{y} \in S$ and $c = 2$, $d = 0$. Then

$$c\mathbf{x} + d\mathbf{y} = \begin{bmatrix} 2 \\ 2 \end{bmatrix} \notin S,$$

because $2 \neq 2^2$. Thus rule (ii) fails. Consequently, S is not a subspace. □

It is important that if a general rule (a 'for all' statement) fails to give a specific example where it goes wrong, as was done in the last example. We call this a **counter example**.

Example 2.1.5. Let $S = \left\{ \mathbf{x} \in \mathbb{R}^3 : x_1 + 2x_2 - 4x_3 = 0 \right\}$. Is S a subspace of $\mathbb{R}^3$?

Clearly $\mathbf{0} \in S$, since $0 + 2 \cdot 0 - 4 \cdot 0 = 0$. Next, suppose that $\mathbf{x}, \mathbf{y} \in S$. Then $x_1 + 2x_2 - 4x_3 = 0$ and $y_1 + 2y_2 - 4y_3 = 0$. If we let $c, d \in \mathbb{R}$, then

$$c\mathbf{x} + d\mathbf{y} = \begin{bmatrix} cx_1 + dy_1 \\ cx_2 + dy_2 \\ cx_3 + dy_3 \end{bmatrix}.$$

To check whether $c\mathbf{x} + d\mathbf{y} \in S$ we calculate

$$cx_1 + dy_1 + 2(cx_2 + dy_2) - 4(cx_3 + dy_3) =$$

$$c(x_1 + 2x_2 - 4x_3) + d(y_1 + 2y_2 - 4y_3) = c \cdot 0 + d \cdot 0 = 0.$$

Thus $c\mathbf{x} + d\mathbf{y} \in S$, and we can conclude that S is a subspace of $\mathbb{R}^3$. $\qquad\square$

It is not hard to see that the subspace in Example 2.1.3 is Span $\left\{ \begin{bmatrix} 2 \\ 1 \end{bmatrix} \right\}$, and that the subspace in Example 2.1.5 is Span $\left\{ \begin{bmatrix} -2 \\ 1 \\ 0 \end{bmatrix}, \begin{bmatrix} 4 \\ 0 \\ 1 \end{bmatrix} \right\}$. In fact, as the next result shows, any span is a subspace.

Theorem 2.1.6. *For $\mathbf{v}_1, \ldots, \mathbf{v}_p \in \mathbb{R}^n$, we have that $S = \mathrm{Span}\,\{\mathbf{v}_1, \ldots, \mathbf{v}_p\}$ is a subspace of $\mathbb{R}^n$.*

Proof. (i) Clearly $\mathbf{0} \in S$ as $c_1 = \cdots = c_p = 0$ gives $\sum_{i=1}^{p} c_i \mathbf{v}_i = \mathbf{0}$; thus $S \neq \emptyset$. (ii) For $\mathbf{u} = c_1 \mathbf{v}_1 + \cdots + c_p \mathbf{v}_p$, $\mathbf{v} = d_1 \mathbf{v}_1 + \cdots + d_p \mathbf{v}_p \in S$,

$$\alpha\mathbf{u} + \beta\mathbf{v} = \alpha(c_1 \mathbf{v}_1 + \cdots + c_p \mathbf{v}_p) + \beta(d_1 \mathbf{v}_1 + \cdots + d_p \mathbf{v}_p) =$$

$$(\alpha c_1 + \beta d_1)\mathbf{v}_1 + \cdots + (\alpha c_p + \beta d_p)\mathbf{v}_p,$$

and thus $\mathbf{u}, \mathbf{v} \in S$ and $\alpha, \beta \in \mathbb{R}$ imply $\alpha\mathbf{u} + \beta\mathbf{v} \in S$. $\qquad\square$

Given two subspaces U and W of $\mathbb{R}^n$, we introduce

$$U + W := \{\mathbf{v} \in \mathbb{R}^n \ : \ \text{there exist } \mathbf{u} \in U \text{ and } \mathbf{w} \in W \text{ so that } \mathbf{v} = \mathbf{u} + \mathbf{w}\},$$

$$U \cap W := \{\mathbf{v} \in \mathbb{R}^n \ : \ \mathbf{v} \in U \text{ and } \mathbf{v} \in W\}.$$

Proposition 2.1.7. *Given two subspaces U and W of $\mathbb{R}^n$, then $U + W$ and $U \cap W$ are also subspaces of $\mathbb{R}^n$.*

Proof. Clearly $\mathbf{0} = \mathbf{0} + \mathbf{0} \in U + W$ as $\mathbf{0} \in U$ and $\mathbf{0} \in W$. Let $\mathbf{v}, \hat{\mathbf{v}} \in U + W$ and $c \in \mathbb{R}$. Then there exist $\mathbf{u}, \hat{\mathbf{u}} \in U$ and $\mathbf{w}, \hat{\mathbf{w}} \in W$ so that $\mathbf{v} = \mathbf{u} + \mathbf{w}$ and $\hat{\mathbf{v}} = \hat{\mathbf{u}} + \hat{\mathbf{w}}$. Then $\mathbf{v} + \hat{\mathbf{v}} = (\mathbf{u} + \mathbf{w}) + (\hat{\mathbf{u}} + \hat{\mathbf{w}}) = (\mathbf{u} + \hat{\mathbf{u}}) + (\mathbf{w} + \hat{\mathbf{w}}) \in U + W$, since $\mathbf{u} + \hat{\mathbf{u}} \in U$ and $\mathbf{w} + \hat{\mathbf{w}} \in W$. Also $c\mathbf{v} = c(\mathbf{u} + \mathbf{w}) = c\mathbf{u} + c\mathbf{w} \in U + W$ as $c\mathbf{u} \in U$ and $c\mathbf{w} \in W$. This proves that $U + W$ is a subspace.

As $\mathbf{0} \in U$ and $\mathbf{0} \in W$, we have that $\mathbf{0} \in U \cap W$. Next, let $\mathbf{v}, \hat{\mathbf{v}} \in U \cap W$ and $c \in \mathbb{R}$. Then $\mathbf{v}, \hat{\mathbf{v}} \in U$, and since U is a subspace, we have $\mathbf{v} + \hat{\mathbf{v}} \in U$. Similarly, $\mathbf{v} + \hat{\mathbf{v}} \in W$. Thus $\mathbf{v} + \hat{\mathbf{v}} \in U \cap W$. Finally, since $\mathbf{v} \in U$ and U is a subspace, $c\mathbf{v} \in U$. Similarly, $c\mathbf{v} \in W$. Thus $c\mathbf{v} \in U \cap W$.   □

When $U \cap W = \{\mathbf{0}\}$, then we refer to $U + W$ as a **direct sum** of U and W, and write $U \dotplus W$.

2.2   Column Space, Row Space and Null Space of a Matrix

We define the **column space** of a matrix A to be the span of the columns of A. We use the notation Col A. For example,

$$\mathrm{Col} \begin{bmatrix} 1 & 3 \\ 4 & -2 \\ -3 & 6 \end{bmatrix} = \mathrm{Span} \left\{ \begin{bmatrix} 1 \\ 4 \\ -3 \end{bmatrix}, \begin{bmatrix} 3 \\ -2 \\ 6 \end{bmatrix} \right\} \subseteq \mathbb{R}^3.$$

As the span of vectors form a subspace, we obtain the following result.

Corollary 2.2.1. *If A is $m \times n$, then Col A is a subspace of $\mathbb{R}^m$.*

We define the **row space** of a matrix A to be the span of the rows of A. We use the notation Row A. For example,

$$\mathrm{Row} \begin{bmatrix} 1 & 3 \\ 4 & -2 \\ -3 & 6 \end{bmatrix} = \mathrm{Span} \left\{ \begin{bmatrix} 1 & 3 \end{bmatrix}, \begin{bmatrix} 4 & -2 \end{bmatrix}, \begin{bmatrix} -3 & 6 \end{bmatrix} \right\} \subseteq \mathbb{R}^{1 \times 2}.$$

In the same way the span of column vectors form a subspace, the span of row vectors form a subspace. We thus obtain:

Corollary 2.2.2. *If A is $m \times n$, then Row A is a subspace of $\mathbb{R}^{1 \times n}$.*

We define the **null space** of a $m \times n$ matrix A by

$$\mathrm{Nul}\, A = \{\mathbf{x} \in \mathbb{R}^n : A\mathbf{x} = \mathbf{0}\}.$$

Let us do an example.

Example 2.2.3. Let us determine the null space of $A = \begin{bmatrix} 1 & 3 & 4 & 7 \\ 3 & 9 & 7 & 6 \end{bmatrix}$. We start with row reduction of the augmented matrix:

$$\begin{bmatrix} 1 & 3 & 4 & 7 & | & 0 \\ 3 & 9 & 7 & 6 & | & 0 \end{bmatrix} \rightarrow \cdots \rightarrow \begin{bmatrix} \boxed{1} & 3 & 0 & -5 & | & 0 \\ 0 & 0 & \boxed{1} & 3 & | & 0 \end{bmatrix}.$$

Thus we find

$$\begin{cases} x_1 &=& -3x_2 + 5x_4 \\ x_2 &=& x_2 \quad \text{(free variable)} \\ x_3 &=& -3x_4 \\ x_4 &=& x_4 \quad \text{(free variable)} \end{cases} \quad \text{or} \quad \begin{bmatrix} x_1 \\ x_2 \\ x_3 \\ x_4 \end{bmatrix} = x_2 \begin{bmatrix} -3 \\ 1 \\ 0 \\ 0 \end{bmatrix} + x_4 \begin{bmatrix} 5 \\ 0 \\ -3 \\ 1 \end{bmatrix}.$$

Thus

$$\text{Nul } A = \text{Span} \left\{ \begin{bmatrix} -3 \\ 1 \\ 0 \\ 0 \end{bmatrix}, \begin{bmatrix} 5 \\ 0 \\ -3 \\ 1 \end{bmatrix} \right\}.$$

□

Notice that in the row reduction, the augmented part will always stay all zeros. Thus we often do not write the augmented part (and just remember in our head that it is there). That is what we were referring to in Interpretation 3 of Section 1.1.

As another example, $\text{Nul } \begin{bmatrix} 1 & 1 \\ 1 & 1 \\ 1 & 1 \end{bmatrix} = \left\{ x_2 \begin{bmatrix} -1 \\ 1 \end{bmatrix} : x_2 \in \mathbb{R} \right\} = \text{Span} \left\{ \begin{bmatrix} -1 \\ 1 \end{bmatrix} \right\}.$

In the above examples the null space of an $m \times n$ matrix is the span of vectors in $\mathbb{R}^n$, and thus a subspace of $\mathbb{R}^n$. This is true in general.

Theorem 2.2.4. *If A is $m \times n$, then* Nul A *is a subspace of* $\mathbb{R}^n$.

Proof. (i) First note that $\mathbf{0} \in \text{Nul } A$, since $A\mathbf{0} = \mathbf{0}$. In particular, Nul $A \neq \emptyset$.

(ii) Let $\mathbf{x}, \mathbf{y} \in \text{Nul } A$ and $\alpha, \beta \in \mathbb{R}$ be arbitrary. Then $A\mathbf{x} = \mathbf{0}$ and $A\mathbf{y} = \mathbf{0}$. But then

$$A(\alpha \mathbf{x} + \beta \mathbf{y}) = \alpha A\mathbf{x} + \beta A\mathbf{y} = \alpha \mathbf{0} + \beta \mathbf{0} = \mathbf{0},$$

and thus $\alpha \mathbf{x} + \beta \mathbf{y} \in \text{Nul } A$. □

It can happen that the null space of a matrix only consists of the zero vector $\mathbf{0}$. Let us look at such an example.

Example 2.2.5. Let us determine the null space of $A = \begin{bmatrix} 1 & 0 & 1 \\ 2 & 1 & 1 \\ 3 & 0 & -2 \\ 1 & 2 & 3 \end{bmatrix}$.

We start with row reduction of A:

$$\begin{bmatrix} \boxed{1} & 0 & 1 \\ 2 & 1 & 1 \\ 3 & 0 & -2 \\ 1 & 2 & 3 \end{bmatrix} \rightarrow \begin{bmatrix} 1 & 0 & 1 \\ 0 & \boxed{1} & -1 \\ 0 & 0 & -5 \\ 0 & 2 & 2 \end{bmatrix} \rightarrow \begin{bmatrix} \boxed{1} & 0 & 1 \\ 0 & \boxed{1} & -1 \\ 0 & 0 & \boxed{-5} \\ 0 & 0 & 4 \end{bmatrix}.$$

We can continue the row reduction, but we do not have to: we can already see what is happening. The third row gives that $-5x_3 = 0$ (remember that the augmented part is $\mathbf{0}$), thus $x_3 = 0$. The second row states $x_2 - x_3 = 0$, and thus $x_2 = x_3 = 0$. The first row gives $x_1 + x_3 = 0$, thus $x_1 = -x_3 = 0$. So the only solution of $A\mathbf{x} = \mathbf{0}$ is

$$\mathbf{x} = \begin{bmatrix} x_1 \\ x_2 \\ x_3 \end{bmatrix} = \begin{bmatrix} 0 \\ 0 \\ 0 \end{bmatrix} = \mathbf{0} \in \mathbb{R}^3.$$

Thus Nul $A = \{\mathbf{0}\}$. $\qquad\qquad\square$

Why does it work out this way? Because there is a pivot in every column of the echelon form of A. As we will see in the next section, we will say in case when Nul $A = \{\mathbf{0}\}$ that the columns of A are linearly independent.

The column space, the row space and the null space are important subspaces associated with a matrix. Let us see how the process of row reduction affects these spaces.

Theorem 2.2.6. *Let A be a $m \times n$ matrix, and let B be obtained from A by elementary row operations. Then*

(i) Row $A =$ Row B,

(ii) Nul $A =$ Nul B,

but

(iii) Col $A \neq$ Col B, in general.

Proof. (i) When we do elementary row operations, we either switch rows, replace row(i) by row(i) $+ \alpha$ row(k), $i \neq k$, or multiply a row by a nonzero scalar. None of these operations change the row space.

(ii) In Chapter 1 we have seen that elementary row reductions do not change the set of solutions of the system. For the null space we solve the system $A\mathbf{x} = \mathbf{0}$, which has the same solution set as $B\mathbf{x} = \mathbf{0}$.

(iii) Here it suffices to come up with an example. Let us take $A = \begin{bmatrix} 1 \\ 1 \end{bmatrix}$. Then Col $A = \text{Span}\left\{ \begin{bmatrix} 1 \\ 1 \end{bmatrix} \right\}$. Row reducing the matrix A, we get the matrix $B = \begin{bmatrix} 1 \\ 0 \end{bmatrix}$. Then Col $B = \text{Span}\left\{ \begin{bmatrix} 1 \\ 0 \end{bmatrix} \right\}$. Clearly , Col $A \neq$ Col B. $\qquad\square$

While row reducing changes the column space, there is still some use in doing the row reductions when we are interested in the column space. Let us consider the following matrix and its reduced row echelon form

$$A = \begin{bmatrix} \mathbf{a}_1 & \cdots & \mathbf{a}_4 \end{bmatrix} = \begin{bmatrix} 1 & 3 & 4 & 7 \\ 3 & 9 & 7 & 6 \\ 1 & 3 & 4 & 7 \end{bmatrix} \rightarrow \cdots \rightarrow \begin{bmatrix} \boxed{1} & 3 & 0 & -5 \\ 0 & 0 & \boxed{1} & 3 \\ 0 & 0 & 0 & 0 \end{bmatrix}.$$

From the reduced row echelon form we can see that $\mathbf{a}_2 = 3\mathbf{a}_1$ and $\mathbf{a}_4 = -5\mathbf{a}_1 + 3\mathbf{a}_3$. Thus Col A equals

$$\text{Span}\{\mathbf{a}_1, \mathbf{a}_2, \mathbf{a}_3, \mathbf{a}_4\} = \text{Span}\{\mathbf{a}_1, 3\mathbf{a}_1, \mathbf{a}_3, -5\mathbf{a}_1 + 3\mathbf{a}_3\} = \text{Span}\{\mathbf{a}_1, \mathbf{a}_3\},$$

since $\mathbf{a}_2 = 3\mathbf{a}_1$ and $\mathbf{a}_4 = -5\mathbf{a}_1 + 3\mathbf{a}_3$ are in $\text{Span}\{\mathbf{a}_1, \mathbf{a}_3\}$. We now find that

$$\text{Col } A = \text{Span}\left\{ \begin{bmatrix} 1 \\ 3 \\ 1 \end{bmatrix}, \begin{bmatrix} 4 \\ 7 \\ 4 \end{bmatrix} \right\}.$$

So, if we are interested in finding the fewest number of vectors that span the column space, the reduced row echelon form is helpful. We will be using this when we are talking about finding a basis for the column space.

2.3   Linear Independence

Definition 2.3.1. A set of vectors $\{\mathbf{v}_1, \ldots, \mathbf{v}_k\}$ in $\mathbb{R}^n$ is said to be **linearly independent** if the vector equation

$$x_1\mathbf{v}_1 + \cdots + x_k\mathbf{v}_k = \mathbf{0}$$

only has the solution $x_1 = x_2 = \cdots = x_k = 0$ (the **trivial solution**). If a set of vectors is **not** linearly independent, then the set of vectors is called **linearly dependent**.

Let us do some examples.

Example 2.3.2. Is

$$\left\{ \begin{bmatrix} 1 \\ -5 \\ 3 \\ 1 \end{bmatrix}, \begin{bmatrix} 1 \\ 4 \\ -3 \\ 0 \end{bmatrix}, \begin{bmatrix} 1 \\ -2 \\ 0 \\ 3 \end{bmatrix} \right\}$$

linearly independent or dependent?
Let us row reduce the matrix

$$A = \begin{bmatrix} \boxed{1} & 1 & 1 \\ -5 & 4 & -2 \\ 3 & -3 & 0 \\ 1 & 0 & 3 \end{bmatrix} \rightarrow \begin{bmatrix} \boxed{1} & 1 & 1 \\ 0 & 9 & 3 \\ 0 & -6 & -3 \\ 0 & -1 & 2 \end{bmatrix} \rightarrow \begin{bmatrix} \boxed{1} & 1 & 1 \\ 0 & \boxed{1} & -2 \\ 0 & 0 & \boxed{-15} \\ 0 & 0 & 21 \end{bmatrix}.$$

We are interested in solving $A\mathbf{x} = \mathbf{0}$. The third row gives that $-15x_3 = 0$, thus $x_3 = 0$. The second row states $x_2 - 2x_3 = 0$, and thus $x_2 = 2x_3 = 0$. The first row gives $x_1 + x_2 + x_3 = 0$, thus $x_1 = -x_2 - x_3 = 0$. So the only solution of $A\mathbf{x} = \mathbf{0}$ is $\mathbf{x} = \mathbf{0}$. Thus the columns of A are linearly independent. $\square$

Example 2.3.3. Is

$$\left\{ \begin{bmatrix} 0 \\ 2 \\ 2 \end{bmatrix}, \begin{bmatrix} 1 \\ -2 \\ 0 \end{bmatrix}, \begin{bmatrix} 3 \\ -4 \\ 2 \end{bmatrix} \right\}$$

linearly independent or dependent?
Let us row reduce the matrix

$$A = \begin{bmatrix} 0 & 1 & 3 \\ 2 & -2 & -4 \\ 2 & 0 & 2 \end{bmatrix} \rightarrow \begin{bmatrix} \boxed{2} & 0 & 2 \\ 0 & -2 & -6 \\ 0 & 1 & 3 \end{bmatrix} \rightarrow \begin{bmatrix} \boxed{1} & 0 & 1 \\ 0 & \boxed{1} & 3 \\ 0 & 0 & 0 \end{bmatrix}.$$

We see that $\mathbf{a}_3 = \mathbf{a}_1 + 3\mathbf{a}_2$. Thus the columns of A are linearly dependent. Indeed,

$$x_1 \begin{bmatrix} 0 \\ 2 \\ 2 \end{bmatrix} + x_2 \begin{bmatrix} 1 \\ -2 \\ 0 \end{bmatrix} + x_3 \begin{bmatrix} 3 \\ -4 \\ 2 \end{bmatrix} = \mathbf{0}$$

when $x_1 = 1$, $x_2 = 3$, and $x_3 = -1$. Thus the vector equation has a non-trivial solution, and the vectors are linearly dependent. $\square$

The difference in the above examples is that in Example 2.3.2 all the columns in the matrix A are pivot columns, while in Example 2.3.3 not all columns in A are pivot columns. This is a general principle. We have the following result.

Theorem 2.3.4. *Let* $\{a_1, \ldots, a_k\}$ *be vectors in* $\mathbb{R}^n$. *Put* $A = \begin{bmatrix} a_1 & \cdots & a_k \end{bmatrix}$. *The following are equivalent.*

(i) The set $\{a_1, \ldots, a_k\}$ *is linearly independent.*

(ii) The homogeneous equation $Ax = 0$ *only has the solution* $x = 0$.

(iii) Nul $A = \{0\}$.

(iv) Every column of A *is a pivot column. That is, in the row (reduced) echelon form of* A *every column has a pivot.*

Proof. (i)$\leftrightarrow$(ii)$\leftrightarrow$(iii). The equation $Ax = 0$ is the same as the equation

$$x_1 a_1 + \cdots + x_k a_k = 0. \tag{2.1}$$

Thus (2.1) having the only solution $x_1 = \cdots = x_k = 0$ is the same as $Ax = 0$ only having the solution $x = 0$. Next, Nul A consists of all solution of $Ax = 0$, and thus Nul $A = \{0\}$ is exactly the statement that $Ax = 0$ only has the solution $x = 0$.

(ii)$\rightarrow$(iv). If $Ax = 0$ only has the solution $x = 0$, then there can not be any free variables, and thus all the columns of A are pivot columns.

(iv)$\rightarrow$(ii). If all the columns of A are pivot columns, then there are no free variables and thus $Ax = 0$ has the unique solution $x = 0$. $\quad\square$

Corollary 2.3.5. *Consider a set of vectors* $\{v_1, \ldots, v_k\}$ *in* $\mathbb{R}^n$. *If* $k > n$, *then* $\{v_1, \ldots, v_k\}$ *is necessarily linearly dependent.*

Proof. If $k > n$, the matrix $\begin{bmatrix} v_1 & \cdots & v_k \end{bmatrix}$ has more columns than rows, so not all columns are pivot columns. $\quad\square$

We can refine the statement of Theorem 2.3.4 as follows.

Theorem 2.3.6. *Any set of pivot columns of a matrix is linearly independent.*

Proof. Let B be a matrix consisting columns $i_1, \ldots, i_k$ of A, which are pivot columns of A. Let us perform a row operations on A and get the matrix C, and perform the same row operations on B and get the matrix D. Then Proposition 1.2.4 yields that D is obtained from C by taking columns $i_1, \ldots, i_k$ from C. But then it follows that the reduced row echelon form of B has a pivot in every column. Thus by Theorem 2.3.4 the columns of B are linearly independent. $\quad\square$

2.4　Basis

We have already seen that the span of a set of vectors is a subspace. In this section we are interested in finding a minimal number of vectors that span a given subspace. The key to this is to add the condition of linear independence. This leads to the notion of a basis.

Definition 2.4.1. A set of vectors $\{v_1, \ldots, v_p\}$ is a **basis** for a subspace H if

1. the vectors span H, i.e., $H = \text{Span}\{v_1, \ldots, v_p\}$, and

2. the set $\{v_1, \ldots, v_p\}$ is linearly independent.

Let us revisit an earlier example.

Example 2.4.2. Let

$$A = \begin{bmatrix} a_1 & \cdots & a_4 \end{bmatrix} = \begin{bmatrix} 1 & 3 & 4 & 7 \\ 3 & 9 & 7 & 6 \\ 1 & 3 & 4 & 7 \end{bmatrix}.$$

The reduced row echelon form of A is

$$\begin{bmatrix} \boxed{1} & 3 & 0 & -5 \\ 0 & 0 & \boxed{1} & 3 \\ 0 & 0 & 0 & 0 \end{bmatrix}.$$

From the reduced row echelon form we can see that $a_2 = 3a_1$ and $a_4 = -5a_1 + 3a_3$. Thus Col A equals

$$\text{Span}\{a_1, a_2, a_3, a_4\} = \text{Span}\{a_1, 3a_1, a_3, -5a_1 + 3a_3\} = \text{Span}\{a_1, a_3\},$$

since $a_2 = 3a_1$ and $a_4 = -5a_1 + 3a_3$ are in $\text{Span}\{a_1, a_3\}$. We now find that

$$\text{Col } A = \text{Span}\left\{ \begin{bmatrix} 1 \\ 3 \\ 1 \end{bmatrix}, \begin{bmatrix} 4 \\ 7 \\ 4 \end{bmatrix} \right\}.$$

These two vectors are linearly independent due to Theorem 2.3.6, as they are pivot columns of A. Thus

$$\left\{ \begin{bmatrix} 1 \\ 3 \\ 1 \end{bmatrix}, \begin{bmatrix} 4 \\ 7 \\ 4 \end{bmatrix} \right\}$$

is a basis for Col A.

Next, solving $A\mathbf{x} = \mathbf{0}$ we obtain

$$\text{Nul } A = \text{Span} \left\{ \begin{bmatrix} -3 \\ 1 \\ 0 \\ 0 \end{bmatrix}, \begin{bmatrix} 5 \\ 0 \\ -3 \\ 1 \end{bmatrix} \right\}.$$

These two vectors are linearly independent, as

$$c_1 \begin{bmatrix} -3 \\ 1 \\ 0 \\ 0 \end{bmatrix} + c_2 \begin{bmatrix} 5 \\ 0 \\ -3 \\ 1 \end{bmatrix} = \mathbf{0} \quad \Rightarrow \quad \begin{cases} \cdots \\ c_1 = 0, \\ \cdots \\ c_2 = 0. \end{cases}$$

Thus

$$\left\{ \begin{bmatrix} -3 \\ 1 \\ 0 \\ 0 \end{bmatrix}, \begin{bmatrix} 5 \\ 0 \\ -3 \\ 1 \end{bmatrix} \right\}$$

is a basis for Nul A.

Finally, from the reduced row echelon form we can immediately see that

$$\left\{ \begin{bmatrix} 1 & 3 & 0 & -5 \end{bmatrix}, \begin{bmatrix} 0 & 0 & 1 & 3 \end{bmatrix} \right\}$$

is a basis for Row A. Indeed by Theorem 2.2.6(i) they span the rowspace of A, and the equation

$$c_1 \begin{bmatrix} 1 & 3 & 0 & -5 \end{bmatrix} + c_2 \begin{bmatrix} 0 & 0 & 1 & 3 \end{bmatrix} = \begin{bmatrix} 0 & 0 & 0 & 0 \end{bmatrix}$$

yields in the first and third component that $c_1 = 0$ and $c_2 = 0$. Thus these two row vectors are linearly independent.   □

A subspace W has in general many different bases. However, each basis of W always has the same number of vectors. We will prove this soon. First we need the following result.

Proposition 2.4.3. *Let $\mathcal{B} = \{\mathbf{v}_1, \ldots, \mathbf{v}_n\}$ be a basis for the subspace W, and let $\mathcal{C} = \{\mathbf{w}_1, \ldots, \mathbf{w}_m\}$ be a set of vectors in W with $m > n$. Then $\mathcal{C}$ is linearly dependent.*

Proof. As $\mathcal{B}$ is a basis, we can express each $\mathbf{w}_j$ as a linear combination of elements of $\mathcal{B}$:

$$\mathbf{w}_j = a_{1j}\mathbf{v}_1 + \cdots + a_{nj}\mathbf{v}_n, \ j = 1, \ldots, m.$$

The matrix $A = (a_{ij})_{i=1, j=1}^{n, \ m}$ has more columns than rows (and thus a non-pivot column), so the equation $A\mathbf{c} = \mathbf{0}$ has a nontrivial solution $\mathbf{c} = \begin{bmatrix} c_1 \\ \vdots \\ c_m \end{bmatrix} \neq \mathbf{0}$. But then it follows that

$$\sum_{j=1}^{m} c_j \mathbf{w}_j = \sum_{j=1}^{m} [c_j \sum_{i=1}^{n} a_{ij} \mathbf{v}_i)] = \sum_{i=1}^{n} (\sum_{j=1}^{m} a_{ij} c_j) \mathbf{v}_i = \sum_{i=1}^{n} 0 \mathbf{v}_i = \mathbf{0}.$$

Thus a nontrivial linear combination of elements of $\mathcal{C}$ equals $\mathbf{0}$, and thus $\mathcal{C}$ is linearly dependent. $\qquad\square$

Theorem 2.4.4. *Let $\mathcal{B} = \{\mathbf{v}_1, \ldots, \mathbf{v}_n\}$ and $\mathcal{C} = \{\mathbf{w}_1, \ldots, \mathbf{w}_m\}$ be bases for the subspace W. Then $n = m$.*

Proof. Suppose that $n \neq m$. If $m > n$ it follows by Proposition 2.4.3 that $\mathcal{C}$ is linearly dependent. But then $\mathcal{C}$ is not a basis. Contradiction. Similarly, if $n > m$ we obtain a contradiction. Thus $m = n$. $\qquad\square$

Definition 2.4.5. For a subspace W, we define its **dimension** (notation: $\dim W$) to be the number of elements in a basis for W. The 'trivial' subspace $H = \{\mathbf{0}\}$ has, by definition, dimension 0.

Corollary 2.4.6. *Suppose that $\widehat{W} \subseteq W$ for subspaces $\widehat{W}$ and W both with dimension n. Then $W = \widehat{W}$.*

Proof. If $\widehat{W} \neq W$, then there exists a $\mathbf{w} \in W$ with $w \notin \widehat{W}$. Let $\mathcal{B} = \{\mathbf{v}_1, \ldots, \mathbf{v}_n\}$ be a basis for $\widehat{W}$. But now $\{\mathbf{w}, \mathbf{v}_1, \ldots, \mathbf{v}_n\} \subset W$ is linearly independent (see Exercise 2.6.8). But since, $\dim W = n$, due to Proposition 2.4.3 there can not be $n+1$ linearly independent vectors in W. Contradiction. Thus $\widehat{W} = W$. $\qquad\square$

Example 2.4.7. The vectors

$$\mathbf{e}_1 = \begin{bmatrix} 1 \\ 0 \\ \vdots \\ 0 \end{bmatrix}, \mathbf{e}_2 = \begin{bmatrix} 0 \\ 1 \\ \vdots \\ 0 \end{bmatrix}, \ldots, \mathbf{e}_n = \begin{bmatrix} 0 \\ 0 \\ \vdots \\ 1 \end{bmatrix},$$

form a basis for $\mathbb{R}^n$. Thus $\dim \mathbb{R}^n = n$. We call this the standard basis for $\mathbb{R}^n$. $\qquad\square$

Example 2.4.2 is representative of what is true in general regarding the dimension and bases of the column, null, and row space of a matrix. The result is as follows.

Theorem 2.4.8. *Let A be an $m \times n$ matrix. Then*

   (i) The pivot columns of A form a basis for Col A.

   (ii) dim Col A *is equal to the number of pivot columns of A.*

   (iii) The dimension of the null space of A equals the number of free variables in the equation $A\mathbf{x} = \mathbf{0}$; that is, dim Nul A is equal to the number of non-pivot columns of A.

   (iv) dim Col $A +$ dim Nul $A = n =$ *the number of columns of A.*

   (v) The nonzero columns in the reduced row echelon form of A form a basis for Row A.

   (vi) dim Row $A =$ dim Col A.

Proof. (i) The non-pivot columns of A are linear combinations of the pivot columns, so the span of the columns of A is not changed by removing the non-pivot columns. By Theorem 2.3.6 the pivot columns of A are linearly independent, showing that they form a basis for Col A.

(ii) follows directly from (i).

(iii) In solving the equation $A\mathbf{x} = \mathbf{0}$, the free variables correspond to the non-pivot columns, which are columns $j_1, \ldots, j_r$, say. Then the general solution is given by $\mathbf{x} = \sum_{s=1}^{r} x_{j_s} \mathbf{f}_s$ for certain vectors $f_1, \ldots f_r$. The vector f_s has a 1 in position j_s and zeros in the positions j_l for $l \neq s$. Thus it is easy to see that $\{f_1, \ldots f_r\}$ is linearly independent. But then $\{f_1, \ldots f_r\}$ is a basis for Nul A. Thus dim Nul $A = r =$ the number of non-pivot columns of A.

(iv) follows as every column of A is either a pivot column or a non-pivot column.

(v) By Theorem 2.2.6(i) the nonzero rows of the reduced row echelon form of A span Row A. These rows are easily seen to be linearly independent since in the pivot columns exactly one of them has a 1 and the others have a zero. Thus they form a basis for Row A.

(vi) Since the number of pivots equals both dim Col A and dim Row A, equality follows. $\qquad\square$

We define the **rank** of a matrix A as the dimension of the column space of A:

$$\text{rank } A := \dim \text{Col } A.$$

Theorem 2.4.8(iv) gives the following.

Corollary 2.4.9. (The rank theorem)

> rank A + dim Nul A = the number of columns of A.

The dimension of the null space is sometimes also called the **nullity**; that is

$$\text{nullity } (A) := \dim \text{Nul } A.$$

Thus one may restate the rank theorem as

rank A + nullity A = the number of columns of A.

Let us next do an example involving the sum and intersection of two subspaces.

Example 2.4.10. Consider the following subspaces of $\mathbb{R}^n$:

$$U = \text{Span}\left\{ \begin{bmatrix} 1 \\ 0 \\ 2 \\ 1 \end{bmatrix}, \begin{bmatrix} 1 \\ 1 \\ 1 \\ 1 \end{bmatrix} \right\}, W = \text{Span}\left\{ \begin{bmatrix} 4 \\ 2 \\ 2 \\ 0 \end{bmatrix}, \begin{bmatrix} 2 \\ 0 \\ 2 \\ 0 \end{bmatrix} \right\}.$$

Find bases for $U \cap W$ and $U + W$.

Vectors in $U \cap W$ are of the form

$$x_1 \begin{bmatrix} 1 \\ 0 \\ 2 \\ 1 \end{bmatrix} + x_2 \begin{bmatrix} 1 \\ 1 \\ 1 \\ 1 \end{bmatrix} = x_3 \begin{bmatrix} 4 \\ 2 \\ 2 \\ 0 \end{bmatrix} + x_4 \begin{bmatrix} 2 \\ 0 \\ 2 \\ 0 \end{bmatrix}. \tag{2.2}$$

Setting up the homogeneous system of linear equations, and subsequently row reducing, we get

$$\begin{bmatrix} 1 & 1 & -4 & -2 \\ 0 & 1 & -2 & 0 \\ 2 & 1 & -2 & -2 \\ 1 & 1 & 0 & 0 \end{bmatrix} \rightarrow \begin{bmatrix} 1 & 1 & -4 & -2 \\ 0 & 1 & -2 & 0 \\ 0 & -1 & 6 & 2 \\ 0 & 0 & 4 & 2 \end{bmatrix} \rightarrow \begin{bmatrix} 1 & 1 & -4 & -2 \\ 0 & 1 & -2 & 0 \\ 0 & 0 & 4 & 2 \\ 0 & 0 & 0 & 0 \end{bmatrix}.$$

This gives that x_4 is free and $x_3 = -\frac{x_4}{2}$. Plugging this into the right-hand side of (2.2) gives

$$-\frac{x_4}{2}\begin{bmatrix} 4 \\ 2 \\ 2 \\ 0 \end{bmatrix} + x_4 \begin{bmatrix} 2 \\ 0 \\ 2 \\ 0 \end{bmatrix} = x_4 \begin{bmatrix} 0 \\ -1 \\ 1 \\ 0 \end{bmatrix}$$

as a typical element of $U \cap W$. So

$$\left\{ \begin{bmatrix} 0 \\ -1 \\ 1 \\ 0 \end{bmatrix} \right\}$$

is a basis for $U \cap W$.

Notice that

$$U + W = \text{Span} \left\{ \begin{bmatrix} 1 \\ 0 \\ 2 \\ 1 \end{bmatrix}, \begin{bmatrix} 1 \\ 1 \\ 1 \\ 1 \end{bmatrix}, \begin{bmatrix} 4 \\ 2 \\ 2 \\ 0 \end{bmatrix}, \begin{bmatrix} 2 \\ 0 \\ 2 \\ 0 \end{bmatrix} \right\}.$$

From the row reductions above, we see that the fourth vector is a linear combination of the first three, while the first three are linearly independent. Thus a basis for $U + W$ is

$$\left\{ \begin{bmatrix} 1 \\ 0 \\ 2 \\ 1 \end{bmatrix}, \begin{bmatrix} 1 \\ 1 \\ 1 \\ 1 \end{bmatrix}, \begin{bmatrix} 4 \\ 2 \\ 2 \\ 0 \end{bmatrix} \right\}.$$

Notice that

$$\dim(U + W) = 3 = 2 + 2 - 1 = \dim U + \dim W - \dim(U \cap W).$$

In Theorem 5.3.14 we will see that this holds in general.

$\square$

2.5 Coordinate Systems

If you have a subspace H of $\mathbb{R}^m$ with $\dim H = n$, then you can actually view H as a copy of $\mathbb{R}^n$. For example a plane in $\mathbb{R}^3$ can be viewed as a copy of $\mathbb{R}^2$.

The process requires settling on a basis $\mathcal{B}$ for the subspace H. The statement is as follows.

Theorem 2.5.1. *Let $\mathcal{B} = \{v_1, \ldots, v_n\}$ be a basis for the subspace H. Then for each $v \in H$ there exists unique $c_1, \ldots, c_n \in \mathbb{R}$ so that*

$$v = c_1 v_1 + \cdots + c_n v_n. \tag{2.3}$$

Proof. Let $v \in H$. As Span $\mathcal{B} = H$, we have that $v = c_1 v_1 + \cdots + c_n v_n$ for some $c_1, \ldots, c_n \in \mathbb{R}$. Suppose that we also have $v = d_1 v_1 + \cdots + d_n v_n$ for some $d_1, \ldots, d_n \in \mathbb{R}$. Then

$$0 = v - v = \sum_{j=1}^{n} c_j v_j - \sum_{j=1}^{n} d_j v_j = (c_1 - d_1)v_1 + \cdots + (c_n - d_n)v_n.$$

As $\{v_1, \ldots, v_n\}$ is linearly independent, we must have $c_1 - d_1 = 0, \ldots, c_n - d_n = 0$. This yields $c_1 = d_1, \ldots, c_n = d_n$, giving uniqueness. $\square$

When (2.3) holds, we say that $c_1, \ldots, c_n$ are the **coordinates of v relative to the basis** $\mathcal{B}$, and we write

$$[v]_\mathcal{B} = \begin{bmatrix} c_1 \\ \vdots \\ c_n \end{bmatrix}.$$

Thus, when $\mathcal{B} = \{v_1, \ldots, v_n\}$ we have

$$v = c_1 v_1 + \cdots + c_n v_n \Leftrightarrow [v]_\mathcal{B} = \begin{bmatrix} c_1 \\ \vdots \\ c_n \end{bmatrix}. \tag{2.4}$$

Example 2.5.2. The vectors $v_1 = \begin{bmatrix} 1 \\ 1 \end{bmatrix}$, $v_2 = \begin{bmatrix} 1 \\ -1 \end{bmatrix}$ form a basis $\mathcal{B}$ of $\mathbb{R}^2$. Let $v = \begin{bmatrix} 7 \\ 1 \end{bmatrix}$. Find $[v]_\mathcal{B} = \begin{bmatrix} c_1 \\ c_2 \end{bmatrix}$.

Thus we need to solve the system

$$c_1 \begin{bmatrix} 1 \\ 1 \end{bmatrix} + c_2 \begin{bmatrix} 1 \\ -1 \end{bmatrix} = \begin{bmatrix} 7 \\ 1 \end{bmatrix}.$$

We do this by row reducing the augmented matrix

$$\begin{bmatrix} 1 & 1 & | & 7 \\ 1 & -1 & | & 1 \end{bmatrix} \to \begin{bmatrix} 1 & 1 & | & 7 \\ 0 & -2 & | & -6 \end{bmatrix} \to \begin{bmatrix} 1 & 1 & | & 7 \\ 0 & 1 & | & 3 \end{bmatrix} \to \begin{bmatrix} 1 & 0 & | & 4 \\ 0 & 1 & | & 3 \end{bmatrix},$$

which gives that $[\mathbf{v}]_{\mathcal{B}} = \begin{bmatrix} 4 \\ 3 \end{bmatrix}$. You can see below that to get to $\mathbf{v}$ one makes 4 steps in the direction of $\mathbf{v}_1$ and then 3 steps in the direction of $\mathbf{v}_2$.

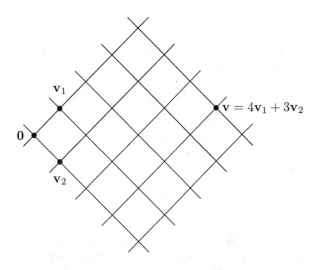

Figure 2.1: Illustration of a basis in $\mathbb{R}^2$.

□

Example 2.5.3. Let $H = \text{Span } \mathcal{B}$ and $\mathbf{v}$ be given by

$$\mathcal{B} = \left\{ \begin{bmatrix} 1 \\ 1 \\ 1 \\ 0 \end{bmatrix}, \begin{bmatrix} 1 \\ 2 \\ 3 \\ 0 \end{bmatrix}, \begin{bmatrix} 1 \\ 3 \\ 6 \\ 1 \end{bmatrix} \right\}, \mathbf{v} = \begin{bmatrix} 3 \\ 5 \\ 8 \\ 1 \end{bmatrix}. \text{ Find } [\mathbf{v}]_{\mathcal{B}} = \begin{bmatrix} c_1 \\ c_2 \\ c_3 \end{bmatrix} = ?$$

We need to solve for c_1, c_2, c_3 in

$$c_1 \begin{bmatrix} 1 \\ 1 \\ 1 \\ 0 \end{bmatrix} + c_2 \begin{bmatrix} 1 \\ 2 \\ 3 \\ 0 \end{bmatrix} + c_3 \begin{bmatrix} 1 \\ 3 \\ 6 \\ 1 \end{bmatrix} = \begin{bmatrix} 3 \\ 5 \\ 8 \\ 1 \end{bmatrix}.$$

Setting up the augmented matrix and row reducing gives

$$\begin{bmatrix} 1 & 1 & 1 & 3 \\ 1 & 2 & 3 & 5 \\ 1 & 3 & 6 & 8 \\ 0 & 0 & 1 & 1 \end{bmatrix} \rightarrow \begin{bmatrix} 1 & 1 & 1 & 3 \\ 0 & 1 & 2 & 2 \\ 0 & 2 & 5 & 5 \\ 0 & 0 & 1 & 1 \end{bmatrix} \rightarrow$$

$$\begin{bmatrix} 1 & 1 & 1 & 3 \\ 0 & 1 & 2 & 2 \\ 0 & 0 & 1 & 1 \\ 0 & 0 & 1 & 1 \end{bmatrix} \rightarrow \begin{bmatrix} 1 & 0 & 0 & 2 \\ 0 & 1 & 0 & 0 \\ 0 & 0 & 1 & 1 \\ 0 & 0 & 0 & 0 \end{bmatrix}. \text{ Thus } [\mathbf{v}]_{\mathcal{B}} = \begin{bmatrix} 2 \\ 0 \\ 1 \end{bmatrix}.$$

$\square$

Clearly, when $\mathbf{v} = c_1\mathbf{v}_1 + \cdots + c_n\mathbf{v}_n$, $\mathbf{w} = d_1\mathbf{v}_1 + \cdots + d_n\mathbf{v}_n$, then $\mathbf{v} + \mathbf{w} = \sum_{j=1}^{n}(c_j + d_j)\mathbf{v}_j$, and thus

$$[\mathbf{v} + \mathbf{w}]_{\mathcal{B}} = \begin{bmatrix} c_1 + d_1 \\ \vdots \\ c_n + d_n \end{bmatrix} = [\mathbf{v}]_{\mathcal{B}} + [\mathbf{w}]_{\mathcal{B}}.$$

Similarly,

$$[\alpha\mathbf{v}]_{\mathcal{B}} = \begin{bmatrix} \alpha c_1 \\ \vdots \\ \alpha c_n \end{bmatrix} = \alpha[\mathbf{v}]_{\mathcal{B}}.$$

We conclude that adding two vectors in the n dimensional subspace H corresponds to adding their corresponding coordinate vectors (which are both with respect to the basis $\mathcal{B}$), and that multiplying a vector in H by a scalar corresponds to multiplying the corresponding coordinate vector by the scalar. Thus, if we represent every element in H by its coordinate vector with respect to the basis $\mathcal{B}$, we can treat the subspace H as a copy of $\mathbb{R}^n$.

2.6   Exercises

Exercise 2.6.1. Draw the following subsets of $\mathbb{R}^2$ and explain why they are not subspaces of $\mathbb{R}^2$.

(a) $\left\{ \begin{bmatrix} x_1 \\ x_2 \end{bmatrix} : x_1 x_2 \geq 0 \right\}$.

(b) $\left\{ \begin{bmatrix} x_1 \\ x_2 \end{bmatrix} : x_1 + 2x_2 \leq 0 \right\}$.

(c) $\left\{ \begin{bmatrix} x_1 \\ x_2 \end{bmatrix} : |x_1 + x_2| \leq 1 \right\}$.

Exercise 2.6.2. For the following sets H prove or disprove that it is a subspace. When H is a subspace, find a basis for it.

(a) $H = \left\{ \begin{bmatrix} x_1 \\ x_2 \\ x_3 \end{bmatrix} : 2x_1 - 3x_2 + 5x_3 = 0 \right\} \subset \mathbb{R}^3$.

(b) $H = \left\{ \begin{bmatrix} x_1 \\ x_2 \end{bmatrix} : x_1 + x_2^2 = 0 \right\} \subset \mathbb{R}^2$.

(c) $H = \left\{ \begin{bmatrix} a - b \\ b - c \\ c - a \end{bmatrix} : a, b, c \in \mathbb{R} \right\} \subset \mathbb{R}^3$.

(d) $H = \left\{ \begin{bmatrix} x_1 \\ x_2 \\ x_3 \end{bmatrix} : x_1 + x_2 = 0 = x_2 + x_3 \right\} \subset \mathbb{R}^3$.

(e) $H = \left\{ \begin{bmatrix} x_1 \\ x_2 \\ x_3 \end{bmatrix} : x_1 + x_2 + x_3 = 5 \right\} \subset \mathbb{R}^3$.

Exercise 2.6.3. Let U and W be subspaces of $\mathbb{R}^n$.

(a) Give an example of subspaces U and W so that their union $U \cup W$ is not a subspace.

(b) Show that $U \cup W$ is a subspace if and only if $U \subseteq W$ or $W \subseteq U$.

Exercise 2.6.4. For the following sets of vectors in $\mathbb{R}^4$, determine whether they are linearly independent.

(a) $\begin{bmatrix} 2 \\ 2 \\ 0 \\ -1 \end{bmatrix} , \begin{bmatrix} 3 \\ 0 \\ 2 \\ 1 \end{bmatrix} , \begin{bmatrix} -2 \\ 4 \\ 0 \\ 1 \end{bmatrix} , \begin{bmatrix} -1 \\ 2 \\ 1 \\ 0 \end{bmatrix}$.

(b) $\begin{bmatrix} 5 \\ -1 \\ 1 \\ 2 \end{bmatrix} , \begin{bmatrix} 1 \\ 7 \\ 3 \\ -4 \end{bmatrix} , \begin{bmatrix} 3 \\ -15 \\ -5 \\ 10 \end{bmatrix}$. *

(c) $\begin{bmatrix} 1 \\ 0 \\ 1 \\ 0 \end{bmatrix} , \begin{bmatrix} 1 \\ 1 \\ -1 \\ 0 \end{bmatrix} , \begin{bmatrix} 1 \\ 0 \\ 1 \\ 2 \end{bmatrix} , \begin{bmatrix} 1 \\ 1 \\ 0 \\ 0 \end{bmatrix} , \begin{bmatrix} 1 \\ 3 \\ -2 \\ 0 \end{bmatrix}$.

---

*You can check www.wolframalpha.com for the answer by entering 'Are (5, -1, 1, 2), (1, 7, 3, -4), and (3, -15, -5, 10) linearly independent?' and hit 'return'.

(d) $\begin{bmatrix} 1 \\ 0 \\ 3 \\ 1 \end{bmatrix}, \begin{bmatrix} 0 \\ 5 \\ 2 \\ 3 \end{bmatrix}, \begin{bmatrix} 0 \\ 0 \\ 1 \\ 0 \end{bmatrix}, \begin{bmatrix} 2 \\ 5 \\ 0 \\ 5 \end{bmatrix}.$

(e) $\begin{bmatrix} 1 \\ -1 \\ 2 \\ 0 \end{bmatrix}, \begin{bmatrix} -2 \\ -4 \\ 3 \\ 2 \end{bmatrix}, \begin{bmatrix} 3 \\ 3 \\ -1 \\ -2 \end{bmatrix}.$

Exercise 2.6.5. For which values of b is $\left\{ \begin{bmatrix} 1 \\ b \end{bmatrix}, \begin{bmatrix} b+3 \\ 3b+4 \end{bmatrix} \right\}$ linearly independent?

Exercise 2.6.6. If possible, provide a 4×6 matrix whose columns are linearly independent. If this is not possible, explain why not.

Exercise 2.6.7. Suppose that $\{v_1, v_2, v_3, v_4\}$ is linearly independent. For the following sets, determine whether they are linearly independent or not.

(a) $\{v_1, v_2 + v_3, v_3, v_4\}$.

(b) $\{v_1 + v_2, v_2 + v_3, v_3 + v_4, v_4 + v_1\}$.

(c) $\{v_1 + v_2, v_2 - v_3, v_3 + v_4, v_4 - v_1, v_3 - v_1\}$.

(d) $\{2v_1 + v_2, v_1 + 2v_3, v_4 + 2v_1\}$.

Exercise 2.6.8. Suppose that $\{v_1, \ldots, v_k\}$ is linearly independent. Show that $\{v_1, \ldots, v_k, v_{k+1}\}$ is linearly independent if and only if $v_{k+1} \notin \text{Span}\{v_1, \ldots, v_k\}$.

Exercise 2.6.9. Is $\left\{ \begin{bmatrix} 1 \\ 2 \\ 0 \end{bmatrix}, \begin{bmatrix} 3 \\ 5 \\ 3 \end{bmatrix}, \begin{bmatrix} 0 \\ 1 \\ 2 \end{bmatrix} \right\}$ a basis for $\mathbb{R}^3$?

Exercise 2.6.10. For which value of c is the following set of vectors a basis for $\mathbb{R}^3$?
$$\left\{ \begin{bmatrix} 1 \\ 1 \\ 1 \end{bmatrix}, \begin{bmatrix} 1 \\ 2 \\ c \end{bmatrix}, \begin{bmatrix} 0 \\ 4 \\ 6 \end{bmatrix} \right\}.$$

Exercise 2.6.11. Find bases for the column space, row space and null space of the following matrices. Provide also the rank of these matrices. For some matrices the reduced row echelon form is given.

(a) $\begin{bmatrix} 1 & 1 & 2 \\ 1 & 0 & 3 \\ 1 & 2 & 1 \end{bmatrix}.$

(b) $\begin{bmatrix} 0 & 0 & 3 & -6 & 6 \\ 3 & 1 & -7 & 8 & -5 \\ 3 & 1 & -9 & 12 & -9 \end{bmatrix}$.

(c) $\begin{bmatrix} 1 & 1 & 0 & 1 & 0 \\ 2 & 2 & 0 & 2 & 0 \\ 0 & 0 & 1 & 2 & 0 \\ 0 & 0 & 2 & 4 & 5 \end{bmatrix} \rightarrow \cdots \rightarrow \begin{bmatrix} 1 & 1 & 0 & 1 & 0 \\ 0 & 0 & 1 & 2 & 0 \\ 0 & 0 & 0 & 0 & 1 \\ 0 & 0 & 0 & 0 & 0 \end{bmatrix}$.

(d) $\begin{bmatrix} 1 & -1 & 1 & 1 \\ -1 & 1 & 1 & -1 \\ 1 & -1 & -1 & 1 \\ -1 & 1 & -1 & -1 \end{bmatrix}$.

(e) $\begin{bmatrix} 1 & 4 & 1 & 5 \\ 2 & 8 & 3 & 11 \\ 2 & 8 & 4 & 12 \end{bmatrix}$.

(f) $\begin{bmatrix} 1 & -1 & -3 & -12 \\ 0 & 4 & 5 & 19 \\ 1 & 3 & 2 & 7 \\ -1 & 1 & 3 & 12 \\ 3 & 8 & 9 & 29 \end{bmatrix} \rightarrow \cdots \rightarrow \begin{bmatrix} 1 & 0 & 0 & -2 \\ 0 & 1 & 0 & 1 \\ 0 & 0 & 1 & 3 \\ 0 & 0 & 0 & 0 \\ 0 & 0 & 0 & 0 \end{bmatrix}$.

(g) $\begin{bmatrix} 2 & 4 & 0 & 6 \\ 1 & 2 & 0 & 3 \\ -1 & -2 & 0 & -4 \end{bmatrix}$.[†]

(h) $\begin{bmatrix} 1 & 0 & 6 \\ 3 & 4 & 9 \\ 1 & 4 & -3 \end{bmatrix}$.

(i) $\begin{bmatrix} 1 & 6 & -3 \\ 0 & 1 & 2 \\ 0 & 0 & 0 \end{bmatrix}$.[‡]

Exercise 2.6.12. For Exercise 2.6.11 determine the dimensions of the column, null and row spaces, and check that the rank theorem is satisfied in all cases.

Exercise 2.6.13. Show that for a 5×7 matrix A we have that Col $A = \mathbb{R}^5$ if and only if dim Nul $A = 2$.

Exercise 2.6.14. Construct a 2×3 rank 1 matrix A so that $\mathbf{e}_2 \in$ Nul A. What is the dimension of Nul A?

[†]You can check the row space basis by going to sagecell.sagemath.org and entering 'A=matrix(RR, [[2,4,0,6], [1,2,0,3], [-1,-2,0,4]])' and 'A.row_space()' and hit 'Evaluate'. It may give you a different basis than you have found.

[‡]You can check the null space basis by going to live.sympy.org and entering 'A=Matrix(3, 3, [1, 6, -3, 0, 1, 2, 0, 0, 0])' and 'A.nullspace()'.

Exercise 2.6.15. In this exercise we want to establish the uniqueness of the reduced row echelon form, and observe that the row echelon form is not unique. Let A be an $m \times n$ matrix.

(a) Observe that column k will be a pivot column if and only if $\mathbf{a}_k \notin$ Span $\{\mathbf{a}_1, \ldots, \mathbf{a}_{k-1}\}$ (use Proposition 1.3.6).

(b) Show that in the reduced row echelon form of A the jth column with a pivot (counting from the left) will be equal to $\mathbf{e}_j$.

(c) Show that if $\mathbf{a}_l$ is not a pivot column of A, then it can be uniquely written as $\mathbf{a}_l = \sum_{r=1}^{s} c_r \mathbf{a}_{j_r}$, where $\mathbf{a}_{j_1}, \ldots, \mathbf{a}_{j_s}$, $j_1 < \cdots < j_s$, are the pivot columns to the left of $\mathbf{a}_l$.

(d) Using the notation of part (c), show that in the reduced row echelon form of A the lth column equals $\sum_{r=1}^{s} c_r \mathbf{e}_r$.

(e) Conclude that the reduced row echelon form of A is uniquely determined by A.

(f) Explain why the row echelon form of a nonzero matrix A is not unique (Hint: One can multiply a nonzero row by 2 and still maintain the echelon form).

Exercise 2.6.16. For the following bases $\mathcal{B}$ and vectors $\mathbf{v}$, find the coordinate vector $[\mathbf{v}]_\mathcal{B}$.

(a) $\mathcal{B} = \left\{ \begin{bmatrix} 1 \\ 1 \\ 1 \end{bmatrix}, \begin{bmatrix} 1 \\ 2 \\ 3 \end{bmatrix}, \begin{bmatrix} 1 \\ 3 \\ 6 \end{bmatrix} \right\}$, and $\mathbf{v} = \begin{bmatrix} 6 \\ 5 \\ 4 \end{bmatrix}$.

(b) $\mathcal{B} = \left\{ \begin{bmatrix} 1 \\ 3 \end{bmatrix}, \begin{bmatrix} 2 \\ 5 \end{bmatrix} \right\}$, and $\mathbf{v} = \begin{bmatrix} 0 \\ 3 \end{bmatrix}$.

Exercise 2.6.17. Let $H \subset \mathbb{R}^4$ be a subspace with basis

$$\mathcal{B} = \left\{ \begin{bmatrix} 1 \\ 0 \\ 2 \\ 0 \end{bmatrix}, \begin{bmatrix} 0 \\ 1 \\ 2 \\ 3 \end{bmatrix}, \begin{bmatrix} 1 \\ 0 \\ -2 \\ 4 \end{bmatrix} \right\}.$$

(a) Determine dim H.

(b) If $\mathbf{v} \in H$ has the coordinate vector $[\mathbf{v}]_\mathcal{B} = \begin{bmatrix} 1 \\ -1 \\ 2 \end{bmatrix}$, then $\mathbf{v}$ is which vector in $\mathbb{R}^4$?

(c) Let $\mathbf{w} = \begin{bmatrix} 1 \\ -1 \\ 0 \\ -3 \end{bmatrix}$. Check whether $\mathbf{w} \in H$, and if so, determine $[\mathbf{w}]_\mathcal{B}$.

Exercise 2.6.18. Let the subspace H have a basis $\mathcal{B} = \{\mathbf{v}_1, \mathbf{v}_2, \mathbf{v}_3, \mathbf{v}_4\}$. For $\mathbf{w} = 3\mathbf{v}_1 + 10\mathbf{v}_3$, determine $[\mathbf{w}]_\mathcal{B}$.

Exercise 2.6.19. For the following subspaces U and W, determine $\dim U$, $\dim W$, $\dim(U + W)$, and $\dim(U \cap W)$.

(a) $U = \text{Span} \left\{ \begin{bmatrix} 1 \\ -1 \\ 2 \end{bmatrix} \right\}$, $W = \text{Span} \left\{ \begin{bmatrix} 0 \\ 3 \\ 2 \end{bmatrix} \right\} \subseteq \mathbb{R}^3$.

(b) $U = \text{Span} \left\{ \begin{bmatrix} 1 \\ -1 \\ 2 \end{bmatrix} \right\}$, $W = \text{Span} \left\{ \begin{bmatrix} 2 \\ -2 \\ 4 \end{bmatrix} \right\} \subseteq \mathbb{R}^3$.

(c) $U = \text{Span} \left\{ \begin{bmatrix} 1 \\ 0 \\ 3 \\ 2 \end{bmatrix}, \begin{bmatrix} 0 \\ 2 \\ -1 \\ 2 \end{bmatrix} \right\}$, $W = \text{Span} \left\{ \begin{bmatrix} 2 \\ 0 \\ -3 \\ 0 \end{bmatrix}, \begin{bmatrix} 1 \\ 2 \\ 2 \\ 2 \end{bmatrix} \right\} \subseteq \mathbb{R}^4$.

(d) $U = \text{Span} \left\{ \begin{bmatrix} 1 \\ 0 \\ 3 \\ 2 \end{bmatrix}, \begin{bmatrix} 0 \\ 2 \\ -1 \\ 2 \end{bmatrix} \right\}$, $W = \text{Span} \left\{ \begin{bmatrix} 2 \\ 0 \\ -3 \\ 0 \end{bmatrix}, \begin{bmatrix} 1 \\ 2 \\ 2 \\ 4 \end{bmatrix} \right\} \subseteq \mathbb{R}^4$.

Exercise 2.6.20. True or False? Justify each answer.

(i) If $\{\mathbf{v}_1, \mathbf{v}_2, \mathbf{v}_3\}$ is linearly independent, then $\{\mathbf{v}_1, \mathbf{v}_2\}$ is also linearly independent.

(ii) If the set $\{\mathbf{v}_1, \mathbf{v}_2\}$ is linearly independent and the set $\{\mathbf{u}_1, \mathbf{u}_2\}$ is linearly independent, then the set $\{\mathbf{v}_1, \mathbf{v}_2, \mathbf{u}_1, \mathbf{u}_2\}$ is linearly independent.

(iii) Let $\mathcal{B} = \left\{ \begin{bmatrix} 1 \\ 2 \end{bmatrix}, \begin{bmatrix} 2 \\ -1 \end{bmatrix} \right\}$, which is a basis for $\mathbb{R}^2$, and let $\mathbf{x} = \begin{bmatrix} -5 \\ 5 \end{bmatrix}$. Is $[\mathbf{x}]_\mathcal{B} = \begin{bmatrix} 1 \\ -3 \end{bmatrix}$?

(iv) For a matrix A we have $\dim \text{ Row } A = \dim \text{ Col } A$.

(v) The system $A\mathbf{x} = \mathbf{0}$ always has infinitely many solutions.

(vi) If $A = \begin{bmatrix} 1 & 1 & 3 \\ 1 & 2 & 3 \end{bmatrix}$, then a basis for Col A is $\left\{ \begin{bmatrix} 1 \\ 1 \end{bmatrix} \right\}$.

(vii) If $A = \begin{bmatrix} 1 & 2 & 3 \\ 1 & 2 & 3 \end{bmatrix}$, then a basis for Col A is $\left\{ \begin{bmatrix} 1 \\ 0 \end{bmatrix} \right\}$.

(viii) If A is an 2×3 matrix, then there could be a vector that is both in Col A and in Nul A.

(ix) If A is an 3×3 matrix, then there could be a vector that is both in Col A and in Nul A.

(x) If the null space of a $m \times n$ matrix A has dimension n, then the zero vector is the only vector in the column space of A.

Exercise 2.6.21. Minimal rank completion is a technique used by companies to try to guess their customers' preferences for the companies' products. In this case we have a matrix with some unknowns in it (indicated by a ?); we call this a **partial matrix**. An example of a partial matrix is

$$\mathcal{A} = \begin{bmatrix} -1 & ? & 0 & ? & 2 \\ ? & 0 & ? & 1 & -1 \end{bmatrix}.$$

As an example, let us use Netflix where the company tries to guess your movie preferences based on movies you have watched and possibly rated. The matrix above could represent two customers (represented by rows) and five movies (represented by columns). The numbers indicate how the customers rated the movies compared to an average rating. Thus customer 1 rated movie 1 below average (1 below average) and movie 5 above average (2 above average), etc. Customer 1 did not watch movie 2 and movie 4. Customer 2 did not watch movie 1 and movie 3. How to guess whether they will like the movies they have not watched yet?

Netflix put out a challenge in 2006 to win one million dollars for those who could improve on Netflix's algorithm by 10% (for details, look for 'Netflix challenge' or 'Netflix prize'). Part of the prize winning idea was to use minimal rank completions. Given a partial matrix $\mathcal{A}$, we call A a **completion** of $\mathcal{A}$ if A coincides with $\mathcal{A}$ on the known entries. We call A a **minimal rank completion** of $\mathcal{A}$ if among all completions of $\mathcal{A}$, the matrix A has the lowest possible rank. For the above partial matrix $\mathcal{A}$ any completion will have rank ≥ 1, and we can actually make a rank 1 completion A by taking

$$\begin{array}{c} \begin{array}{ccccc} \text{m1} & \text{m2} & \text{m3} & \text{m4} & \text{m5} \end{array} \\ A = \begin{bmatrix} -1 & 0 & 0 & -2 & 2 \\ \frac{1}{2} & 0 & 0 & 1 & -1 \end{bmatrix} \begin{array}{l} \text{customer1} \\ \text{customer2} \end{array} . \end{array}$$

The matrix A is a minimal rank completion of $\mathcal{A}$. Based on this completion we would suggest to customer 2 to watch movie 1, and customer 1 may like movie 2 better than movie 4 (but maybe not strong enough to bother with a recommendation). In this example, we had of course very little to go on. In practice there are millions of customers and a huge number of movies, and also we may have other data on customers (how long/short they watched movies, what type of movies they mostly watch, their approximate geographical location (relevant?; number crunching may tell), etc.). The simple idea presented here is of course just part of the story. If you would like more details, look for **svd++**. In Chapter 8 we will explain the singular value decomposition (svd), which is one of the ingredients of the algorithms used.

(a) Find a minimal rank completion for $\begin{bmatrix} 1 & ? \\ 1 & 1 \end{bmatrix}$.

(b) Show that any minimal rank completion of $\begin{bmatrix} 1 & ? \\ 0 & 1 \end{bmatrix}$ will have rank 2.

(c) Find a minimal rank completion for

$$\begin{bmatrix} -1 & ? & 2 & ? & 0 & -1 \\ ? & 2 & 0 & ? & 1 & 1 \\ 0 & ? & 2 & 1 & ? & -2 \\ ? & 3 & 2 & -1 & ? & -1 \end{bmatrix}.$$

(Hint: Try to make all columns linear combinations of columns 3 and 6.)

(d) If the partial matrix in (c) represents four customers partially rating 6 movies, which movie (if any) would you recommend to each of the customers?

Exercise 2.6.22. Matrices can be used to represent graphs. In this exercise, we introduce the **incidence matrix** of a graph. A graph has vertices $v_1, \ldots, v_m$ and edges (x, y), where $x, y \in \{v_1, \ldots, v_m\}$. Here we do not allow edges of the form (x, x). We depict a graph by representing the vertices as small circles, and edges as lines between the vertices. Figure 2.2 is an example of a **directed graph**. The graph is directed as the edges are arrows.

The graph in Figure 2.2 can be summarized as $G = (V, E)$, where

$$V = \{\text{set of vertices}\} = \{1, 2, 3, 4\},$$

$$E = \{\text{set of edges}\} = \{(1,3), (1,4), (2,1), (3,1), (3,2), (3,4), (4,2), (4,3)\}.$$

Notice that we numbered the edges e1, e2, ..., e8. For each edge (x, y), the starting point is x and the endpoint is y. As an example, edge e2 corresponds to $(1, 4)$.

The incidence matrix A of a graph $G = (V, E)$ is an $m \times n$ matrix, where

$$m = \text{the number of vertices}, \quad n = \text{the number of edges},$$

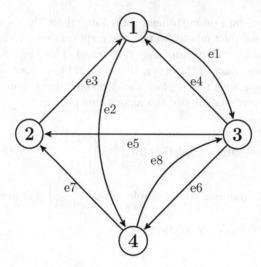

Figure 2.2: A directed graph with 4 vertices and 8 edges.

and if (x, y) is the jth edge, then we put -1 in position (x, j) and 1 in position (y, j). For the above example, we obtain

$$
A =
\begin{array}{c}

\begin{array}{cccccccc}
\text{e1} & \text{e2} & \text{e3} & \text{e4} & \text{e5} & \text{e6} & \text{e7} & \text{e8}
\end{array} \\
\left[
\begin{array}{cccccccc}
-1 & -1 & 1 & 1 & 0 & 0 & 0 & 0 \\
0 & 0 & -1 & 0 & 1 & 0 & 1 & 0 \\
1 & 0 & 0 & -1 & -1 & -1 & 0 & 1 \\
0 & 1 & 0 & 0 & 0 & 1 & -1 & -1
\end{array}
\right]
\begin{array}{c}
\text{v1} \\ \text{v2} \\ \text{v3} \\ \text{v4}
\end{array}
\end{array}
\qquad (2.5)
$$

(a) Find the incidence matrix of the graph in Figure 2.3.

(b) Show that the sum of the rows in an incidence matrix is always $\mathbf{0}$. Use this to prove that the rank of an $m \times n$ incidence matrix is at most $m - 1$.

The interesting thing about the incidence matrix is that the linear dependance between the columns gives us information about the loops in the graph. We first need some more terminology.

A **walk** of length k in a graph is a subset of k edges of the form $(x_1, x_2), (x_2, x_3), \ldots, (x_k, x_{k+1})$. We call x_1 the starting point of the walk, and x_{k+1} the endpoint. A **loop** is a walk where the starting point and the endpoint are the same.

Notice that in the incidence matrix A in (2.5) we have that $\mathbf{a}_3 - \mathbf{a}_4 + \mathbf{a}_5 = \mathbf{0}$. In the corresponding graph this means that e3, $-$e4, e5 form a loop (where $-$e4 means that we take edge 4 in the reverse direction). Similarly, $\mathbf{a}_6 + \mathbf{a}_8 = \mathbf{0}$. This means that e6, e8 form a loop. It is not too hard to figure out why it

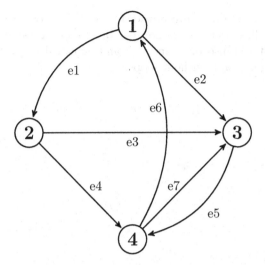

Figure 2.3: A directed graph with 4 vertices and 7 edges.

works out this way; indeed, if we look at $\mathbf{a}_6 + \mathbf{a}_8 = \mathbf{0}$, it means that for vertex 1 (which corresponds to the first row) there are as many -1's as there are 1's among edges e6, e8, and thus vertex 1 is as many times a starting point as it is an endpoint in the walk. The same goes for the other vertices.

(c) Illustrate this phenomena in the incidence matrix you have found in part (a). (For instance, use the loop e1, e4, e7, $-$e2.)

If we now find a basis for the columns space of A, it means that we are identifying a maximal subset of the columns so that they are linearly independent. For the graph that means that we are trying to find a maximal subset of the edges, so that there are no loops. This is called a **spanning forest** for the graph. For instance, for A in (2.5) we find that

$$
\{\mathbf{a}_1, \mathbf{a}_2, \mathbf{a}_3\} = \left\{ \begin{bmatrix} -1 \\ 0 \\ 1 \\ 0 \end{bmatrix}, \begin{bmatrix} -1 \\ 0 \\ 0 \\ 1 \end{bmatrix}, \begin{bmatrix} 1 \\ -1 \\ 0 \\ 0 \end{bmatrix} \right\}
$$

is a basis for Col$A$, and thus e1, e2, e3 form a spanning forest for the corresponding graph.

(d) Find a spanning forest for the graph in part (a).

By the way, we call a graph **connected** if between any two vertices there is a walk (where we allow going in the reverse direction of an edge). In that case, a spanning forest will also be connected, and be called a **spanning tree**. A **tree** is a connected graph without loops.

3

Matrix Algebra

CONTENTS

3.1   Matrix Addition and Multiplication

We denote the set of $m \times n$ matrices with real entries as $\mathbb{R}^{m \times n}$. An $m \times n$ matrix A is typically denoted as

$$A = (a_{ij})_{i=1,j=1}^{m} {}^{n} = \begin{bmatrix} a_{11} & \cdots & a_{1n} \\ \vdots & & \vdots \\ a_{m1} & \cdots & a_{mn} \end{bmatrix} = \begin{bmatrix} \mathbf{a}_1 & \cdots & \mathbf{a}_n \end{bmatrix}.$$

Thus a_{ij} is the entry in the ith row and the jth column. Also, as before we denote the jth column of A as $\mathbf{a}_j$. Occasionally, we will use the ith row of A, which we denote as $\text{row}_i(A)$. Thus

$$\text{row}_i(A) = \begin{bmatrix} a_{i1} & \cdots & a_{i,n} \end{bmatrix} \in \mathbb{R}^{1 \times n}.$$

Sometimes we will also use the notation $\text{col}_j(A)$ for the jth column of A. This is convenient when the matrix is not denoted by a single capital letter.

We define **addition** of two matrices of the same size by

$$A + B = (a_{ij})_{i=1,j=1}^{m} {}^{n} + (b_{ij})_{i=1,j=1}^{m} {}^{n} = (a_{ij} + b_{ij})_{i=1,j=1}^{m} {}^{n}.$$

We define **scalar multiplication** of a matrix by

$$cA = c(a_{ij})_{i=1,j=1}^{m} {}^{n} = (ca_{ij})_{i=1,j=1}^{m} {}^{n}.$$

For example,

$$\begin{bmatrix} 1 & -1 \\ 2 & 0 \\ 3 & -4 \end{bmatrix} + \begin{bmatrix} 10 & 3 \\ 7 & -3 \\ 0 & 5 \end{bmatrix} = \begin{bmatrix} 11 & 2 \\ 9 & -3 \\ 3 & 1 \end{bmatrix}, \ 3\begin{bmatrix} 6 & -1 \\ 2 & 0 \end{bmatrix} = \begin{bmatrix} 18 & -3 \\ 6 & 0 \end{bmatrix}.$$

The above operations are similar to those on vectors. In fact, we can view a vector in $\mathbb{R}^m$ as an $m \times 1$ matrix. In fact, we identify $\mathbb{R}^{m \times 1}$ with $\mathbb{R}^m$.

We let $0_{m \times n}$ be the $m \times n$ matrix of all zeros. Sometimes we just write 0 for a matrix of all zeros, in which case the size of the matrix should be clear from the context. Thus, if we write $A + 0$, then A and 0 are understood to be matrices of the same size.

The rules we had for addition and scalar multiplication of vectors carry over to matrices, as the next result states.

Proposition 3.1.1. *Let $A, B, C \in \mathbb{R}^{m \times n}$ and $c, d \in \mathbb{R}$. Then*

$$A + B = B + A, A + (B + C) = (A + B) + C, 0 + A = A, A + (-A) = 0,$$

$$c(A + B) = cA + cB, (c + d)A = cA + dA, c(dA) = (cd)A, 1A = A.$$

Due to these rules $\mathbb{R}^{m \times n}$ is a vector space over $\mathbb{R}$, as we will see in Chapter 5.

We have already seen that we can multiply an $m \times n$ matrix A with a vector $\mathbf{x} \in \mathbb{R}^n$ to obtain the product vector $A\mathbf{x} \in \mathbb{R}^m$. We will extend this operation to allow for a matrix product AB as follows.

Definition 3.1.2. We define **matrix multiplication** as follows: if A is an $m \times n$ matrix and $B = \begin{bmatrix} \mathbf{b}_1 & \cdots & \mathbf{b}_k \end{bmatrix}$ an $n \times k$ matrix, we define AB via

$$AB := \begin{bmatrix} A\mathbf{b}_1 & \cdots & A\mathbf{b}_k \end{bmatrix} \in \mathbb{R}^{m \times k}.$$

Thus, for a matrix product AB to make sense, we need that

number of columns of A = number of rows of B.

Here is an example:

$$\begin{bmatrix} 1 & -1 \\ 2 & 0 \\ 3 & -4 \\ 0 & 1 \end{bmatrix} \begin{bmatrix} 6 & 1 \\ 2 & 0 \end{bmatrix} = \begin{bmatrix} 6\begin{bmatrix} 1 \\ 2 \\ 3 \\ 0 \end{bmatrix} + 2\begin{bmatrix} -1 \\ 0 \\ -4 \\ 1 \end{bmatrix} & 1\begin{bmatrix} 1 \\ 2 \\ 3 \\ 0 \end{bmatrix} + 0\begin{bmatrix} -1 \\ 0 \\ -4 \\ 1 \end{bmatrix} \end{bmatrix} = \begin{bmatrix} 4 & 1 \\ 12 & 2 \\ 10 & 3 \\ 2 & 0 \end{bmatrix}.$$

If we look at what happens at the entry level when we take a matrix product, we get the following.

$$C = AB \quad \Leftrightarrow \quad c_{rs} = \sum_{j=1}^{n} a_{rj}b_{js}, \ 1 \le r \le m, 1 \le s \le k.$$

In other words,

$$c_{rs} = \text{row}_r(A) \, \text{col}_s(B) = \begin{bmatrix} a_{r1} & \cdots & a_{rn} \end{bmatrix} \begin{bmatrix} b_{1s} \\ \vdots \\ b_{ns} \end{bmatrix}.$$

Here are some more examples:

$$\begin{bmatrix} 3 & -2 & 0 \\ -1 & 5 & 1 \end{bmatrix} \begin{bmatrix} -2 & 3 & 0 \\ 1 & -4 & 1 \\ 7 & 0 & 0 \end{bmatrix} = \begin{bmatrix} -8 & 17 & -2 \\ 14 & -23 & 5 \end{bmatrix},$$

$$\begin{bmatrix} -2 & 3 & 0 \\ 1 & -4 & 1 \\ 7 & 0 & 0 \end{bmatrix} \begin{bmatrix} 3 & -2 & 0 \\ -1 & 5 & 1 \end{bmatrix} \quad \text{is undefined,}$$

$$\begin{bmatrix} 1 \\ 2 \\ 3 \\ 4 \end{bmatrix} \begin{bmatrix} 1 & -1 & 2 & -2 \end{bmatrix} = \begin{bmatrix} 1 & -1 & 2 & -2 \\ 2 & -2 & 4 & -4 \\ 3 & -3 & 6 & -6 \\ 4 & -4 & 8 & -8 \end{bmatrix},$$

$$\begin{bmatrix} 1 & -1 & 2 & -2 \end{bmatrix} \begin{bmatrix} 1 \\ 2 \\ 3 \\ 4 \end{bmatrix} = \begin{bmatrix} -3 \end{bmatrix} = -3.$$

In the last example we arrive at a 1×1 matrix, which we just identify with its single entry. In other words, for a 1×1 matrix we sometimes do not write the square brackets around the single number.

It should be noticed that in general matrix multiplication is **not** commutative:

$$\boxed{AB \ne BA.}$$

Indeed, if we take $A = \begin{bmatrix} 1 & 0 \\ 0 & 0 \end{bmatrix}$ and $B = \begin{bmatrix} 0 & 1 \\ 0 & 0 \end{bmatrix}$, then

$$AB = \begin{bmatrix} 0 & 1 \\ 0 & 0 \end{bmatrix} \ne \begin{bmatrix} 0 & 0 \\ 0 & 0 \end{bmatrix} = BA.$$

It can also happen that AB is defined but that BA is undefined, as you can see above. Finally, let us give another example to show that it is possible that $AB = 0$ while $A \neq 0$ and $B \neq 0$:

$$\begin{bmatrix} 1 & 0 \\ 7 & 0 \end{bmatrix} \begin{bmatrix} 0 & 0 & 0 \\ 1 & 2 & 5 \end{bmatrix} = \begin{bmatrix} 0 & 0 & 0 \\ 0 & 0 & 0 \end{bmatrix}.$$

We have the following rules.

Proposition 3.1.3. *Let* $A, D \in \mathbb{R}^{m \times n}$, $B, E \in \mathbb{R}^{n \times k}$, $C \in \mathbb{R}^{k \times l}$, $F, G \in \mathbb{R}^{r \times n}$ *and* $c \in \mathbb{R}$. *Then*

$$A(BC) = (AB)C, A(B + E) = AB + AE, (F + G)A = FA + GA,$$

$$c(AB) = (cA)B = A(cB), \quad I_m A = A = AI_n.$$

Partial proof. Let us prove the associativity rule $A(BC) = (AB)C$, where A is $m \times n$, B is $n \times k$ and C is $k \times l$. When M is some product of matrices, we write $[M]_{ij}$ for the (i, j) entry of the matrix M. We have

$$[A(BC)]_{ij} = \sum_{r=1}^{n} a_{ir}[BC]_{rj} = \sum_{r=1}^{n} a_{ir}\left(\sum_{s=1}^{k} b_{rs}c_{sj}\right) = \sum_{r=1}^{n}\sum_{s=1}^{k} a_{ir}b_{rs}c_{sj}$$

$$= \sum_{s=1}^{k}\sum_{r=1}^{n} a_{ir}b_{rs}c_{sj} = \sum_{s=1}^{k}\left(\sum_{r=1}^{n} a_{ir}b_{rs}\right)c_{sj} = \sum_{s=1}^{k}[AB]_{is}c_{sj} = [(AB)C]_{ij}.$$

Let us also prove that $I_m A = A$. This follows as $I_m \mathbf{a}_j = \mathbf{a}_j$, $j = 1, \ldots, m$.

To show $AI_n = A$, we write (as we did before), $I_n = \begin{bmatrix} \mathbf{e}_1 & \cdots & \mathbf{e}_n \end{bmatrix}$. Next notice that

$$A\mathbf{e}_j = 0 \, \mathbf{a}_1 + \cdots + 0 \, \mathbf{a}_{j-1} + 1 \, \mathbf{a}_j + 0 \, \mathbf{a}_{j+1} + \cdots + 0 \, \mathbf{a}_n = \mathbf{a}_j.$$

But then

$$AI_n = \begin{bmatrix} A\mathbf{e}_1 & \cdots & A\mathbf{e}_n \end{bmatrix} = \begin{bmatrix} \mathbf{a}_1 & \cdots & \mathbf{a}_n \end{bmatrix} = A$$

follows. The remaining items we leave to the reader to prove.   □

When we defined the matrix product AB, we viewed B in terms of its columns and defined the jth column of the product as $A\mathbf{b}_j$. There is also a point of view via the rows. For this you need the following observation:

$$\text{row}_i(AB) = \text{row}_i(A)B. \tag{3.1}$$

Thus, if we had started out by defining the product of a row vector with a matrix (with the matrix appearing on the right of the row vector), we could have defined the matrix product via (3.1). Either way, the resulting definition of the matrix product is the same. It is useful to observe that

- the rows of AB are linear combinations of the rows of B,

- the columns of AB are linear combinations of the columns of A.

In the next section, we introduce the transpose of a matrix, which gives a way to transition between these different viewpoints.

3.2   Transpose

Definition 3.2.1. If $A = (a_{ij})_{i=1,j=1}^{m}{}_{,j=1}^{n}$, then the **transpose** A^T is the $n \times m$ matrix defined by $[A^T]_{ij} = a_{ji}$.

For example,

$$\begin{bmatrix} 1 & -1 \\ 2 & 0 \\ 3 & -4 \end{bmatrix}^T = \begin{bmatrix} 1 & 2 & 3 \\ -1 & 0 & -4 \end{bmatrix}.$$

Notice that the jth column in A^T comes from the jth row of A, and that the ith row of A^T comes from the ith column of A. To be more precise,

$$\text{row}_j(A^T) = (\text{col}_j(A))^T, \ \text{col}_i(A^T) = (\text{row}_i(A))^T.$$

Proposition 3.2.2. *Let* $A, B \in \mathbb{R}^{m \times n}$ *and* $c \in \mathbb{R}$*. Then*

$$(A^T)^T = A, (A+B)^T = A^T + B^T, (cA)^T = cA^T, I_n^T = I_n, (AB)^T = B^T A^T.$$

Partial proof. We only prove the last statement, as it is the most involved one. We have

$$[B^T A^T]_{ij} = \sum_{r=1}^{n} [B^T]_{ir} [A^T]_{rj} = \sum_{i=1}^{n} b_{ri} a_{jr} = \sum_{i=1}^{n} a_{jr} b_{ri} = [AB]_{ji} = [(AB)^T]_{ij}.$$

□

It is important to remember that when you take the transpose of a product, such as $(AB)^T$, and want to write it as a product of transposes, you have to reverse the order of the matrices (B^T first and then A^T).

We will see the importance of the transpose when we introduce the dot product (in Chapter 8).

3.3   Inverse

Definition 3.3.1. We say that an $n \times n$ matrix A is **invertible** if there exists a $n \times n$ matrix B so that $AB = I_n = BA$.

The following result shows that if such a matrix B exists, it is unique.

Lemma 3.3.2. *Let* $A \in \mathbb{R}^{n \times n}$. *Suppose that* $B, C \in \mathbb{R}^n$ *exist so that* $AB = I_n = BA$ *and* $AC = I_n = CA$. *Then* $B = C$.

Proof. We have $B = BI_n = B(AC) = (BA)C = I_nC = C$. $\square$

If $AB = I_n = BA$, then we say that B is the **inverse** of A and we write $B = A^{-1}$. Thus

$$AA^{-1} = I_n = A^{-1}A.$$

If we let $\mathbf{e}_i$ be the ith column of the identity matrix, then solving $AB = I_n$ means we are trying to find $B = \begin{bmatrix} \mathbf{b}_1 & \cdots & \mathbf{b}_n \end{bmatrix}$ so that

$$A\mathbf{b}_1 = \mathbf{e}_1, A\mathbf{b}_2 = \mathbf{e}_2, \ldots, A\mathbf{b}_n = \mathbf{e}_n.$$

Let us try this out on an example.

Example 3.3.3. Find the inverse of $A = \begin{bmatrix} 1 & 2 \\ 3 & 4 \end{bmatrix}$.

To compute the first column of the inverse, we solve the system represented by the augmented matrix

$$\left[\begin{array}{cc|c} 1 & 2 & 1 \\ 3 & 4 & 0 \end{array}\right] \rightarrow \left[\begin{array}{cc|c} 1 & 2 & 1 \\ 0 & -2 & -3 \end{array}\right] \rightarrow \left[\begin{array}{cc|c} 1 & 2 & 1 \\ 0 & 1 & \frac{3}{2} \end{array}\right] \rightarrow \left[\begin{array}{cc|c} 1 & 0 & -2 \\ 0 & 1 & \frac{3}{2} \end{array}\right].$$

To compute the second column of the inverse, we do

$$\left[\begin{array}{cc|c} 1 & 2 & 0 \\ 3 & 4 & 1 \end{array}\right] \rightarrow \left[\begin{array}{cc|c} 1 & 2 & 0 \\ 0 & -2 & 1 \end{array}\right] \rightarrow \left[\begin{array}{cc|c} 1 & 2 & 0 \\ 0 & 1 & -\frac{1}{2} \end{array}\right] \rightarrow \left[\begin{array}{cc|c} 1 & 0 & 1 \\ 0 & 1 & -\frac{1}{2} \end{array}\right].$$

Thus we find

$$A^{-1} = \begin{bmatrix} -2 & 1 \\ \frac{3}{2} & -\frac{1}{2} \end{bmatrix}.$$

Let us check our answer:

$$\begin{bmatrix} 1 & 2 \\ 3 & 4 \end{bmatrix}\begin{bmatrix} -2 & 1 \\ \frac{3}{2} & -\frac{1}{2} \end{bmatrix} = \begin{bmatrix} 1 & 0 \\ 0 & 1 \end{bmatrix} = \begin{bmatrix} -2 & 1 \\ \frac{3}{2} & -\frac{1}{2} \end{bmatrix}\begin{bmatrix} 1 & 2 \\ 3 & 4 \end{bmatrix}.$$

It works!

Note that there was redundancy in what we did: we row reduced the same coefficient matrix twice. We can do better by combining the two systems (remember Interpretation 2 in Section 1.1!), as follows:

$$\left[\begin{array}{cc|cc} 1 & 2 & 1 & 0 \\ 3 & 4 & 0 & 1 \end{array}\right] \rightarrow \left[\begin{array}{cc|cc} 1 & 2 & 1 & 0 \\ 0 & -2 & -3 & 1 \end{array}\right] \rightarrow \left[\begin{array}{cc|cc} 1 & 0 & -2 & 1 \\ 0 & 1 & \frac{3}{2} & -\frac{1}{2} \end{array}\right].$$

□

This observation leads to the general algorithm for finding the inverse of an $n \times n$ matrix:

$$\textbf{Algorithm}: \quad \begin{bmatrix} A & | & I_n \end{bmatrix} \rightarrow \cdots \rightarrow \begin{bmatrix} I_n & | & A^{-1} \end{bmatrix}.$$

This works when Col $A = \mathbb{R}^n$. Indeed, in that case $A\mathbf{x} = \mathbf{e}_j$ can always be solved. Col $A = \mathbb{R}^n$ implies that the reduced row echelon form of A has a pivot in every row, which combined with A being square gives that its reduced row echelon form is I_n. It is also true that when A is invertible, we necessarily need to have that Col $A = \mathbb{R}^n$, as we will see. First let us do another example.

Example 3.3.4. Find the inverse of $A = \begin{bmatrix} 0 & -1 & 1 \\ 2 & 0 & 2 \\ 3 & 4 & 1 \end{bmatrix}$. We row reduce

$$\left[\begin{array}{ccc|ccc} 0 & -1 & 1 & 1 & 0 & 0 \\ 2 & 0 & 2 & 0 & 1 & 0 \\ 3 & 4 & 1 & 0 & 0 & 1 \end{array}\right] \rightarrow \left[\begin{array}{ccc|ccc} 1 & 0 & 1 & 0 & \frac{1}{2} & 0 \\ 0 & -1 & 1 & 1 & 0 & 0 \\ 0 & 4 & -2 & 0 & -\frac{3}{2} & 1 \end{array}\right] \rightarrow$$

$$\left[\begin{array}{ccc|ccc} 1 & 0 & 1 & 0 & \frac{1}{2} & 0 \\ 0 & 1 & -1 & -1 & 0 & 0 \\ 0 & 0 & 2 & 4 & -\frac{3}{2} & 1 \end{array}\right] \rightarrow \left[\begin{array}{ccc|ccc} 1 & 0 & 0 & -2 & \frac{5}{4} & -\frac{1}{2} \\ 0 & 1 & 0 & 1 & -\frac{3}{4} & \frac{1}{2} \\ 0 & 0 & 1 & 2 & -\frac{3}{4} & \frac{1}{2} \end{array}\right].$$

We indeed find that

$$A^{-1} = \begin{bmatrix} -2 & \frac{5}{4} & -\frac{1}{2} \\ 1 & -\frac{3}{4} & \frac{1}{2} \\ 2 & -\frac{3}{4} & \frac{1}{2} \end{bmatrix}.$$

□

The rules for inverses include the following.

Theorem 3.3.5. *Let A and B be $n \times n$ invertible matrices. Then*

(i) The solution to $A\mathbf{x} = \mathbf{b}$ is $\mathbf{x} = A^{-1}\mathbf{b}$.

(ii) $(A^{-1})^{-1} = A$.

(iii) $(AB)^{-1} = B^{-1}A^{-1}$.

(iv) $(A^T)^{-1} = (A^{-1})^T$.

Proof. For (i) observe that $A\mathbf{x} = \mathbf{b}$ implies $A^{-1}(A\mathbf{x}) = A^{-1}\mathbf{b}$. Using associativity, we get $A^{-1}(A\mathbf{x}) = (A^{-1}A)\mathbf{x} = I_n\mathbf{x} = \mathbf{x}$. Thus $\mathbf{x} = A^{-1}\mathbf{b}$ follows.

(ii) follows immediately from the equations $AA^{-1} = I_n = A^{-1}A$.

(iii) Observe that $ABB^{-1}A^{-1} = A(BB^{-1})A^{-1} = AI_nA^{-1} = AA^{-1} = I$. Similarly, $B^{-1}A^{-1}AB = B^{-1}I_nB = I_n$. Thus $(AB)^{-1} = B^{-1}A^{-1}$ follows.

(iv) We have $(A^{-1})^T A^T = (AA^{-1})^T = I_n^T = I_n$ and $A^T(A^{-1})^T = (A^{-1}A)^T = I_n^T = I_n$. This yields $(A^T)^{-1} = (A^{-1})^T$.   □

Similar to our remark regarding transposes, it is important to remember when you take the inverse of a product, such as $(AB)^{-1}$, and want to write it as a product of inverses, you have to reverse the order of the matrices (B^{-1} first and then A^{-1}) .

When we get to determinants in the next chapter, we will derive formulas for the inverse of an $n \times n$ matrix. The one that is easiest to remember is when $n = 2$, which we now present.

Theorem 3.3.6. *Let $A = \begin{bmatrix} a & b \\ c & d \end{bmatrix}$. Then A is invertible if and only if $ad - bc \neq 0$, and in that case*

$$A^{-1} = \frac{1}{ad - bc} \begin{bmatrix} d & -b \\ -c & a \end{bmatrix}. \tag{3.2}$$

Proof. Notice that

$$\begin{bmatrix} a & b \\ c & d \end{bmatrix} \begin{bmatrix} d & -b \\ -c & a \end{bmatrix} = (ad - bc)I_2 = \begin{bmatrix} d & -b \\ -c & a \end{bmatrix} \begin{bmatrix} a & b \\ c & d \end{bmatrix}. \tag{3.3}$$

If $ad - bc \neq 0$, we can divide equation (3.3) by $ad - bc$ and obtain (3.2).

Next, suppose $ad - bc = 0$. Then

$$A \begin{bmatrix} d & -b \\ -c & a \end{bmatrix} = 0.$$

If A has an inverse, we get that

$$\begin{bmatrix} d & -b \\ -c & a \end{bmatrix} = A^{-1}A \begin{bmatrix} d & -b \\ -c & a \end{bmatrix} = A^{-1}0 = 0.$$

This gives that $a = b = c = d = 0$, which in turn yields $A = 0$. But then for every matrix B we have $BA = 0$, and thus A can not be invertible. Consequently, $ad - bc = 0$ implies that A is not invertible.   □

3.4 Elementary Matrices

In line with the three elementary row operations, we introduce three types of elementary matrices. In a square matrix we refer to the entries (i, i) as the **diagonal entries**, and to the entries $(i, j), i \neq j$, as the **off-diagonal entries**.

- **Type I**: $E \in \mathbb{R}^{n \times n}$ is of type I if for some choice of $i, j \in \{1, \ldots, n\}, i < j$, we have that $[E]_{kk} = 1, k \neq i, j$, $[E]_{ij} = [E]_{ji} = 1$, and all other entries in E are 0. For example,

$$E_1 = E_{1 \leftrightarrow 2}^{(1)} = \begin{bmatrix} 0 & 1 & 0 \\ 1 & 0 & 0 \\ 0 & 0 & 1 \end{bmatrix}.$$

  is of type I (here $i = 1$, $j = 2$).

- **Type II**: $E \in \mathbb{R}^{n \times n}$ is of type II if all its diagonal entries equal 1, and exactly one off-diagonal entry is nonzero. For example,

$$E_2 = E_{(1,3,c)}^{(2)} = \begin{bmatrix} 1 & 0 & c \\ 0 & 1 & 0 \\ 0 & 0 & 1 \end{bmatrix}$$

  is a type II elementary matrix.

- **Type III**: $E \in \mathbb{R}^{n \times n}$ is of type III if for some choice of $i \in \{1, \ldots, n\}$, we have that $[E]_{kk} = 1, k \neq i$, $[E]_{ii} = c(\neq 0)$, and all other entries in E are 0. For example,

$$E_3 = E_{(3,c)}^{(3)} = \begin{bmatrix} 1 & 0 & 0 \\ 0 & 1 & 0 \\ 0 & 0 & c \end{bmatrix}, c \neq 0,$$

  is of type III.

We used the notation from Section 1.2. It is not hard to convince yourself that each of these three types of matrices are invertible, and that the inverses are of the same type. For the three examples above, we find

$$E_1^{-1} = \begin{bmatrix} 0 & 1 & 0 \\ 1 & 0 & 0 \\ 0 & 0 & 1 \end{bmatrix} = E_{1 \leftrightarrow 2}^{(1)}, E_2^{-1} = \begin{bmatrix} 1 & 0 & -c \\ 0 & 1 & 0 \\ 0 & 0 & 1 \end{bmatrix} = E_{(1,3,-c)}^{(2)},$$

$$E_3^{-1} = \begin{bmatrix} 1 & 0 & 0 \\ 0 & 1 & 0 \\ 0 & 0 & \frac{1}{c} \end{bmatrix} = E_{(3,1/c)}^{(3)}.$$

The next proposition shows that the elementary row operations correspond exactly with multiplying a matrix on the left with an elementary matrix.

Proposition 3.4.1. *Let $A \in \mathbb{R}^{m \times n}$. If $E \in \mathbb{R}^{m \times m}$ is an elementary matrix then EA is obtained from A by performing an elementary row operation. More specifically,*

- *If E is type I and 1's appear in off-diagonal entries (i, j) and (j, i), then EA is obtained from A by switching rows i and j.*

- *If E is type II and the nonzero off-diagonal entry (k, l) has value c, then EA is obtained from A by replacing $\text{row}_k(A)$ by $\text{row}_k(A) + c \cdot \text{row}_l(A)$.*

- *If E is type III and the diagonal entry (k, k) has value $c \neq 0$, then EA is obtained from A by multiplying $\text{row}_k(A)$ by c.*

Proof. This is easily done by inspection. $\qquad\qquad\qquad\qquad\qquad\square$

Theorem 3.4.2. *Let A be $n \times n$. The following statements are equivalent.*

(i) *A is invertible.*

(ii) *A is a product of elementary matrices.*

(iii) *A is row equivalent to the $n \times n$ identity matrix.*

(iv) *A has n pivot positions.*

(v) *The equation $A\mathbf{x} = \mathbf{0}$ has only the solution $\mathbf{x} = \mathbf{0}$; that is, Nul $A = \{\mathbf{0}\}$.*

(vi) *The columns of A form a linearly independent set.*

(vii) *The equation $A\mathbf{x} = \mathbf{b}$ has at least one solution for every $\mathbf{b} \in \mathbb{R}^n$.*

(viii) *The columns of A span $\mathbb{R}^n$; that is, Col $A = \mathbb{R}^n$.*

(ix) *There is a $n \times n$ matrix C so that $CA = I_n$.*

(x) *There is a $n \times n$ matrix D so that $AD = I_n$.*

(xi) *A^T is invertible.*

Proof. (i)→(ix): Since A is invertible, we have $A^{-1}A = I$. So we can chose $C = A^{-1}$.

(ix)→(v): Suppose $A\mathbf{x} = \mathbf{0}$. Multiplying both sides on the left with C gives $CA\mathbf{x} = C\mathbf{0}$. Since $CA = I_n$, we thus get $I_n\mathbf{x} = \mathbf{x} = \mathbf{0}$.

(v)↔(vi): This is due to the definition of linear independence.

(vi)→(iv): Since the columns of A form a linearly independent set, there is a pivot in each column of the echelon form of A. Thus A has n pivot positions.

(iv)→(iii): Since A has n pivot positions, and A is $n \times n$, the reduced row echelon form of A is I_n.

(iii)→(ii): Since the reduced row echelon form of A is I_n, there exist elementary matrices $E_1, \ldots, E_k$ so that $E_k E_{k-1} \cdots E_1 A = I$. Multiplying on both sides with $E_1^{-1} E_2^{-1} \cdots E_k^{-1}$, gives $A = E_1^{-1} E_2^{-1} \cdots E_k^{-1}$. Since the inverse of an elementary matrix is an elementary matrix, we obtain (ii).

(ii)→(i): If $A = E_1 \cdots E_k$, then since each E_j is invertible, we get by repeated use of Theorem 3.3.5(iii) that $A^{-1} = E_k^{-1} \cdots E_1^{-1}$. Thus A is invertible.

(i)↔(xi): This follows from Theorem 3.3.5(iv).

(i)→(x): Since A is invertible $AA^{-1} = I_n$. So we can chose $D = A^{-1}$.

(x)→(viii): Let $\mathbf{b} \in \mathbb{R}^n$. Then $AD\mathbf{b} = I_n\mathbf{b}$. Thus with $\mathbf{x} = D\mathbf{b}$, we have $A\mathbf{x} = \mathbf{b}$. Thus $\mathbf{b} \in \text{Col } A$. This holds for every $\mathbf{b} \in \mathbb{R}^n$, and thus $\text{Col } A = \mathbb{R}^n$.

(viii)↔(vii): This follows directly.

(vii)→(iv): Since $\text{Col } A = \mathbb{R}^n$, every row in the row echelon form has a pivot. Since A has n rows, this gives that A has n pivot positions. □

Note that C and D in Theorem 3.4.2(ix) and (x) are the same as $C = CI_n = C(AD) = (CA)D = I_nD = D$ and equal A^{-1} (by definition of the inverse). Thus, indeed, to find the inverse of A it suffices to solve the system $AB = I_n$ as indicated in the algorithm in the previous section.

3.5 Block Matrices

Definition 3.5.1. A block matrix (also called **partitioned matrix**) has the form

$$A = \begin{bmatrix} A_{11} & \cdots & A_{1n} \\ \vdots & & \vdots \\ A_{m1} & \cdots & A_{mn} \end{bmatrix}, \tag{3.4}$$

where A_{ij} are matrices themselves, of size $\mu_i \times \nu_j$, say.

For example, the 3×6 matrix

$$
A = \left[\begin{array}{ccc|cc|c}
1 & 2 & 3 & -1 & 7 & 8 \\
-3 & 6 & 5 & -4 & 2 & 9 \\
\hline
3 & 8 & 9 & -10 & -5 & -2
\end{array}\right]
$$

can be viewed as a 2×3 block matrix

$$
A = \left[\begin{array}{ccc}
A_{11} & A_{12} & A_{13} \\
A_{21} & A_{22} & A_{23}
\end{array}\right],
$$

where $A_{11} = \left[\begin{array}{ccc} 1 & 2 & 3 \\ -3 & 6 & 5 \end{array}\right], A_{12} = \left[\begin{array}{cc} -1 & 7 \\ -4 & 2 \end{array}\right], A_{13} = \left[\begin{array}{c} 8 \\ 9 \end{array}\right],$

$$
A_{21} = \left[\begin{array}{ccc} 3 & 8 & 9 \end{array}\right], A_{22} = \left[\begin{array}{cc} -10 & -5 \end{array}\right], A_{23} = \left[\begin{array}{c} -2 \end{array}\right].
$$

If we have A as in (3.4) and

$$
B = \left[\begin{array}{ccc}
B_{11} & \cdots & B_{1k} \\
\vdots & & \vdots \\
B_{n1} & \cdots & B_{nk}
\end{array}\right],
$$

where B_{ij} are matrices of size $\nu_i \times \kappa_j$, then the product AB can be computed as

$$
AB = \left[\begin{array}{ccc}
\sum_{r=1}^{n} A_{1r}B_{r1} & \cdots & \sum_{r=1}^{n} A_{1r}B_{rk} \\
\vdots & & \vdots \\
\sum_{r=1}^{n} A_{mr}B_{r1} & \cdots & \sum_{r=1}^{n} A_{mr}B_{rk}
\end{array}\right].
$$

Notice that the order in the multiplication is important: in the products $A_{ir}B_{rj}$, the A's are always on the left and the B's on the right. As an example, if $A = \left[\begin{array}{cc} A_{11} & A_{12} \\ A_{21} & A_{22} \end{array}\right]$ and $B = \left[\begin{array}{cc} B_{11} & B_{12} \\ B_{21} & B_{22} \end{array}\right]$, then

$$
AB = \left[\begin{array}{cc}
A_{11}B_{11} + A_{12}B_{21} & A_{11}B_{12} + A_{12}B_{22} \\
A_{21}B_{11} + A_{22}B_{21} & A_{21}B_{12} + A_{22}B_{22}
\end{array}\right].
$$

We can also add block matrices provided the sizes of the blocks match. If A is as in (3.4) and

$$
C = \left[\begin{array}{ccc}
C_{11} & \cdots & C_{1n} \\
\vdots & & \vdots \\
C_{m1} & \cdots & C_{mn}
\end{array}\right],
$$

where C_{ij} is of size $\mu_i \times \nu_j$, then $A + C$ can be computed as

$$
A + C = \left[\begin{array}{ccc}
A_{11} + C_{11} & \cdots & A_{1n} + C_{1n} \\
\vdots & & \vdots \\
A_{m1} + C_{m1} & \cdots & A_{mn} + C_{mn}
\end{array}\right].
$$

Note that when we write $B = \begin{bmatrix} \mathbf{b}_1 & \cdots & \mathbf{b}_k \end{bmatrix}$, we in effect view B as a block matrix. Then $AB = \begin{bmatrix} A\mathbf{b}_1 & \cdots & A\mathbf{b}_k \end{bmatrix}$ is a block matrix product.

The block matrix point of view is at times useful. Let us give some examples. We start with a formula for an invertible 2×2 block lower triangular matrix.

Theorem 3.5.2. *If A_{11} is $k \times k$ invertible, and A_{22} is $m \times m$ invertible, then*

$$\begin{bmatrix} A_{11} & 0 \\ A_{21} & A_{22} \end{bmatrix}^{-1} = \begin{bmatrix} A_{11}^{-1} & 0 \\ -A_{22}^{-1}A_{21}A_{11}^{-1} & A_{22}^{-1} \end{bmatrix}. \tag{3.5}$$

As a special case,

$$\begin{bmatrix} A_{11} & 0 \\ 0 & A_{22} \end{bmatrix}^{-1} = \begin{bmatrix} A_{11}^{-1} & 0 \\ 0 & A_{22}^{-1} \end{bmatrix}. \tag{3.6}$$

Proof.

$$\begin{bmatrix} A_{11} & 0 \\ A_{21} & A_{22} \end{bmatrix} \begin{bmatrix} A_{11}^{-1} & 0 \\ -A_{22}^{-1}A_{21}A_{11}^{-1} & A_{22}^{-1} \end{bmatrix} =$$

$$\begin{bmatrix} A_{11}A_{11}^{-1} & 0 \\ A_{21}A_{11}^{-1} - A_{22}A_{22}^{-1}A_{21}A_{11}^{-1} & A_{22}A_{22}^{-1} \end{bmatrix} = \begin{bmatrix} I_k & 0 \\ 0 & I_m \end{bmatrix}.$$

Similarly,

$$\begin{bmatrix} A_{11}^{-1} & 0 \\ -A_{22}^{-1}A_{21}A_{11}^{-1} & A_{22}^{-1} \end{bmatrix} \begin{bmatrix} A_{11} & 0 \\ A_{21} & A_{22} \end{bmatrix} = \begin{bmatrix} I_k & 0 \\ 0 & I_m \end{bmatrix}.$$

This proves (3.5).

Equation (3.6) follows directly from (3.5) by letting $A_{21} = 0$. $\qquad \square$

Let us do a block matrix sample problem.

Example 3.5.3. Let A, B, C and D be matrices of the same square size, and assume that A is invertible. Solve for matrices X, Y, Z, W (in terms of A, B, C and D) so that

$$\begin{bmatrix} A & B \\ C & D \end{bmatrix} \begin{bmatrix} Z & I \\ I & 0 \end{bmatrix} = \begin{bmatrix} 0 & X \\ Y & W \end{bmatrix}.$$

Answer. Let us multiply out the block matrix product

$$\begin{bmatrix} AZ + B & A \\ CZ + D & C \end{bmatrix} = \begin{bmatrix} 0 & X \\ Y & W \end{bmatrix}.$$

From the $(1,1)$ block entry we get $AZ + B = 0$, thus $AZ = -B$. Multiplying with A^{-1} on the left on both sides, we get $A^{-1}AZ = -A^{-1}B$, and thus $Z = -A^{-1}B$. From the $(2,1)$ entry we get $A = X$, and from the $(2,2)$ entry we find $C = W$. The $(1, 2)$ entry yields $Y = CZ + D$. Substituting $Z = -A^{-1}B$, we obtain $Y = -CA^{-1}B + D$. Thus the solution is

$$W = C, X = A, Y = -CA^{-1}B + D, Z = -A^{-1}B.$$

$\square$

Notice in the above example, when we solve for Z, we do

$$AZ = -B \Rightarrow A^{-1}(AZ) = A^{-1}(-B) \Rightarrow (A^{-1}A)Z = -A^{-1}B \Rightarrow Z = -A^{-1}B.$$

If we would have multiplied with A^{-1} on the right we would have obtained $AZA^{-1} = -BA^{-1}$, which we can not further simplify to extract Z. It is thus important that when manipulating these types of equations you should carefully keep track of the multiplication order.

In the above example we could simply manipulate matrix equations as it was given to us that A was invertible. In the example below we do not have that luxury, and we will have to go back to solving systems of linear equations.

Example 3.5.4. Find all solutions X to the matrix equation $AX = B$, where

$$A = \begin{bmatrix} 1 & -1 & 2 \\ 3 & 0 & 8 \end{bmatrix}, B = \begin{bmatrix} 1 & -4 \\ 2 & 3 \end{bmatrix}.$$

Answer. Writing $X = \begin{bmatrix} \mathbf{x} & \mathbf{y} \end{bmatrix}$ and $B = \begin{bmatrix} \mathbf{b}_1 & \mathbf{b}_2 \end{bmatrix}$, we obtain the equations $A\mathbf{x} = \mathbf{b}_1$ and $A\mathbf{y} = \mathbf{b}_2$. As these equations have the same coefficient matrix we combine these systems of linear equations in one augmented matrix:

$$\left[\begin{array}{ccc|cc} 1 & -1 & 2 & 1 & -4 \\ 3 & 0 & 8 & 2 & 3 \end{array} \right] \rightarrow \left[\begin{array}{ccc|cc} 1 & -1 & 2 & 1 & -4 \\ 0 & 3 & 2 & -1 & 15 \end{array} \right] \rightarrow$$

$$\left[\begin{array}{ccc|cc} 1 & -1 & 2 & 1 & -4 \\ 0 & 1 & \frac{2}{3} & -\frac{1}{3} & 5 \end{array} \right] \rightarrow \left[\begin{array}{ccc|cc} 1 & 0 & \frac{8}{3} & \frac{2}{3} & 1 \\ 0 & 1 & \frac{2}{3} & -\frac{1}{3} & 5 \end{array} \right].$$

Thus in the system $A\mathbf{x} = \mathbf{b}_1$ we have a free variable x_3 and in the system $A\mathbf{y} = \mathbf{b}_2$ we have a free variable y_3. We find

$$\begin{bmatrix} x_1 \\ x_2 \\ x_3 \end{bmatrix} = \begin{bmatrix} \frac{2}{3} \\ -\frac{1}{3} \\ 0 \end{bmatrix} + x_3 \begin{bmatrix} -\frac{8}{3} \\ -\frac{2}{3} \\ 1 \end{bmatrix}, \begin{bmatrix} y_1 \\ y_2 \\ y_3 \end{bmatrix} = \begin{bmatrix} 1 \\ 5 \\ 0 \end{bmatrix} + y_3 \begin{bmatrix} -\frac{8}{3} \\ -\frac{2}{3} \\ 1 \end{bmatrix}.$$

The final solution is now

$$X = \begin{bmatrix} \frac{2}{3} & 1 \\ -\frac{1}{3} & 5 \\ 0 & 0 \end{bmatrix} + \begin{bmatrix} -\frac{8}{3} \\ -\frac{2}{3} \\ 1 \end{bmatrix} \begin{bmatrix} x_3 & y_3 \end{bmatrix}, \text{ with } x_3 \text{ and } y_3 \text{ free.}$$

$\square$

3.6 Lower and Upper Triangular Matrices and LU Factorization

A matrix $L = (l_{ij})_{i,j=1}^n$ is **lower triangular** if $l_{ij} = 0$ when $i < j$. If, in addition, $l_{ii} = 1$ for $i = 1, \ldots n$, then L is called a **unit lower triangular** matrix. For instance

$$L_1 = \begin{bmatrix} 1 & 0 & 0 \\ -8 & 1 & 0 \\ 2 & -5 & 1 \end{bmatrix}, L_2 = \begin{bmatrix} 0 & 0 & 0 \\ -10 & 9 & 0 \\ 3 & 4 & 11 \end{bmatrix},$$

are lower triangular matrices, with L_1 a unit one. A matrix $U = (u_{ij})_{i,j=1}^n$ is **upper triangular** if $u_{ij} = 0$ when $i > j$. If, in addition, $u_{ii} = 1$ for $i = 1, \ldots n$, then U is called a **unit upper triangular** matrix. It is easy to see that when L is (unit) lower triangular, then L^T is (unit) upper triangular. Similarly, if U is (unit) upper triangular, then U^T is (unit) lower triangular. We say that $D = (d_{ij})_{i,j=1}^n$ is a **diagonal matrix** if $d_{ij} = 0$ for $i \neq j$. A matrix is a diagonal matrix if and only if it is both lower and upper triangular.

Lemma 3.6.1. *If L_1 and L_2 are $n \times n$ lower triangular matrices, then so is their product $L_1 L_2$.*
If U_1 and U_2 are $n \times n$ upper triangular matrices, then so is their product $U_1 U_2$.

Proof. Write $L_1 = (l_{ij})_{i,j=1}^n$ and $L_2 = (\lambda_{ij})_{i,j=1}^n$. Then for $i < j$

$$[L_1 L_2]_{ij} = \sum_{r=1}^n l_{ir} \lambda_{rj} = \sum_{r=1}^{j-1} l_{ir} \lambda_{rj} + \sum_{r=j}^n l_{ir} \lambda_{rj} = \sum_{r=1}^{j-1} l_{ir} \cdot 0 + \sum_{r=j}^n 0 \cdot \lambda_{rj} = 0.$$

The proof for the product of upper triangular matrices is similar, or one can use that $(U_1 U_2)^T = U_2^T U_1^T$ is the product of two lower triangular matrices. $\square$

If we compute the inverse of L_1 above, we obtain

$$\begin{bmatrix} 1 & 0 & 0 \\ -8 & 1 & 0 \\ 2 & -5 & 1 \end{bmatrix}^{-1} = \begin{bmatrix} 1 & 0 & 0 \\ 8 & 1 & 0 \\ 38 & 5 & 1 \end{bmatrix},$$

which is also unit lower triangular. This is true in general as we state now.

Lemma 3.6.2. *Let $L = (l_{ij})_{i,j=1}^n$ be lower triangular. Then L is invertible if and only $l_{ii} \neq 0$, $i = 1, \ldots, n$. In that case, L^{-1} is lower triangular as well, and has diagonal entries $\frac{1}{l_{ii}}$, $i = 1, \ldots, n$. In particular, if L is unit lower triangular, then so is L^{-1}.*
Similarly, an upper triangular $U = (u_{ij})_{i,j=1}^n$ is invertible if and only if $u_{ii} \neq 0$ for $i = 1, \ldots n$. In that case, U^{-1} is upper triangular as well, and has diagonal entries $\frac{1}{u_{ii}}$, $i = 1, \ldots, n$. In particular, if U is unit lower triangular, then so is U^{-1}.

Proof. First let us assume that L is lower triangular with $l_{ii} \neq 0$, $i = 1, \ldots n$. When one does row reduction

$$\begin{bmatrix} L & | & I_n \end{bmatrix} \to \cdots \to \begin{bmatrix} I_n & | & L^{-1} \end{bmatrix},$$

one can choose l_{ii}, $i = 1, \ldots n$, as the pivots and only use Operation 2 (replacing row_i by $\mathrm{row}_i + \alpha\,\mathrm{row}_k$) with $i > k$, until the coefficient matrix is diagonal (with diagonal entries l_{ii}, $i = 1, \ldots n$). After this one performs Operation 3 (multiply row_i with $\frac{1}{l_{ii}}$), to make the coefficient part equal to I_n. One then finds that L^{-1} is lower triangular with diagonal entries $\frac{1}{l_{ii}}$, $i = 1, \ldots, n$.

If $l_{ii} = 0$ for some i, choose i to be the lowest index where this happens. Then it is easy to see that $\mathrm{row}_i(L)$ is a linear combinations of rows $\mathrm{row}_1(L), \ldots, \mathrm{row}_{i-1}(L)$, and one obtains a row of all zeroes when one performs row reduction. This yields that L is not invertible.

For the statement on upper triangular matrices, one can apply the transpose and make use of the results on lower triangular matrices. $\square$

A matrix A is said to allow an **LU factorization** if we can write $A = LU$, with L lower triangular and U upper triangular. We shall focus on the case when L is a unit lower triangular matrix. Recall that doing elementary row operations corresponds to multiplying on the left with elementary matrices. If we choose these to be lower triangular unit matrices $L_1, \ldots, L_k$ we obtain

$$A \to L_1 A \to L_2 L_1 A \to \cdots \to L_k \cdots L_1 A.$$

We hope to achieve that $L_k \cdots L_1 A = U$ is upper triangular. Then $A = (L_1^{-1} \cdots L_k^{-1})U$ gives an LU factorization. To compute $L = L_1^{-1} \cdots L_k^{-1}$ easily, we use the following technical result.

Proposition 3.6.3. *Let $L_r = (l_{ij}^{(r)})_{i,j=1}^n$, $r = 1, \ldots, n-1$, be unit lower triangular matrices so that $l_{ij}^{(r)} = 0$ when $i > j \neq r$. Then $L = L_1^{-1} \cdots L_{n-1}^{-1}$ is the unit lower triangular matrix where the entries below the diagonal are* *given by*

$$L = (l_{ij})_{i,j=1}^n, \quad l_{ir} = -l_{ir}^{(r)}, i > r. \tag{3.7}$$

As an example, Proposition 3.6.3 yields for instance that

$$\begin{bmatrix} 1 & 0 & 0 \\ -8 & 1 & 0 \\ 2 & 0 & 1 \end{bmatrix}^{-1} \begin{bmatrix} 1 & 0 & 0 \\ 0 & 1 & 0 \\ 0 & -5 & 1 \end{bmatrix}^{-1} = \begin{bmatrix} 1 & 0 & 0 \\ 8 & 1 & 0 \\ -2 & 5 & 1 \end{bmatrix}.$$

It is easy to check, since

$$\begin{bmatrix} 1 & 0 & 0 \\ -8 & 1 & 0 \\ 2 & 0 & 1 \end{bmatrix} \begin{bmatrix} 1 & 0 & 0 \\ 8 & 1 & 0 \\ -2 & 5 & 1 \end{bmatrix} = \begin{bmatrix} 1 & 0 & 0 \\ 0 & 1 & 0 \\ 0 & 5 & 1 \end{bmatrix}$$

and

$$\begin{bmatrix} 1 & 0 & 0 \\ 0 & 1 & 0 \\ 0 & -5 & 1 \end{bmatrix} \begin{bmatrix} 1 & 0 & 0 \\ 0 & 1 & 0 \\ 0 & 5 & 1 \end{bmatrix} = \begin{bmatrix} 1 & 0 & 0 \\ 0 & 1 & 0 \\ 0 & 0 & 1 \end{bmatrix}.$$

The proof of Proposition 3.6.3 follows essentially the observations illustrated in this example.

Proof. Let L_r be as in the statement of the Proposition, and introduce the unit lower triangular matrix L via (3.7). Then $L_1 L$ is easily seen to be the matrix obtained from L by making the off diagonal entries in column 1 equal to 0. Next $L_2 L_1 L$ is easily seen to be the matrix obtained from $L_1 L$ by making the off diagonal entries in column 2 equal to 0. Continuing this way, $L_k L_{k-1} \cdots L_1 L$ is obtained from $L_{k-1} \cdots L_1 L$ by making the off-diagonal entries in column k equal to 0. But then it follows that $L_{n-1} \cdots L_1 L$ is the identity matrix. $\qquad\square$

Let us use these ideas to find an LU factorization of

$$\begin{bmatrix} 4 & 3 & -5 \\ -4 & -5 & 7 \\ 8 & 0 & -2 \end{bmatrix}.$$

First we will use the $(1,1)$ entry as a pivot and make zeros below it. We will keep track of the operations, so that we can easily make the L. We (i) replace row$_2$ by row$_2 + \textcircled{1}$row$_1$, and (ii) row$_3$ by row$_3\textcircled{-2}$ row$_1$:

$$\begin{bmatrix} \boxed{4} & 3 & -5 \\ -4 & -5 & 7 \\ 8 & 0 & -2 \end{bmatrix} \rightarrow \begin{bmatrix} 4 & 3 & -5 \\ 0 & -2 & 2 \\ 0 & -6 & 8 \end{bmatrix}, \quad L = \begin{bmatrix} 1 & 0 & 0 \\ \textcircled{-1} & 1 & 0 \\ \textcircled{2} & * & 1 \end{bmatrix}.$$

Notice that the negative of the circled numbers end up in the matrix L. Next, we make the $(2,2)$ entry the pivot and make zeros below it by replacing row$_3$ by row$_3 \left(-3\right)$row$_2$. This gives

$$
\begin{bmatrix} 4 & 3 & -5 \\ 0 & -2 & 2 \\ 0 & -6 & 8 \end{bmatrix} \rightarrow \begin{bmatrix} 4 & 3 & -5 \\ 0 & -2 & 2 \\ 0 & 0 & 2 \end{bmatrix} = U, \ L = \begin{bmatrix} 1 & 0 & 0 \\ -1 & 1 & 0 \\ 2 & 3 & 1 \end{bmatrix}.
$$

We find that A has been row reduced to the upper triangular matrix U, which yields that we have found an LU factorization of A:

$$
A = \begin{bmatrix} 4 & 3 & -5 \\ -4 & -5 & 7 \\ 8 & 0 & -2 \end{bmatrix} = \begin{bmatrix} 1 & 0 & 0 \\ -1 & 1 & 0 \\ 2 & 3 & 1 \end{bmatrix} \begin{bmatrix} 4 & 3 & -5 \\ 0 & -2 & 2 \\ 0 & 0 & 2 \end{bmatrix} = LU. \tag{3.8}
$$

This process works as long as we can keep picking diagonal entries as the next pivot. For instance, for the matrices

$$
\begin{bmatrix} 0 & 1 & -4 \\ 1 & 2 & 6 \\ 5 & 1 & -1 \end{bmatrix} \text{ and } \begin{bmatrix} 2 & 4 & -5 \\ 1 & 2 & -3 \\ 5 & -2 & 1 \end{bmatrix}
$$

the process will break down. In the next chapter (see Proposition 4.2.6), we will use determinants to give necessary and sufficient conditions for a matrix A to have an LU factorization $A = LU$, with L and U invertible.

Below is the pseudo code for LU factorization.

Algorithm 3 LU factorization

1: **procedure** LU(A)   ▷ LU factorization of $n \times n$ matrix A, if possible
2:     $L \leftarrow I_n, U \leftarrow A, i \leftarrow 1$
3:     **while** $i \leq n$ **do**   ▷ We have L, U invertible if $i > n$
4:        **if** $u_{ii} \neq 0$ **then**
5:           **for** $s = i+1, \ldots, n$ **do** $U \leftarrow E^{(2)}_{(s,i,-u_{si}/u_{ii})} U$ and $l_{si} \leftarrow u_{si}/u_{ii}$
6:           $i \leftarrow i+1$
7:        **else**
8:           TERMINATE   ▷ No invertible LU factorization exists
9:     **return** L, U ▷ $A = LU$; if $i > n$, L unit lower, U invertible upper

---

If $A = LU$, with L and U invertible, then the equation $A\mathbf{x} = \mathbf{b}$ can be solved in two steps:

$$
A\mathbf{x} = \mathbf{b} \iff \begin{cases} L\mathbf{y} = \mathbf{b}, \\ U\mathbf{x} = \mathbf{y}. \end{cases}
$$

Indeed, $\mathbf{b} = L\mathbf{y} = L(U\mathbf{x}) = (LU)\mathbf{x} = A\mathbf{x}$.

The advantage is that solving systems of linear equations with triangular matrices is easier than general systems. Let us illustrate this using the factorization (3.8). Consider the equation

$$\begin{bmatrix} 4 & 3 & -5 \\ -4 & -5 & 7 \\ 8 & 0 & -2 \end{bmatrix} \begin{bmatrix} x_1 \\ x_2 \\ x_3 \end{bmatrix} = \begin{bmatrix} 2 \\ -2 \\ 6 \end{bmatrix} = \mathbf{b}.$$

We first solve $L\mathbf{y} = \mathbf{b}$:

$$\begin{bmatrix} 1 & 0 & 0 \\ -1 & 1 & 0 \\ 2 & 3 & 1 \end{bmatrix} \begin{bmatrix} y_1 \\ y_2 \\ y_3 \end{bmatrix} = \begin{bmatrix} 2 \\ -2 \\ 6 \end{bmatrix},$$

which gives $y_1 = 2$, $-y_1 + y_2 = -2$ and $2y_1 + 3y_2 + y_3 = 6$. And thus $y_1 = 2$, $y_2 = -2 + y_1 = 0$, and $y_3 = 6 - 2y_1 - 3y_2 = 2$. Next we solve $U\mathbf{x} = \mathbf{y}$:

$$\begin{bmatrix} 4 & 3 & -5 \\ 0 & -2 & 2 \\ 0 & 0 & 2 \end{bmatrix} \begin{bmatrix} x_1 \\ x_2 \\ x_3 \end{bmatrix} = \begin{bmatrix} 2 \\ 0 \\ 2 \end{bmatrix}.$$

This gives $2x_3 = 2$, $-2x_2 + 2x_3 = 0$, and $4x_1 + 3x_2 - 5x_3 = 2$. Thus we find $x_3 = 1$, $x_2 = -\frac{1}{2}(-2x_3) = 1$, $x_1 = \frac{1}{4}(2 - 3x_2 + 5x_3) = 1$. And indeed $x_1 = x_2 = x_3 = 1$ is the solution as

$$\begin{bmatrix} 4 & 3 & -5 \\ -4 & -5 & 7 \\ 8 & 0 & -2 \end{bmatrix} \begin{bmatrix} 1 \\ 1 \\ 1 \end{bmatrix} = \begin{bmatrix} 2 \\ -2 \\ 6 \end{bmatrix}.$$

3.7 Exercises

Exercise 3.7.1. Compute the following products.

(a) $\begin{bmatrix} 1 & -1 \\ 3 & -2 \\ 5 & -1 \end{bmatrix} \begin{bmatrix} 0 & 1 \\ 2 & -1 \end{bmatrix}$.

(b) $\begin{bmatrix} 1 & 1 & 0 \\ 2 & 1 & 1 \end{bmatrix} \begin{bmatrix} 1 & 0 & 2 \\ 1 & 2 & 1 \\ 2 & 0 & 1 \end{bmatrix}$. *

---

*In live.sympy.org you can enter 'A=Matrix(2, 3, [1, 1, 0, 2, 1, 1])' and 'B=Matrix(3, 3, [1, 0, 2, 1, 2, 1, 2, 0, 1])' and 'A*B' to get the answer.

(c) $\begin{bmatrix} 2 & 2 \\ 2 & -10 \end{bmatrix} \begin{bmatrix} 5 & 6 \\ 1 & 2 \end{bmatrix}$.

Exercise 3.7.2. Let $A = \begin{bmatrix} 1 & 2 & 4 \\ 3 & -1 & 7 \\ -6 & 2 & -3 \end{bmatrix}$.

(a) Let $e_1 = \begin{bmatrix} 1 \\ 0 \\ 0 \end{bmatrix}$. Compute Ae_1 and $e_1^T A$. How would you describe the result of these multiplications?

(b) Let e_k be the kth standard basis vector of $\mathbb{R}^n$, and A a $n \times n$ matrix. How would you describe the result of the multiplications Ae_k and $e_k^T A$?

(c) How would you describe the result of the multiplication $e_k^T Ae_l$?

Exercise 3.7.3. Give an example that shows that matrix multiplication for 3×3 matrices is not commutative (in other words: give 3×3 matrices A and B such that $AB \neq BA$).

Exercise 3.7.4. Prove the rule $A(B+C) = AB + AC$, where $A \in \mathbb{R}^{m \times n}$ and $B, C \in \mathbb{R}^{n \times k}$.

Exercise 3.7.5. Let A be an $m \times k$ matrix and B a $k \times n$ matrix. Let $C = AB$.

(a) To compute the $(i, j)th$ entry of C, we need to compute

$$c_{ij} = a_{i1}b_{1j} + \cdots + a_{ik}b_{kj}.$$

How many multiplications are needed to compute c_{ij}?

(b) How many multiplications are needed to compute the matrix C?

Let D be a $n \times l$ matrix. One can compute the product in two ways, namely using the order given via $(AB)D$ or given via $A(BD)$.

(c) How many multiplications are needed to compute the matrix $(AB)D$? (Thus, first compute how many multiplications are needed to compute $C = AB$ and then add the number of multiplications needed to compute CD.)

(d) How many multiplications are needed to compute the matrix $A(BD)$?

When $m = 15$, $k = 40$, $n = 6$ and $l = 70$, then the answer under (c) is 9900, while the answer under (d) is 58800. Thus, when multiplying matrices, the order in which you multiply affects the efficiency. To find the most efficient way is the so-called 'Matrix Chain Ordering Problem'.

Exercise 3.7.6. The **trace** of a square matrix is defined to be the sum of its diagonal entries. Thus $\operatorname{tr}[(a_{ij})_{i,j=1}^n] = a_{11} + \cdots + a_{nn} = \sum_{j=1}^n a_{jj}$.

(a) Show that if $A \in \mathbb{R}^{n \times n}$, then $\operatorname{tr}(A^T) = \operatorname{tr}(A)$.

(b) Show that if $A \in \mathbb{R}^{m \times n}$ and $B \in \mathbb{R}^{n \times m}$, then $\operatorname{tr}(AB) = \operatorname{tr}(BA)$.

(c) Show that if $A \in \mathbb{R}^{m \times n}$, $B \in \mathbb{R}^{n \times k}$, and $C \in \mathbb{R}^{k \times m}$, then $\operatorname{tr}(ABC) = \operatorname{tr}(CAB) = \operatorname{tr}(BCA)$.

(d) Give an example of matrices $A, B, C \in \mathbb{R}^{n \times n}$ so that $\operatorname{tr}(ABC) \neq \operatorname{tr}(BAC)$.

Exercise 3.7.7. Let $A = \begin{bmatrix} 0 & 1 & -1 \\ 1 & 2 & 2 \end{bmatrix}$, $B = \begin{bmatrix} -2 & 4 \\ -2 & 6 \end{bmatrix}$. Compute the following matrices. If the expression is not well-defined, explain why this is the case.

(a) AB.

(b) $B^T B + A A^T$.

(c) B^{-1}.

(d) $(B^T)^{-1}$.

Exercise 3.7.8. Let $A = \begin{bmatrix} 1 & -1 \\ 2 & 2 \end{bmatrix}$, $B = \begin{bmatrix} 1 & -2 & 4 \\ 3 & -2 & 6 \end{bmatrix}$. Compute the following matrices. If the expression is not well-defined, explain why this is the case.

(a) AB.

(b) $BB^T + A$.

(c) A^{-1}.

(d) B^{-1}.

Exercise 3.7.9. Find the inverses of the following matrices, if possible.

(a) $\begin{bmatrix} 1 & 2 & -5 \\ 0 & 1 & 3 \\ 0 & 2 & 7 \end{bmatrix}$.

(b) $\begin{bmatrix} 1 & -1 & 0 \\ 2 & -1 & 0 \\ 1 & 1 & 2 \end{bmatrix}$.[†]

[†]To check in sagecell.sagemath.org, enter 'A = matrix([[1,-1,0], [2,-1,0], [1,1,2]])' and 'A.inverse()'.

(c) $\begin{bmatrix} 2 & 3 & 1 \\ 1 & 4 & 3 \\ 1 & 1 & 0 \end{bmatrix}$.

Exercise 3.7.10. Find the inverse of

$$\begin{bmatrix} 1 & 3 & 0 & 0 & 0 \\ 2 & 5 & 0 & 0 & 0 \\ 0 & 0 & 1 & 2 & 0 \\ 0 & 0 & 1 & 1 & 0 \\ 0 & 0 & 0 & 0 & 3 \end{bmatrix}$$

by viewing it as a partitioned matrix.

Exercise 3.7.11. Show the following for matrices $A \in \mathbb{R}^{m \times n}$, $B \in \mathbb{R}^{n \times k}$, and $C \in \mathbb{R}^{k \times m}$.

(a) $\text{Col}(AB) \subseteq \text{Col}A$.

(b) $\text{Nul}A \subseteq \text{Nul}(CA)$.

(c) $AB = 0$ if and only if $\text{Col}B \subseteq \text{Nul}A$.

Exercise 3.7.12. Let A, B and C be invertible $n \times n$ matrices. Solve for X in the matrix equation

$$A^{-1} \left(X^T + C \right) B = I.$$

Exercise 3.7.13. Let A, B, C and D be invertible matrices of the same size. Solve for matrices X, Y, Z, W (in terms of A, B, C and D) so that

$$\begin{bmatrix} A & B \\ C & D \end{bmatrix} \begin{bmatrix} Z & I \\ I & 0 \end{bmatrix} = \begin{bmatrix} 0 & X \\ Y & W \end{bmatrix}.$$

Exercise 3.7.14. Given is the equality

$$\begin{bmatrix} A & B \\ C & D \end{bmatrix} \begin{bmatrix} X & Y \\ I & 0 \end{bmatrix} = \begin{bmatrix} E & F \\ Z & W \end{bmatrix}.$$

Express X, Y, Z, W in terms of A, B, C, D, E, F. You may assume that all matrices are square and invertible.

Exercise 3.7.15. Let

$$T = \begin{bmatrix} A & B \\ C & D \end{bmatrix} \tag{3.9}$$

be a block matrix, and suppose that D is invertible. Define the **Schur complement** S of D in T by $S = A - BD^{-1}C$. Show that

$$\text{rank } T = \text{rank}(A - BD^{-1}C) + \text{rank } D.$$

Hint: Multiply (3.9) on the left with P and on the right with Q, where

$$P = \begin{bmatrix} I & -BD^{-1} \\ 0 & I \end{bmatrix}, Q = \begin{bmatrix} I & 0 \\ -D^{-1}C & I \end{bmatrix}.$$

Exercise 3.7.16. Let $A \in \mathbb{R}^{m \times n}$ and $B \in \mathbb{R}^{n \times m}$.

(a) Suppose that $I_m - AB$ is invertible, and put $C = I_n + B(I_m - AB)^{-1}A$. Show that $(I_n - BA)C = I_n$ and $C(I_n - BA) = I_n$.

(b) Prove that $I_m - AB$ is invertible if and only if $I_n - BA$ is invertible.

Exercise 3.7.17. Let $A \in \mathbb{R}^{n \times n}, B \in \mathbb{R}^{n \times m}$, and $C \in \mathbb{R}^{m \times m}$, and suppose that A and C are invertible. Show that

$$\begin{bmatrix} A & B \\ 0 & C \end{bmatrix}^{-1} = \begin{bmatrix} A^{-1} & -A^{-1}BC^{-1} \\ 0 & C^{-1} \end{bmatrix}.$$

(Hint: Simply multiply the two block matrices in both ways, and determine that the products equal I_{n+m}.)

Exercise 3.7.18. Let $A \in \mathbb{R}^{m \times n}, B \in \mathbb{R}^{m \times k}$, and $C \in \mathbb{R}^{n \times k}$. Show that

$$\begin{bmatrix} I_m & A & B \\ 0 & I_n & C \\ 0 & 0 & I_k \end{bmatrix}^{-1} = \begin{bmatrix} I_m & -A & AC - B \\ 0 & I_n & -C \\ 0 & 0 & I_k \end{bmatrix}.$$

(Hint: Simply multiply the two block matrices in both ways, and determine that the products equal I_{m+n+k}.)

Exercise 3.7.19. Let $A = \begin{bmatrix} B & C \end{bmatrix}$, where $B \in \mathbb{R}^{m \times n}$ and $C \in \mathbb{R}^{m \times k}$. Show that

$$\mathrm{rank} B \leq \mathrm{rank} A, \ \mathrm{rank} C \leq \mathrm{rank} A.$$

Exercise 3.7.20. Let $A = \begin{bmatrix} B \\ C \end{bmatrix}$, where $B \in \mathbb{R}^{m \times n}$ and $C \in \mathbb{R}^{k \times n}$. Show that

$$\mathrm{rank} B \leq \mathrm{rank} A, \ \mathrm{rank} C \leq \mathrm{rank} A.$$

(Hint: Take transposes and use Exercise 3.7.19.)

Exercise 3.7.21. In this exercise we use notions introduced in Exercise 2.6.21. Let

$$\mathcal{A} = \begin{bmatrix} B & ? \\ C & D \end{bmatrix}$$

be a partial block matrix, where the $(1,2)$ block is unknown. Show that the rank of a minimal rank completion of $\mathcal{A}$ is equal to

$$\mathrm{rank} \begin{bmatrix} B \\ C \end{bmatrix} + \mathrm{rank} \begin{bmatrix} C & D \end{bmatrix} - \mathrm{rank}\, C.$$

(Hint: Consider the pivot columns of $\begin{bmatrix} C & D \end{bmatrix}$, and notice that those appearing in D will remain linearly independent of the columns $\begin{bmatrix} B \\ C \end{bmatrix}$ regardless of how these columns are completed. Once these pivot columns appearing in D are completed, all other columns can be completed to be linear combinations of the other columns.)

Exercise 3.7.22. Let $A \in \mathbb{R}^{m \times n}$ and $B \in \mathbb{R}^{n \times m}$. If $AB = I_m$, we say that B is a **right inverse** of A. We also say that A is a **left inverse** of B.

(a) Show that if A has a right inverse, then $\mathrm{Col} A = \mathbb{R}^m$.

(b) Show that if B has a left inverse, then $\mathrm{Nul} B = \{\mathbf{0}\}$.

Exercise 3.7.23. Find all solutions X to the matrix equation $AX = B$, where

$$A = \begin{bmatrix} 1 & -1 & 2 \\ 2 & 0 & 8 \end{bmatrix}, B = \begin{bmatrix} 1 & -4 \\ 2 & 3 \end{bmatrix}.$$

Exercise 3.7.24. Let $A = \begin{bmatrix} 1 & -1 \\ -1 & 2 \end{bmatrix}$, $B = \begin{bmatrix} -2 & 4 & 0 \\ -2 & 6 & -4 \end{bmatrix}$ and $C = \begin{bmatrix} 1 & 3 & -1 \\ 1 & -5 & -2 \end{bmatrix}$. Solve for the matrix X in the equation $AX + C = B$.

Exercise 3.7.25. Find an LU factorization of the following matrices.

(a) $\begin{bmatrix} 1 & -2 \\ 4 & -3 \end{bmatrix}$.‡

(b) $\begin{bmatrix} 1 & 4 \\ 2 & 9 \end{bmatrix}$.

(c) $\begin{bmatrix} 1 & -1 & 0 \\ 2 & -3 & 1 \\ 0 & 1 & 2 \end{bmatrix}$.

(d) $\begin{bmatrix} -2 & 2 & -2 \\ 3 & -1 & 5 \\ \frac{1}{2} & \frac{3}{2} & \frac{1}{2} \end{bmatrix}$.

Exercise 3.7.26. We call a $n \times n$ matrix N **nilpotent** if there exists $k \in \mathbb{N}$ so that $N^k = 0$.

(a) Show that $\begin{bmatrix} 0 & 2 & -3 \\ 0 & 0 & -4 \\ 0 & 0 & 0 \end{bmatrix}$ is nilpotent.

(b) Show that if $N^k = 0$, then $(I - N)^{-1} = I + N + N^2 + \cdots + N^{k-1}$. (Hint: Compute the products $(I - N)(I + N + N^2 + \cdots + N^{k-1})$ and $(I + N + N^2 + \cdots + N^{k-1})(I - N)$.)

---

‡In www.wolframalpha.com you can enter 'LU factorization of [[1,-2], [4,-3]]' to get the answer. Sometimes this command will find you the PLU factorization, where P is a permutation.

(c) Use (a) and (b) to compute $\begin{bmatrix} 1 & -2 & 3 \\ 0 & 1 & 4 \\ 0 & 0 & 1 \end{bmatrix}^{-1}$.

Exercise 3.7.27. The **Leontief input-output model**[§] describes how different industries affect each other. It makes effective use of matrices. Let us explain this on a small example, using the interaction between just two industries: the electric company (E) and the water company (W). Both companies need to make use of their own product as well as the other's product to meet their production for the community.

Let us measure everything in dollars. Suppose that creating one dollar's worth of electricity requires 0.2\$ worth of electricity and 0.1\$ worth of water, and that creating one dollar's worth of water requires 0.3\$ worth of electricity and 0.4\$ worth of water. Ultimately, we want to create \$10 million worth of electricity and \$12 million worth of water for the community (we call this the 'outside demand'). To meet this outside demand, we need to make more electricity and water as part of it gets used in the production (the part that gets used in the production is called the 'inside demand'). This leads to the equation

Total production = Internal demand + Outside demand.

Let us denote by e the total amount of electricity and by w the total amount of water we need to produce. We obtain the following equations (measured in millions of dollars).

$$\begin{cases} e = 0.2e + 0.1w + 10, \\ w = 0.3e + 0.4w + 12. \end{cases}$$

Letting $\mathbf{x} = \begin{bmatrix} e \\ w \end{bmatrix}$ be the vector of unknows, $\mathbf{d} = \begin{bmatrix} 10 \\ 12 \end{bmatrix}$ the vector of outside demand, and C the internal cost matrix

$$C = \begin{bmatrix} 0.2 & 0.1 \\ 0.3 & 0.4 \end{bmatrix},$$

we obtain the equation

$$\mathbf{x} = C\mathbf{x} + \mathbf{d}.$$

To solve this equation, we bring all the terms involving $\mathbf{x}$ to the left side, giving

$$(I - C)\mathbf{x} = \mathbf{x} - C\mathbf{x} = \mathbf{d}.$$

If $I - C$ is invertible, we obtain

$$\mathbf{x} = (I - C)^{-1}\mathbf{d}.$$

In this case, we find $e = 16$ million dollars, and $w = 28$ million dollars.

[§]The Russian-American economist Wassily Leontief won the Nobel Prize in Economics in 1973 for his model.

(a) The following economy is based on three sectors, agriculture (A), energy
    (E) and manufacturing (M). To produce one dollar worth of agriculture
    requires an input of $0.25 from the agriculture sector and $0.35 from the
    energy sector. To produce one dollar worth of energy requires an input
    of $0.15 from the energy sector and $0.3 from the manufacturing sector.
    To produce one dollar worth of manufacturing requires an input of $0.15
    from the agriculture sector, $0.2 from the energy sector, and $0.25 from
    the manufacturing sector. The community's demand is $20 million for
    agriculture, $10 million for energy, and $30 million for manufacturing. Set
    up the equation to find the total production of agriculture, energy and
    manufacturing to meet the community's demand.

(b) Compute the total production agriculture, energy and manufacturing to
    meet the community's demand. (Use your favorite computational soft-
    ware.)

Exercise 3.7.28. In Exercise 2.6.22 we described the incidence matrix of a
directed graph. In this exercise we will introduce the **adjacency matrix** A
of a graph $G = (V, E)$ with n vertices $V = \{1, \ldots n\}$ by letting

$$a_{ij} = 1 \text{ if and only if } (i, j) \text{ is an edge}$$

and $a_{ij} = 0$ otherwise. Thus for the graph

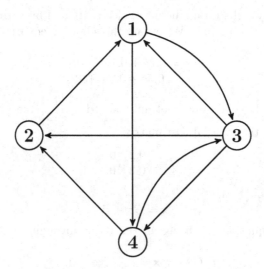

Figure 3.1: A graph and its adjacency matrix.

we obtain

$$A = \begin{bmatrix} 0 & 0 & 1 & 1 \\ 1 & 0 & 0 & 0 \\ 1 & 1 & 0 & 1 \\ 0 & 1 & 1 & 0 \end{bmatrix}.$$

Squaring A, we get

$$A^2 = \begin{bmatrix} 1 & 2 & 1 & 1 \\ 0 & 0 & 1 & 1 \\ 1 & 1 & 2 & 1 \\ 2 & 1 & 0 & 1 \end{bmatrix}.$$

What does A^2 tell us about the graph? Let us denote $A^k = (a_{ij}^{(k)})_{i,j=1}^n$. So, for instance, $a_{12}^{(2)} = 2$; that is, the $(1,2)$ entry of A^2 equals 2. Using how matrices multiply, we know that

$$a_{12}^{(2)} = \sum_{r=1}^n a_{1r} a_{r2}.$$

Thus $a_{12}^{(2)}$ counts how many times both a_{1r} and a_{r2} equal one when $r = 1, \ldots, n$. This counts exactly how many walks of length 2 there are in the graph to go from 1 to 2. Indeed, we can go $1 \to 3 \to 2$ or $1 \to 4 \to 2$. In general we have the rule

$a_{ij}^{(k)}$ = the number of walks of length k in the graph from vertex $\mathbf{i}$ to vertex $\mathbf{j}$.
(3.10)

(a) For the graph above, find the number of walks of length 4 from vertex 1 to 3 (by using matrix multiplication).

(b) Find the adjacency matrix of the following graph

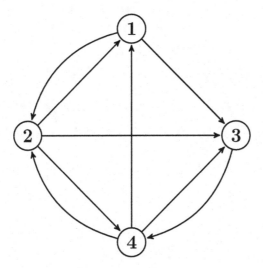

Figure 3.2: An example graph.

(c) In the graph from part (b), find the number of walks of length 3 from vertex 2 to 3.

(d) Prove rule (3.10). (Hint: Use induction on k. Note that for a walk of length $k+1$ from vertex i to vertex j, you need a vertex r so that there is a walk of length k from i to r, and then an edge from r to j.)

Exercise 3.7.29. True or False? Justify each answer.

(i) For matrices A and B, if $A + B$ is well-defined, then so is AB^T.

(ii) If A has two identical columns, and the product AB is well-defined, then AB also has (at least) two identical columns.

(iii) If rows 1 and 2 of the matrix B are the same, and A is a matrix, then rows 1 and 2 of the matrix AB are also the same.

(iv) If A is not an invertible matrix, then the equation $A\mathbf{x} = \mathbf{b}$ does not have a solution.

(v) For matrices A and B, if AB is well-defined, then so is $A^T B^T$.

4

Determinants

CONTENTS

4.1   Definition of the Determinant and Properties

The determinant function assigns a scalar (denoted $\det A$) to a square matrix A. We build up the definition according to the size of the matrix.

- For a 1×1 matrix $A = [a_{11}]$, we define $\det A = a_{11}$.

- For a 2×2 matrix $A = \begin{bmatrix} a_{11} & a_{12} \\ a_{21} & a_{22} \end{bmatrix}$ we define

$$\det A = a_{11}a_{22} - a_{12}a_{21}.$$

For example,

$$\det \begin{bmatrix} 1 & 2 \\ 3 & 4 \end{bmatrix} = 4 - 6 = -2.$$

- For a 3×3 matrix $A = (a_{ij})_{i,j=1}^{3}$, we define

$$\det A = a_{11} \det \begin{bmatrix} a_{22} & a_{23} \\ a_{32} & a_{33} \end{bmatrix} - a_{12} \det \begin{bmatrix} a_{21} & a_{23} \\ a_{31} & a_{33} \end{bmatrix} + a_{13} \det \begin{bmatrix} a_{21} & a_{22} \\ a_{31} & a_{32} \end{bmatrix}$$

$$= (-1)^{1+1} a_{11} \det A_{11} + (-1)^{1+2} a_{12} \det A_{12} + (-1)^{1+3} a_{13} \det A_{13},$$

where

$$A_{rs} = \text{the matrix } A \text{ without row } r \text{ and column } s.$$

For example,

$$\det \begin{bmatrix} 4 & 2 & -3 \\ 1 & -1 & 0 \\ 5 & -2 & 8 \end{bmatrix} = 4\det \begin{bmatrix} -1 & 0 \\ -2 & 8 \end{bmatrix} - 2\det \begin{bmatrix} 1 & 0 \\ 5 & 8 \end{bmatrix} + (-3)\det \begin{bmatrix} 1 & -1 \\ 5 & -2 \end{bmatrix}$$

$$= 4(-8 - 0) - 2(8 - 0) - 3(-2 + 5) = -57.$$

- For an $n \times n$ matrix $A = (a_{ij})_{i,j=1}^{n}$, we define

$$\det A = (-1)^{1+1}a_{11}\det A_{11} + (-1)^{1+2}a_{12}\det A_{12} + \cdots + (-1)^{1+n}a_{1n}A_{1n}.$$

We introduce the (i,j)th **cofactor** of the matrix A by $C_{ij} = (-1)^{i+j}\det A_{ij}$. With that notation we find that

$$\det A = a_{11}C_{11} + a_{12}C_{12} + \cdots + a_{1n}C_{1n} = \sum_{j=1}^{n} a_{1j}C_{1j}. \qquad (4.1)$$

For example,

$$\det \begin{bmatrix} 1 & 4 & 2 & -3 \\ 0 & 1 & -1 & 0 \\ 5 & 0 & -2 & 8 \\ 6 & 9 & -5 & 7 \end{bmatrix} = 1\det \begin{bmatrix} 1 & -1 & 0 \\ 0 & -2 & 8 \\ 9 & -5 & 7 \end{bmatrix} - 4\det \begin{bmatrix} 0 & -1 & 0 \\ 5 & -2 & 8 \\ 6 & -5 & 7 \end{bmatrix} +$$

$$2\det \begin{bmatrix} 0 & 1 & 0 \\ 5 & 0 & 8 \\ 6 & 9 & 7 \end{bmatrix} - (-3)\det \begin{bmatrix} 0 & 1 & -1 \\ 5 & 0 & -2 \\ 6 & 9 & -5 \end{bmatrix} = ((-14 + 40) - (-1)(0 - 72)) -$$

$$4(-(-1)(35 - 48)) + 2(-1(35 - 48)) + 3(-1(-25 + 12) + (-1)45) = -64.$$

The sign pattern $(-1)^{i+j}$ can be remembered easily by having the following checkered pattern in mind:

$$\begin{bmatrix} + & - & + & - & \cdots \\ - & + & - & + & \cdots \\ + & - & + & - & \cdots \\ - & + & - & + & \cdots \\ \vdots & \vdots & \vdots & \vdots & \end{bmatrix}.$$

We next provide some useful properties, which we prove in the next section.

Theorem 4.1.1. *Let A be an $n \times n$ matrix. The elementary row operations have the following effect on the determinant:*

(i) *If B is obtained from the $n \times n$ matrix A by switching two rows of A, then $\det B = -\det A$.*

(ii) *If B is obtained from the $n \times n$ matrix A by adding a multiple of one row to another row of A, then $\det B = \det A$.*

(iii) *If B is obtained from the $n \times n$ matrix A by multiplying a row of A by c, then $\det B = c \det A$.*

If we combine Theorem 4.1.1(i) with (4.1), we obtain the alternative formula ('expand along ith row'):

$$\det A = \sum_{j=1}^{n} a_{ij} C_{ij} = \sum_{j=1}^{n} (-1)^{i+j} a_{ij} \det A_{ij}$$

$$= (-1)^{i+1} a_{i1} \det A_{i1} + (-1)^{i+2} a_{i2} \det A_{i2} + \cdots + (-1)^{i+n} a_{in} A_{in}.$$

For example, if we expand the 3×3 matrix above and expand it along the second row (thus we choose $i = 2$), we get

$$\det \begin{bmatrix} 4 & 2 & -3 \\ 1 & -1 & 0 \\ 5 & -2 & 8 \end{bmatrix} = -1 \det \begin{bmatrix} 2 & -3 \\ -2 & 8 \end{bmatrix} + (-1) \det \begin{bmatrix} 4 & -3 \\ 5 & 8 \end{bmatrix} - 0 \det \begin{bmatrix} 4 & 2 \\ 5 & -2 \end{bmatrix}$$

$$= -(16 - 6) - (32 + 15) - 0(-8 - 10) = -57.$$

The determinant of a lower or upper triangular matrix is easy to compute, as in that case the determinant is the product of the diagonal entries.

Theorem 4.1.2.
If $L = (l_{ij})_{i,j=1}^{n}$ is lower triangular, then $\det L = \prod_{i=1}^{n} l_{ii} = l_{11} \cdots l_{nn}$.
If $U = (u_{ij})_{i,j=1}^{n}$ is upper triangular, then $\det U = \prod_{i=1}^{n} u_{ii} = u_{11} \cdots u_{nn}$.
In particular, $\det I_n = 1$.

If we combine Theorem 4.1.1 with Theorem 4.1.2, we obtain another way to compute the determinant: row reduce the matrix, while carefully keeping track how the determinant changes in each step, until the matrix is triangular (at which point the determinant can be easily calculated). For instance,

$$\begin{bmatrix} 4 & 2 & -3 \\ 1 & -1 & 0 \\ 5 & -2 & 8 \end{bmatrix} \rightarrow \begin{bmatrix} \boxed{1} & -1 & 0 \\ 4 & 2 & -3 \\ 5 & -2 & 8 \end{bmatrix} \rightarrow \begin{bmatrix} \boxed{1} & -1 & 0 \\ 0 & \boxed{6} & -3 \\ 0 & 3 & 8 \end{bmatrix} \rightarrow \begin{bmatrix} \boxed{1} & -1 & 0 \\ 0 & \boxed{6} & -3 \\ 0 & 0 & \boxed{9\tfrac{1}{2}} \end{bmatrix}.$$

In the first step we switched two rows, so the determinant changes sign, and in the other steps we added a multiple of one row to another, which does not change the determinant. We thus find that

$$
\det \begin{bmatrix} 4 & 2 & -3 \\ 1 & -1 & 0 \\ 5 & -2 & 8 \end{bmatrix} = -\det \begin{bmatrix} \boxed{1} & -1 & 0 \\ 0 & \boxed{6} & -3 \\ 0 & 0 & \boxed{9\frac{1}{2}} \end{bmatrix} = -1 \cdot 6 \cdot 9\frac{1}{2} = -57.
$$

The next result states some more useful properties.

Theorem 4.1.3. *For $n \times n$ matrices A and B, we have*

(i) $\det A^T = \det A$.

(ii) $\det(BA) = (\det B)(\det A)$.

(iii) A is invertible if and only if $\det A \neq 0$, and in that case

$$
\det(A^{-1}) = \frac{1}{\det A}.
$$

Using the transpose rule (Theorem 4.1.3(i)), we now also obtain formulas where we expand along the jth column:

$$
\det A = (-1)^{1+j} a_{1j} \det A_{1j} + (-1)^{2+j} a_{2j} \det A_{2j} + \cdots + (-1)^{n+j} a_{nj} A_{nj}.
$$

For example, expanding along the third column (thus $j = 3$), we get

$$
\det \begin{bmatrix} 4 & 2 & -3 \\ 1 & -1 & 0 \\ 5 & -2 & 8 \end{bmatrix} = -3 \det \begin{bmatrix} 1 & -1 \\ 5 & -2 \end{bmatrix} - 0 \det \begin{bmatrix} 4 & 2 \\ 5 & -2 \end{bmatrix} + 8 \det \begin{bmatrix} 4 & 2 \\ 1 & -1 \end{bmatrix} =
$$

$$
-3(-2 + 5) - 0(-8 - 10) + 8(-4 - 2) = -57.
$$

Theorem 4.1.3(iii) shows that invertibility of a matrix can be captured by its determinant being nonzero. It is this property that inspired the name 'determinant'.

4.2   Alternative Definition and Proofs of Properties

To prove the results in the previous section, it is more convenient to define the determinant in another way. Before we come to this alternative definition of the determinant, we first have to discuss permutations. A **permutation** on the set $\underline{n} = \{1, \ldots, n\}$ is a bijective function on $\underline{n}$. For example, $\sigma : \underline{7} \to \underline{7}$ given by

$$
\begin{array}{c|ccccccc}
i & 1 & 2 & 3 & 4 & 5 & 6 & 7 \\
\hline
\sigma(i) & 5 & 3 & 1 & 4 & 6 & 2 & 7
\end{array}
\tag{4.2}
$$

is a permutation (here $\sigma(1) = 5, \sigma(2) = 3$, etc.). The set of all permutations on $\underline{n}$ is denoted as S_n. The composition $\sigma \circ \tau$ of two permutations $\sigma, \tau \in S_n$ is again a permutation on $\underline{n}$. (As an aside, we mention that S_n with the operation $\circ$ is the so-called 'Symmetric group'; a favorite example in an Abstract Algebra course). To any permutation we can assign a sign whose value is $+1$ or -1. One way to define the sign function is by using the polynomial in variables $x_1, \ldots, x_n$ defined by

$$
P(x_1, \ldots, x_n) = \prod_{1 \leq i < j \leq n} (x_i - x_j).
$$

For example, $P(x_1, x_2, x_3) = (x_1 - x_2)(x_1 - x_3)(x_2 - x_3)$. We now define

$$
\text{sign} : S_n \to \{-1, 1\}, \quad \text{sign}(\sigma) = \frac{P(x_{\sigma(1)}, \ldots, x_{\sigma(n)})}{P(x_1, \ldots, x_n)}.
$$

For instance, the permutation

$$
\begin{array}{c|cccc}
i & 1 & 2 & 3 & 4 \\
\hline
\sigma(i) & 1 & 4 & 3 & 2
\end{array}
\tag{4.3}
$$

has sign

$$
\text{sign}(\sigma) = \frac{(x_1 - x_4)(x_1 - x_3)(x_1 - x_2)(x_4 - x_3)(x_4 - x_2)(x_3 - x_2)}{(x_1 - x_2)(x_1 - x_3)(x_1 - x_4)(x_2 - x_3)(x_2 - x_4)(x_3 - x_4)} = -1.
$$

A **transposition** is a permutation which interchanges two numbers and leaves the others fixed. An example is σ in (4.3) as it interchanges 2 and 4, and leaves the others (1 and 3 in this case) alone.

Lemma 4.2.1. *The sign function has the following properties.*

(i) $\text{sign}(\sigma \circ \tau) = \text{sign}(\sigma)\text{sign}(\tau)$.

(ii) $\text{sign}(id) = 1$, *where id is the identity on* $\underline{n}$.

(iii) $\text{sign}(\sigma) = \text{sign}(\sigma^{-1})$.

(iv) If σ *is a transposition, then* $\text{sign}(\sigma) = -1$.

Proof. (i) follows from

$$
\text{sign}(\sigma \circ \tau) = \frac{P(x_{\sigma(\tau((1))}, \ldots, x_{\sigma(\tau(n))})}{P(x_1, \ldots, x_n)}
$$

$$= \frac{P(x_{\sigma(\tau(1))}, \ldots, x_{\sigma(\tau(n))})}{P(x_{\tau(1)}, \ldots, x_{\tau(n)})} \frac{P(x_{\tau(1)}, \ldots, x_{\tau(n)})}{P(x_1, \ldots, x_n)} = \text{sign}(\sigma)\text{sign}(\tau).$$

(ii) is immediate.

(iii) follows from (i) and (ii) since $\sigma \circ \sigma^{-1} = id$.

(iv) If we let σ denote the transposition that switches i and j, with $i < j$, then its sign equals

$$\text{sign}(\sigma) = \frac{(x_j - x_i) \prod_{i<r<j}(x_j - x_r)(x_r - x_i)}{(x_i - x_j) \prod_{i<r<j}(x_i - x_r)(x_r - x_j)},$$

where we omitted the factors that are the same in numerator and denominator. Thus there are $2(j - i - 1) + 1$ factors that switch sign, which is an odd number. Thus $\text{sign}(\sigma) = -1$.   □

It will be convenient to introduce the shorthand

$$(-1)^{\sigma} := \text{sign}(\sigma).$$

Then we have that $(-1)^{\sigma \circ \tau} = (-1)^{\sigma}(-1)^{\tau}$ and $(-1)^{id} = 1$. A permutation σ with $(-1)^{\sigma} = -1$ is usually referred to as an **odd permutation**, as it is the composition of an odd number of transpositions. Similarly, a permutation σ with $(-1)^{\sigma} = 1$ is referred to as an **even permutation**, as it is the composition of an even number of transpositions. There is a lot more to say about permutations*, but we need to get back to determinants.

Definition 4.2.2. We now define for $A = (a_{ij})_{i,j=1}^n$ its **determinant** to be

$$\det A = \sum_{\sigma \in S_n} (-1)^{\sigma} a_{1,\sigma(1)} a_{2,\sigma(2)} \cdots a_{n,\sigma(n)} = \sum_{\sigma \in S_n} \left((-1)^{\sigma} \prod_{i=1}^n a_{i,\sigma(i)} \right).$$

$$(4.4)$$

When $n = 2$ there are 2 permutations (id and the one that switches 1 and 2), and we get

$$\det A = a_{11} a_{22} - a_{12} a_{21}.$$

When $n = 3$ there are $3! = 6$ permutations (id, $1 \leftrightarrow 2$, $1 \leftrightarrow 3$, $2 \leftrightarrow 3$, $1 \to 2 \to 3 \to 1$, $1 \to 3 \to 2 \to 1$), and we get

$$\det A = a_{11} a_{22} a_{33} - a_{12} a_{21} a_{33} - a_{13} a_{22} a_{31} - a_{11} a_{23} a_{32} + a_{12} a_{23} a_{31} + a_{13} a_{21} a_{32}.$$

---

*For example, the permutation (4.2) in cycle notation is (1 5 6 2 3) and is an even one.

We can now start proving the results in the previous section.

Proof of Theorem 4.1.1(i). Let τ be the permutation that switches i and j and leaves the other numbers alone. Then $b_{r,s} = a_{\tau(r),s}$ for every $r \in \underline{n}$. Thus

$$\det B = \sum_{\sigma \in S_n} \left((-1)^\sigma \prod_{r=1}^n b_{r,\sigma(r)} \right) = \sum_{\sigma \in S_n} \left((-1)^\sigma \prod_{r=1}^n a_{\tau(r),\sigma(r)} \right).$$

Letting $s = \tau(r)$ and $\tilde\sigma = \sigma \circ \tau^{-1}$ and using $(-1)^{\tilde\sigma} = (-1)^\tau (-1)^\sigma = -(-1)^\sigma$, we get that $\det B$ equals

$$\sum_{\sigma \in S_n} \left((-1)^\sigma \prod_{s=1}^n a_{s,\sigma(\tau^{-1}(s))} \right) = -\sum_{\tilde\sigma \in S_n} \left((-1)^{\tilde\sigma} \prod_{s=1}^n a_{s,\tilde\sigma(s)} \right) = -\det A.$$

$\square$

Corollary 4.2.3. *If A has two identical rows, then $\det A = 0$.*

Proof. If B is obtained by switching the identical rows, we get $\det B = -\det A$. But $B = A$, and thus $\det A = -\det A$, which yields $\det A = 0$. $\square$

Proof of Theorem 4.1.1(ii). Suppose we have that $b_{is} = a_{is} + c a_{js}$ and $b_{rs} = a_{rs}, r \neq i$, for $s = 1, \dots, n$. Then

$$\det B = \sum_{\sigma \in S_n} \left((-1)^\sigma \prod_{r=1}^n b_{r,\sigma(r)} \right) = \sum_{\sigma \in S_n} \left((-1)^\sigma (a_{i,\sigma(i)} + c a_{j,\sigma(i)}) \prod_{r \neq i} a_{r,\sigma(r)} \right)$$

$$= \sum_{\sigma \in S_n} \left((-1)^\sigma \prod_{r=1}^n a_{r,\sigma(r)} \right) + c \sum_{\sigma \in S_n} \left((-1)^\sigma a_{j,\sigma(i)} \prod_{r \neq i} a_{r,\sigma(r)} \right).$$

The first sum equals $\det A$, while the second sum is the determinant of a matrix where rows i and j are the same. Thus, by Corollary 4.2.3, the second term equals 0, and we find that $\det B = \det A$. $\square$

Proof of Theorem 4.1.1(iii). Let us say that B is obtained from A by multiplying row i by c. Thus $b_{ij} = c a_{ij}$, and $b_{rj} = a_{rj}, r \neq i$, for $j = 1, \dots, n$. But then $\prod_{r=1}^n b_{r,\sigma(r)} = c \prod_{r=1}^n a_{r,\sigma(r)}$ for all $\sigma \in S_n$. Thus

$$\det B = \sum_{\sigma \in S_n} \left((-1)^\sigma \prod_{r=1}^n b_{r,\sigma(r)} \right) = \sum_{\sigma \in S_n} \left((-1)^\sigma c \prod_{r=1}^n a_{r,\sigma(r)} \right) = c \det A.$$

$\square$

Proof of Theorem 4.1.2. Let $L = (l_{ij})_{i,j=1}^n$ be lower triangular. If $\sigma \neq id$, then there is an $r \in \{1, \dots, n\}$ so that $r < \sigma(r)$, which gives that $l_{r,\sigma(r)} = 0$. But then $\prod_{i=1}^n l_{i,\sigma(i)} = 0$. Thus

$$\det L = \sum_{\sigma \in S_n} \left((-1)^\sigma \prod_{i=1}^n l_{i,\sigma(i)} \right) = \sum_{\sigma = id} \left((-1)^\sigma \prod_{i=1}^n l_{i,\sigma(i)} \right) = \prod_{i=1}^n l_{ii}.$$

The upper triangular case works similarly.   □

Corollary 4.2.4. *Let E and A be $n \times n$ matrices. If E is an elementary matrix then*
$$\det(EA) = (\det E)(\det A).$$

Proof. It is not hard to see that if E is an elementary matrix of

- type I then $\det E = -1$.

- type II then $\det E = 1$.

- type III then $\det E = c$.

Now the corollary follows from Theorem 4.1.1.   □

Proof of Theorem 4.1.3(ii). If B is invertible, then B is a product of elementary matrices $B = E_1 E_2 \cdots E_p$. Now use Corollary 4.2.4 repeatedly to conclude that $\det(BA) = \det B \det A$.

If B is not invertible, then B has a row that is a linear combination of other rows. But then BA also has a row that is a linear combination of other rows. But then $\det B = 0$ and $\det BA = 0$, and so the equality follows.   □

Proof of Theorem 4.1.3(iii). If A is invertible, then $AA^{-1} = I_n$. But then, by Theorem 4.1.3(ii) we have $\det A \det A^{-1} = \det I_n = 1$. Thus $\det A \neq 0$ and $\det A^{-1} = \frac{1}{\det A}$ follows.

If A is not invertible, then there exist elementary matrices $E_1, E_2, \dots, E_p$ so that $E_1 E_2 \cdots E_p A$ has a zero row. Then

$$0 = \det(E_1 E_2 \cdots E_p A) = \left(\prod_{j=1}^p \det(E_j) \right) \det A.$$

Since $\det E_j \neq 0$, $j = 1 \dots, p$, we obtain that $\det A = 0$.   □

Proof of Theorem 4.1.3(i). Let $A = (a_{ij})_{i,j=1}^n$. Then $[A^T]_{ij} = a_{ji}$, and thus

$$\det A^T = \sum_{\sigma \in S_n} \left((-1)^\sigma \prod_{i=1}^n a_{\sigma(i),i} \right) = \sum_{\sigma \in S_n} \left((-1)^\sigma \prod_{i=1}^n a_{i,\sigma^{-1}(i)} \right)$$

$$= \sum_{\sigma^{-1} \in S_n} \left((-1)^{\sigma^{-1}} \prod_{i=1}^n a_{i,\sigma^{-1}(i)} \right) = \det A,$$

where we used that $(-1)^{\sigma^{-1}} = (-1)^\sigma$. $\qquad\square$

Let us also check that the definition of the determinant in this section matches with the one from the previous one. We do this by proving the expansion formula along the last row.

Proposition 4.2.5. *For an $n \times n$ matrix A, we have*

$$\det A = \sum_{i=1}^n (-1)^{i+n} a_{ni} \det A_{ni}.$$

Proof. Let B_i be the matrix that coincides with A in the first $n-1$ rows, and has in the nth row only one nonzero entry in the (n, i)th spot equal to a_{ni}. First observe that

$$\det A = \sum_{\sigma \in S_n} (-1)^\sigma \prod_{r=1}^n a_{r,\sigma(r)}$$

$$= \sum_{i=1}^n \sum_{\substack{\sigma \in S_n \\ \sigma(n) = i}} (-1)^\sigma a_{in} \prod_{r=1}^{n-1} a_{r,\sigma(r)} = \sum_{i=1}^n \det B_i.$$

In the matrix B_i we can move the ith column step by step to the right, taking $n-i$ steps to move it to the last column. In each step the determinant switches sign. We now get that

$$\det B_i = (-1)^{n-i} \det \begin{bmatrix} A_{ni} & * \\ 0 & a_{ni} \end{bmatrix},$$

where $*$ indicates entries whose values are not important. For a matrix of the form

$$C = (c_{ij})_{i,j=1}^n = \begin{bmatrix} \tilde{C} & * \\ 0 & c_{nn} \end{bmatrix}$$

one finds that

$$\det C = \sum_{\sigma \in S_n} (-1)^\sigma \prod_{r=1}^n c_{r,\sigma(r)} = \sum_{\substack{\sigma \in S_n \\ \sigma(n) = n}} (-1)^\sigma c_{nn} \prod_{r=1}^{n-1} c_{r,\sigma(r)} = c_{nn} \det \tilde{C}.$$

Thus $\det B_i = (-1)^{n-i} a_{ni} \det A_{ni}$, and using $(-1)^{n-i} = (-1)^{i+n}$ we find that

$$\det A = \sum_{i=1}^{n} (-1)^{n-i} a_{ni} \det A_{ni} = \sum_{i=1}^{n} (-1)^{i+n} a_{ni} \det A_{ni}.$$

$\square$

We can use the results in this section to give a characterization when A has an LU factorization with invertible factors L and U. For a matrix $A = (a_{ij})_{i,j=1}^{n}$ the determinants

$$\det(a_{ij})_{i,j=1}^{k}, k = 1, \ldots, n,$$

are called the **leading principal minors** of the matrix A.

Theorem 4.2.6. *The matrix A has an LU factorization $A = LU$ with L and U invertible if and only if its leading principal minors are nonzero.*

Proof. First suppose that $A = LU$ with L and U invertible. Let $k \in \{1, \ldots, n\}$. Write L and U as block matrices

$$L = \begin{bmatrix} L_{11} & 0 \\ L_{21} & L_{22} \end{bmatrix}, U = \begin{bmatrix} U_{11} & U_{12} \\ 0 & U_{22} \end{bmatrix},$$

where L_{11} and U_{11} are of size $k \times k$. Notice that L_{11} is lower triangular with nonzero diagonal entries. Thus $\det L_{11} \neq 0$. Similarly, $\det U_{11} \neq 0$. Since

$$A = LU = \begin{bmatrix} L_{11} & 0 \\ L_{21} & L_{22} \end{bmatrix} \begin{bmatrix} U_{11} & U_{12} \\ 0 & U_{22} \end{bmatrix} = \begin{bmatrix} L_{11}U_{11} & L_{11}U_{12} \\ L_{21}U_{11} & L_{21}U_{12} + L_{22}U_{22} \end{bmatrix},$$

we get that

$$(a_{ij})_{i,j=1}^{k} = L_{11}U_{11},$$

and thus $\det(a_{ij})_{i,j=1}^{k} = \det L_{11} \det U_{11} \neq 0$.

Conversely, suppose that A has nonzero leading principal minors. When one performs the row reduction procedure as outlined in Section 3.6, one finds that the first pivot is $a_{11} \neq 0$, and that the pivot in the (k, k), $k > 1$, entry is

$$u_{kk} = \frac{\det(a_{ij})_{i,j=1}^{k}}{\det(a_{ij})_{i,j=1}^{k-1}} \neq 0.$$

Thus the row reduction can continue until one reaches the upper triangular U which has nonzero diagonal entries. The matrix L is a unit lower triangular matrix. Thus A has an LU factorization with invertible factors L and U. $\square$

4.3 Cramer's Rule

Let $\mathbf{a}_i$ denote the ith column of A. Now we define

$$A_i(\mathbf{b}) := \begin{bmatrix} \mathbf{a}_1 & \cdots & \mathbf{a}_{i-1} & \mathbf{b} & \mathbf{a}_{i+1} & \cdots & \mathbf{a}_n \end{bmatrix}, \ i = 1, \ldots, n.$$

Thus $A_i(\mathbf{b})$ is the matrix obtained from A by replacing its ith column by $\mathbf{b}$.

Theorem 4.3.1. *(Cramer's rule) Let A be an invertible $n \times n$ matrix. For any $\mathbf{b} \in \mathbb{R}^n$, the unique solution $\mathbf{x} = (x_i)_{i=1}^n$ to the equation $A\mathbf{x} = \mathbf{b}$ has entries given by*

$$x_i = \frac{\det A_i(\mathbf{b})}{\det A}, \ i = 1, \ldots, n. \tag{4.5}$$

Proof. We denote the columns of the $n \times n$ identity matrix I by $\mathbf{e}_1, \ldots, \mathbf{e}_n$. Let us compute

$$A\, I_i(\mathbf{x}) = A \begin{bmatrix} \mathbf{e}_1 & \cdots & \mathbf{e}_{i-1} & \mathbf{x} & \mathbf{e}_{i+1} & \cdots & \mathbf{e}_n \end{bmatrix} =$$

$$\begin{bmatrix} A\mathbf{e}_1 & \cdots & A\mathbf{e}_{i-1} & A\mathbf{x} & A\mathbf{e}_{i+1} & \cdots & A\mathbf{e}_n \end{bmatrix} = A_i(\mathbf{b}).$$

But then, using the multiplicativity of the determinant, we get $\det A \det I_i(\mathbf{x}) = \det A_i(\mathbf{b})$. It is easy to see that $\det I_i(\mathbf{x}) = x_i$, and (4.5) follows. $\square$

Example 4.3.2. Find the solution to the system of linear equations

$$\begin{cases} x_1 + 2x_2 = 0 \\ x_1 + x_2 = 1 \end{cases}.$$

Applying Cramer's rule, we get

$$x_1 = \det \begin{bmatrix} 0 & 2 \\ 1 & 1 \end{bmatrix} / \det \begin{bmatrix} 1 & 2 \\ 1 & 1 \end{bmatrix} = \frac{-2}{-1} = 2,$$

$$x_2 = \det \begin{bmatrix} 1 & 0 \\ 1 & 1 \end{bmatrix} / \det \begin{bmatrix} 1 & 2 \\ 1 & 1 \end{bmatrix} = \frac{1}{-1} = -1.$$

Checking the answer $(2 + 2(-1) = 0, \ 2 + (-1) = 1)$, confirms that the answer is correct. $\square$

Given $A = (a_{ij})_{i,j=1}^n \in \mathbb{R}^{n \times n}$. As before, we let $A_{ij} \in \mathbb{R}^{(n-1) \times (n-1)}$ be the matrix obtained from A by removing the ith row and the jth column, and we put

$$C_{ij} = (-1)^{i+j} \det A_{ij}, \ i, j = 1, \ldots, n,$$

which is the (i,j)th cofactor of A. Given

$$
A = \begin{bmatrix}
a_{11} & a_{12} & \cdots & a_{1n} \\
a_{21} & a_{22} & \cdots & a_{2n} \\
\vdots & \vdots & & \vdots \\
a_{n1} & a_{n2} & \cdots & a_{nn}
\end{bmatrix},
$$

the **adjugate** of A is defined by

$$
\mathrm{adj}(A) = \begin{bmatrix}
C_{11} & C_{21} & \cdots & C_{n1} \\
C_{12} & C_{22} & \cdots & C_{n2} \\
\vdots & \vdots & & \vdots \\
C_{1n} & C_{2n} & \cdots & C_{nn}
\end{bmatrix}. \tag{4.6}
$$

Thus the (i,j)th entry of $\mathrm{adj}(A)$ is C_{ji} (notice the switch in the indices!).

Example 4.3.3. Compute the adjugate of

$$
A = \begin{bmatrix}
1 & 0 & 2 \\
2 & 3 & 1 \\
1 & 4 & 0
\end{bmatrix}.
$$

We get

$$
\mathrm{adj}(A) = \begin{bmatrix}
3 \cdot 0 - 1 \cdot 4 & -0 \cdot 0 + 2 \cdot 4 & 0 \cdot 1 - 2 \cdot 3 \\
-2 \cdot 0 + 1 \cdot 1 & 1 \cdot 0 - 2 \cdot 1 & -1 \cdot 1 + 2 \cdot 2 \\
2 \cdot 4 - 3 \cdot 1 & -1 \cdot 4 + 0 \cdot 1 & 1 \cdot 3 - 0 \cdot 2
\end{bmatrix} =
$$

$$
\begin{bmatrix}
-4 & 8 & -6 \\
1 & -2 & 3 \\
5 & -4 & 3
\end{bmatrix}.
$$

□

The usefulness of the adjugate matrix is given by the following result.

Theorem 4.3.4. *Let A be an $n \times n$ matrix. Then*

$$
A\,\mathrm{adj}(A) = (\det A)I_n = \mathrm{adj}(A)\,A. \tag{4.7}
$$

In particular, if $\det A \neq 0$, then

$$
A^{-1} = \frac{1}{\det A}\mathrm{adj}(A). \tag{4.8}
$$

Proof. As before, we let $\mathbf{a}_i$ denote the ith column of A. Consider $A_i(\mathbf{a}_j)$, which is the matrix A with the ith column replaced by $\mathbf{a}_j$. Thus, when $i \neq j$ we have that $A_i(\mathbf{a}_j)$ has two identical columns (namely the ith and the jth) and thus $\det A_i(\mathbf{a}_j) = 0$, $i \neq j$. When $i = j$, then $A_i(\mathbf{a}_j) = A$, and thus $\det A_i(\mathbf{a}_j) = \det A$. Computing the (i, j)th entry of the product $\mathrm{adj}(A)\, A$, we get

$$(\mathrm{adj}(A)\, A)_{ij} = \sum_{k=1}^{n} C_{ki} a_{kj} = \det A_i(\mathbf{a}_j) = \begin{cases} \det A & \text{if } i = j \\ 0 & \text{if } i \neq j \end{cases},$$

where we expanded $\det A_i(\mathbf{a}_j)$ along the ith column. This proves the second equality in (4.7). The proof of the first equality in (4.7) is similar.

Equation (4.8) follows directly from (4.7). $\qquad\qquad\qquad\qquad\qquad\square$

Notice that if we apply (4.8) to a 2×2 matrix, we obtain the familiar formula

$$\begin{bmatrix} a & b \\ c & d \end{bmatrix}^{-1} = \frac{1}{ad - bc} \begin{bmatrix} d & -b \\ -c & a \end{bmatrix}.$$

Applying (4.8) to the matrix A in Example 4.3.3, we get

$$\begin{bmatrix} 1 & 0 & 2 \\ 2 & 3 & 1 \\ 1 & 4 & 0 \end{bmatrix}^{-1} = \frac{1}{\det A} \mathrm{adj}(A) = \frac{1}{6} \begin{bmatrix} -4 & 8 & -6 \\ 1 & -2 & 3 \\ 5 & -4 & 3 \end{bmatrix}.$$

4.4 Determinants and Volumes

With a matrix $A = \begin{bmatrix} \mathbf{a}_1 & \cdots & \mathbf{a}_n \end{bmatrix}$ we can associate the region

$$P_A = \{x_1 \mathbf{a}_1 + \cdots + x_n \mathbf{a}_n : 0 \leq x_i \leq 1, i = 1, \ldots, n\} \subset \mathbb{R}^n. \qquad (4.9)$$

For $n = 2$ it is the parallelogram depicted in Figure 4.1. In higher dimensions the region is called a **parallelepiped**.

The main result in this section is that the absolute value of the determinant of A corresponds to the volume of the parallelepiped.

Theorem 4.4.1. *For $A = \begin{bmatrix} \mathbf{a}_1 & \cdots & \mathbf{a}_n \end{bmatrix}$ we have*

$$|\det A| = \mathrm{Volume}(P_A),$$

where P_A is the parallelepiped defined in (4.9).

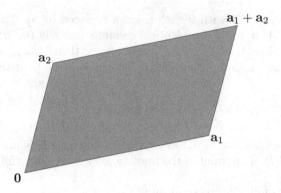

Figure 4.1: The area of this parallelogram equals $|\det A|$, where $A = [\mathbf{a}_1 \ \mathbf{a}_2]$.

In two dimensions 'Volume' should be interpreted as the area. In general, an n-dimensional cube with sides of length $h_1, \ldots, h_n$ has volume $h_1 h_2 \cdots h_n$. We rely on intuition for the volume of other n-dimensional regions. The following 'proof' therefore is actually an intuitive explanation.

Idea of proof. If $\det A = 0$ it means that the columns of A span a subspace of dimension $\leq n-1$. Thus the parallelepiped P_A lies inside a lower dimensional space, and thus the volume is 0.

If $\det A \neq 0$, then A is invertible and thus the product of elementary matrices, say $A = E_1 E_2 \cdots E_k$. We start with P_{I_n} which is the unit cube which has volume $1 = \det I_n$.

Next we look at $E_1 = I_n E_1$. If E_1 is a type I elementary matrix, it means we are just permuting two columns of the identity matrix, which does not change the parallelepiped ($P_{I_n} = P_{E_1}$), and thus Volume(P_{E_1}) $= 1 = |\det(E_1)|$.

If E_1 is a type II elementary matrix, it means that one of the unit vectors $\mathbf{e}_i$ is replaced by $\mathbf{e}_i + c\mathbf{e}_j$, and the other columns are left alone. This corresponds to a 'shear' action, as depicted in Figure 4.2. This does not change the volume. Thus $1 =$ Volume(P_{I_n}) $=$ Volume(P_{E_1}) and thus Volume(P_{E_1}) $= 1 = |\det E_1|$.

If E_1 is a type III elementary matrix, it means that one of the unit vectors $\mathbf{e}_i$ is multiplied by $c \neq 0$. Thus the length of one of the sides of the unit cube gets multiplied by $|c|$ and the other lengths stay the same. Consequently, Volume(P_{E_1}) $= |c| = |\det E_1|$. Note that if c is a negative number, the parallelepiped gets reflected as well, but the reflection does not change the volume.

The same arguments can also be applied if we start with P_B and take an elementary matrix E, and compare Volume(P_B) with Volume(P_{BE}). In the

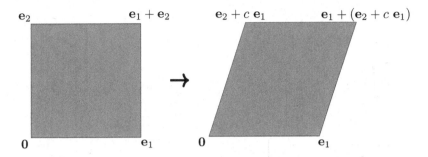

Figure 4.2: Shear action does not change the area/volume.

same manner we get

$$\text{Volume}(P_{BE}) = |\det E|\text{Volume}(P_B).$$

If we use this rule repeatedly on $A = I_n E_1 E_2 \cdots E_k$, we obtain

$$\text{Volume}(P_A) = |\det E_1| \cdots |\det E_k|\text{Volume}(P_{I_n}) = |\det(E_1 \cdots E_k)| = |\det A|.$$

□

The sign of det A can be used to assign an orientation to a collection of vectors. We do not pursue this here. It would fit well in a geometry course.

4.5 Exercises

Exercise 4.5.1. Find the determinant of following matrices.

(a) $\begin{bmatrix} 3 & -2 \\ 4 & 2 \end{bmatrix}$.

(b) $\begin{bmatrix} 2 & 6 \\ 1 & 3 \end{bmatrix}$.

(c) $\begin{bmatrix} 2 & 4 & 1 & 6 \\ 0 & 2 & 0 & 0 \\ -1 & -2 & 4 & 5 \\ 0 & 6 & -3 & 1 \end{bmatrix}$.

(d) $\begin{bmatrix} 2 & 3 & 1 \\ 2 & 1 & 0 \\ -1 & -4 & 5 \end{bmatrix}$. †

(e) $\begin{bmatrix} 2 & 5 & 3 & 0 \\ 1 & 4 & -7 & 2 \\ 1 & 28 & \frac{1}{2} & 0 \\ 0 & 3 & 0 & 0 \end{bmatrix}$.

(f) $\begin{bmatrix} a & b & c \\ a & b+x & c+x \\ a & b+y & c+y+1 \end{bmatrix}$.

Exercise 4.5.2. Suppose that $\det \begin{bmatrix} a & b & c \\ d & e & f \\ g & h & j \end{bmatrix} = 7$. Find the following.

(a) $\det \begin{bmatrix} g-4a & h-4b & j-4c \\ 3a & 3b & 3c \\ d & e & f \end{bmatrix}$.

(b) $\det \begin{bmatrix} g+6a & h+6b & j+6c \\ 2d+a & 2e+b & 2f+c \\ -a & -b & -c \end{bmatrix}$.

Exercise 4.5.3. For what values of a, b, x and y is the matrix

$$\begin{bmatrix} 1 & 1 & 1 \\ a & a & 2a+x \\ b & b+y & b \end{bmatrix}$$

invertible?

Exercise 4.5.4. Let $a, b, c, x, y, z \in \mathbb{R}$. Find the determinants of the following matrices.

(a) $\begin{bmatrix} a-x & a-y & a-z \\ b-x & b-y & b-z \\ c-x & c-y & c-z \end{bmatrix}$.

(b) $\begin{bmatrix} 1+ax & 1+ay & 1+az \\ 1+bx & 1+by & 1+bz \\ 1+cx & 1+cy & 1+cz \end{bmatrix}$.

---

†In sagecell.sagemath.org you can enter 'A = matrix([[2,3,1], [2,1,0], [-1,-4,5]])' and 'A.det()' to check your answer.

(c) $\begin{bmatrix} a & b & c \\ x & y & 0 \\ z & 0 & 0 \end{bmatrix}$.‡

Exercise 4.5.5. For $x_1, \ldots, x_n \in \mathbb{R}$ define

$$V(x_1, \ldots, x_n) = \begin{bmatrix} 1 & x_1 & \cdots & x_1^{n-1} \\ 1 & x_2 & \cdots & x_2^{n-1} \\ \vdots & \vdots & & \vdots \\ 1 & x_n & \cdots & x_n^{n-1} \end{bmatrix}. \tag{4.10}$$

The matrix $V(x_1, \ldots, x_n)$ is called the **Vandermonde matrix**.

The Vandermonde matrix $V(x_1, \ldots, x_n)$ satisfies

$$\det V(x_1, \ldots, x_n) = \prod_{1 \leq j < i \leq n} (x_i - x_j). \tag{4.11}$$

In particular, $V(x_1, \ldots, x_n)$ is invertible as soon as $x_i \neq x_j$ when $i \neq j$.

(a) Prove (4.11) for $n = 2$.

(b) Prove (4.11) for $n = 3$.

(c) Use induction to prove (4.11) for all n.
Hint: Take $V(x_1, \ldots, x_n)$ and subtract row 1 from all the other rows, leaving the determinant unchanged and arriving at the matrix

$$\begin{bmatrix} 1 & x_1 & \cdots & x_1^{n-1} \\ 0 & x_2 - x_1 & \cdots & x_2^{n-1} - x_1^{n-1} \\ \vdots & \vdots & & \vdots \\ 0 & x_n - x_1 & \cdots & x_n^{n-1} - x_1^{n-1} \end{bmatrix}.$$

Next, subtract, in order, x_1 times column $n-1$ from column n, x_1 times column $n-2$ from column $n-1$, and so on, until we subtract x_1 times column 1 from column 2. This again leaves the determinant unchanged, and leads to the matrix

$$\begin{bmatrix} 1 & 0 & 0 & \cdots & 0 & 0 \\ 0 & x_2 - x_1 & (x_2 - x_1)x_2 & \cdots & (x_2 - x_1)x_2^{n-3} & (x_2 - x_1)x_2^{n-2} \\ \vdots & \vdots & \vdots & & \vdots & \vdots \\ 0 & x_n - x_1 & (x_n - x_1)x_n & \cdots & (x_n - x_1)x_n^{n-3} & (x_n - x_1)x_n^{n-2} \end{bmatrix}.$$

---

‡In www.wolframalpha.com you can enter 'det {{a, b, c}, {x, y, 0}, {z, 0, 0}}' to check your answer.

This matrix equals

$$\begin{bmatrix} 1 & 0 & \cdots & 0 \\ 0 & x_2 - x_1 & \cdots & 0 \\ \vdots & \vdots & \ddots & \vdots \\ 0 & 0 & \cdots & x_n - x_1 \end{bmatrix} \begin{bmatrix} 1 & 0 & 0 & \cdots & 0 \\ 0 & 1 & x_2 & \cdots & x_2^{n-2} \\ \vdots & \vdots & \vdots & & \vdots \\ 0 & 1 & x_n & \cdots & x_n^{n-2} \end{bmatrix}.$$

Now use the induction hypothesis.

Exercise 4.5.6. Suppose that we would like to find a polynomial $p(x) = p_0 + p_1 x + p_2 x^2$ so that $p(x_1) = y_1$, $p(x_2) = y_2$ and $p(x_3) = y_3$.

(a) Show that to find the coefficients p_0, p_1, p_2 one needs to solve the equation

$$V(x_1, x_2, x_3) \begin{bmatrix} p_0 \\ p_1 \\ p_2 \end{bmatrix} = \begin{bmatrix} y_1 \\ y_2 \\ y_3 \end{bmatrix},$$

where $V(x_1, \ldots, x_n)$ is defined in (4.10).

(b) Find a polynomial $p(x)$ of degree ≤ 2 so that $p(1) = 3$, $p(2) = 7$, and $p(3) = 13$.

(c) Show that finding a polynomial $p(x) = \sum_{i=0}^{n-1} p_i x^i$ with $p(x_j) = y_j$, $j = 1, \ldots, n$, leads to the equation

$$V(x_1, \ldots, x_n) \begin{bmatrix} p_0 \\ \vdots \\ p_{n-1} \end{bmatrix} = \begin{bmatrix} y_1 \\ \vdots \\ y_n \end{bmatrix}.$$

(d) Use (4.11) to conclude that if $x_1, \ldots, x_n$ in part (c) are all different, then the interpolation problem in (c) has a unique solution.

Exercise 4.5.7. Use Cramer's rule to find the solution to the following systems of linear equations

(a) $\begin{cases} 2x_1 + 2x_2 = 1 \\ x_1 + 2x_2 = 1 \end{cases}$.

(b) $\begin{cases} 2x_1 + 2x_2 = 1 \\ x_1 + 2x_2 = 1 \end{cases}$.

Exercise 4.5.8. Use Cramer's rule to solve for x_3 in the system

$$\begin{cases} 3x_1 - x_2 + x_3 &= 1 \\ 2x_1 - x_2 &= 5 \\ x_1 + x_2 - x_3 &= 0 \end{cases}.$$

Exercise 4.5.9. Use Cramer's rule to solve for x_2 in the system

$$\begin{cases} 2x_1 - 2x_2 - x_3 &= 2 \\ x_1 - x_2 - x_3 &= 4 \\ x_2 - x_3 &= -6 \end{cases}.$$

Exercise 4.5.10. Consider the matrix vector equation $A\mathbf{x} = \mathbf{b}$ given by

$$\begin{bmatrix} 1 & 1 & 2 \\ 1 & \alpha & 0 \\ 1 & 1 & 3 \end{bmatrix} \begin{bmatrix} x_1 \\ x_2 \\ x_3 \end{bmatrix} = \begin{bmatrix} 2 \\ 0 \\ 5 \end{bmatrix}.$$

Determine $\alpha \in \mathbb{R}$ so that A is invertible and $x_1 = x_2$.
(Hint: Use Cramer's rule.)

Exercise 4.5.11. Let $A \in \mathbb{R}^{n \times n}$, $B \in \mathbb{R}^{m \times n}$, and $C \in \mathbb{R}^{m \times m}$. Show that

$$\det \begin{bmatrix} A & 0 \\ B & C \end{bmatrix} = (\det A)(\det C).$$

Exercise 4.5.12. Give an example of matrices $A, B, C, D \in \mathbb{R}^{n \times n}$ so that

$$\det \begin{bmatrix} A & B \\ C & D \end{bmatrix} \neq (\det A)(\det D) - (\det B)(\det C).$$

Exercise 4.5.13. Let

$$T = \begin{bmatrix} A & B \\ C & D \end{bmatrix} \tag{4.12}$$

be a block matrix, and suppose that A is a square matrix and that D is invertible. Show that

$$\det T = \det(A - BD^{-1}C) \det D.$$

Hint: Use the same hint as in Exercise 3.7.15.

Exercise 4.5.14. Compute the adjugate[§] of

$$A = \begin{bmatrix} \frac{1}{2} & 2 & 2 \\ 2 & 3 & 1 \\ 1 & 4 & 0 \end{bmatrix}.$$

Next, find A^{-1}.

Exercise 4.5.15. True or False? Justify each answer.

(i) If A is a 2×2 matrix, then $\det(-A) = \det(A)$.

[§]In live.sympy.org you can enter 'A=Matrix(3, 3, [1/2, 2, 2, 2, 3, 1, 1, 4, 0])' and 'A.adjugate()' to check your answer.

(ii) If for the square matrix A we have that $A^2 = A$, then $\det A = 1$.

(iii) For $n \times n$ matrices A and B, it holds that $\det(A+B) = \det(A) + \det(B)$.

(iv) For a $n \times n$ matrix A, it holds that $\det(A - A) = \det(A) - \det(A)$.

(v) If for the square matrix A we have that $A^{-1} = A$, then $\det A = 1$.

(vi) A square matrix has a nonzero determinant if and only if its columns are linearly independent.

(vii) If all diagonal entries of an $n \times n$-matrix A are 0, then $\det A = 0$.

Exercise 4.5.16. For a $n \times n$ matrix A show that

$$\det(\text{adj}(A)) = (\det A)^{n-1}.$$

Exercise 4.5.17. Let $1_{n \times n}$ be the $n \times n$ matrix with all entries equal to 1. Show that
$$\det(I_n + a 1_{n \times n}) = 1 + an.$$

Exercise 4.5.18. Let

$$A = \begin{bmatrix} 1 & 1 & 1 & \cdots & 1 & 1 \\ 1 & 2 & 2 & \cdots & 2 & 2 \\ 1 & 2 & 3 & \cdots & 3 & 3 \\ \vdots & \vdots & \vdots & & \vdots & \vdots \\ 1 & 2 & 3 & \cdots & n-1 & n-1 \\ 1 & 2 & 3 & \cdots & n-1 & n \end{bmatrix}, B = \begin{bmatrix} 2 & -1 & 0 & \cdots & 0 & 0 \\ -1 & 2 & -1 & \cdots & 0 & 0 \\ 0 & -1 & 2 & \cdots & 0 & 0 \\ \vdots & \vdots & \ddots & \ddots & \ddots & \vdots \\ 0 & 0 & \cdots & -1 & 2 & -1 \\ 0 & 0 & \cdots & 0 & -1 & 1 \end{bmatrix}.$$

(a) Show that $\det A = 1$. (Hint: Start by subtracting row $n-1$ from row n.)

(b) Show that $A^{-1} = B$.

(c) What is $\det B$?

Exercise 4.5.19. Let

$$A_n = \begin{bmatrix} 1 & \frac{1}{2!} & \frac{1}{3!} & \cdots & \frac{1}{(n-1)!} & \frac{1}{n!} \\ 1 & 1 & \frac{1}{2!} & \cdots & \frac{1}{(n-2)!} & \frac{1}{(n-1)!} \\ 0 & 1 & 1 & \ddots & \frac{1}{(n-3)!} & \frac{1}{(n-3)!} \\ \vdots & \vdots & \ddots & \ddots & \ddots & \vdots \\ 0 & 0 & 0 & \ddots & 1 & \frac{1}{2!} \\ 0 & 0 & 0 & \cdots & 1 & 1 \end{bmatrix}.$$

(a) Compute $\det A_2$.

(b) Compute $\det A_3$.

(c) Compute $\det A_4$.

(d) Can you guess a formula for $\det A_n$? Can you prove it?

Exercise 4.5.20. Let

$$
\mathbf{a}_1 = \begin{bmatrix} 2 \\ -1 \\ 0 \end{bmatrix}, \mathbf{a}_2 = \begin{bmatrix} 1 \\ 0 \\ 1 \end{bmatrix}, \mathbf{a}_3 = \begin{bmatrix} 0 \\ 3 \\ 2 \end{bmatrix}.
$$

Find the volume of the parallelepiped with corners

$$
\mathbf{0}, \mathbf{a}_1, \mathbf{a}_2, \mathbf{a}_3, \mathbf{a}_1 + \mathbf{a}_2, \mathbf{a}_1 + \mathbf{a}_3, \mathbf{a}_2 + \mathbf{a}_3, \mathbf{a}_1 + \mathbf{a}_2 + \mathbf{a}_3.
$$

Exercise 4.5.21. Let $A = \begin{bmatrix} \mathbf{a}_1 & \mathbf{a}_2 & \mathbf{a}_3 \end{bmatrix} \in \mathbb{R}^{3\times3}$. Explain using a geometric argument (using Theorem 4.4.1) why $\det A = 0$ if and only if $\{\mathbf{a}_1, \mathbf{a}_2, \mathbf{a}_3\}$ is linearly dependent.

5

Vector Spaces

CONTENTS

5.1   Definition of a Vector Space

So far we have seen $\mathbb{R}^n$ and $\mathbb{R}^{m \times n}$ as examples of vector spaces over $\mathbb{R}$. Next we introduce the general definition of a vector space. The important feature of a vector space is that we have two operations, addition and scalar multiplication, that satisfy several rules. This abstract definition allows one to develop a single theory that captures many different situations. The scalars come in general from a field $\mathbb{F}$. For now one can think of $\mathbb{F}$ as being the real numbers $\mathbb{R}$. In the later chapters we allow $\mathbb{F}$ to be the complex numbers $\mathbb{C}$. The general definition of a field is given in Appendix A.3.

Definition 5.1.1. A **vector space** over $\mathbb{F}$ is a set V along with two operations $+ : V \times V \to V$, $\cdot : \mathbb{F} \times V \to V$ satisfying the following axioms:

1. **Closure of addition**: for all $\mathbf{u}, \mathbf{v} \in V$ we have that $\mathbf{u} + \mathbf{v} \in V$.

2. **Associativity of addition**: for all $\mathbf{u}, \mathbf{v}, \mathbf{w} \in V$ we have that $(\mathbf{u} + \mathbf{v}) + \mathbf{w} = \mathbf{u} + (\mathbf{v} + \mathbf{w})$.

3. **Commutativity of addition**: for all $\mathbf{u}, \mathbf{v} \in V$ we have that $\mathbf{u} + \mathbf{v} = \mathbf{v} + \mathbf{u}$.

4. **Existence of a neutral element for addition**: there exists a $\mathbf{0} \in V$ so that $\mathbf{u} + \mathbf{0} = \mathbf{u} = \mathbf{0} + \mathbf{u}$ for all $\mathbf{u} \in V$.

---

5. **Existence of an additive inverse**: for every $\mathbf{u} \in V$ there exists a $-\mathbf{u} \in V$ so that $\mathbf{u} + (-\mathbf{u}) = \mathbf{0} = (-\mathbf{u}) + \mathbf{u}$.

6. **Closure of scalar multiplication**: for all $c \in \mathbb{F}$ and $\mathbf{u} \in V$ we have that $c\mathbf{u} \in V$.

7. **First distributive law**: for all $c \in \mathbb{F}$ and $\mathbf{u}, \mathbf{v} \in V$ we have that $c(\mathbf{u} + \mathbf{v}) = c\mathbf{u} + c\mathbf{v}$.

8. **Second distributive law**: for all $c, d \in \mathbb{F}$ and $\mathbf{u} \in V$ we have that $(c + d)\mathbf{u} = c\mathbf{u} + d\mathbf{u}$.

9. **Associativity for scalar multiplication**: for all $c, d \in \mathbb{F}$ and $\mathbf{u} \in V$ we have that $c(d\mathbf{u}) = (cd)\mathbf{u}$.

10. **Unit multiplication rule**: for every $\mathbf{u} \in V$ we have that $1\mathbf{u} = \mathbf{u}$.

Technically the closure rules are already stated when we say that addition and scalar multiplication map back into V, but it does not hurt to emphasize the point. These axioms imply several rules that seem 'obvious,' but as all properties in vector spaces have to be traced back to the axioms, we need to reprove these obvious rules. Here are two such examples.

Lemma 5.1.2. *Let V be a vector space over $\mathbb{F}$. Then for all $\mathbf{u} \in V$ we have that*

(i) $0\mathbf{u} = \mathbf{0}$.

(ii) $(-1)\mathbf{u} = -\mathbf{u}$.

Proof. (i) We have

$$0\mathbf{u} + \mathbf{u} = 0\mathbf{u} + 1\mathbf{u} = (0 + 1)\mathbf{u} = 1\mathbf{u} = \mathbf{u}.$$

Adding $-\mathbf{u}$ on both sides (and using associativity) now gives $0\mathbf{u} = \mathbf{0}$.

For (ii) we observe

$$(-1)\mathbf{u} + \mathbf{u} = (-1)\mathbf{u} + 1\mathbf{u} = (-1 + 1)\mathbf{u} = 0\mathbf{u} = \mathbf{0}.$$

Adding $-\mathbf{u}$ on both sides yields (ii). $\qquad\square$

We shall also develop several shorthands. For instance we shall write $\mathbf{u} - \mathbf{v}$ for $\mathbf{u} + (-\mathbf{v})$. Also, when we add several vectors together such as $\mathbf{u} + \mathbf{v} + \mathbf{w}$ or $\mathbf{u}_1 + \cdots + \mathbf{u}_n = \sum_{i=1}^n \mathbf{u}_i$ we typically will not write any parentheses as due to the associativity it does not matter in which order we add them. Indeed, $\mathbf{u} + \mathbf{v} + \mathbf{w}$ can be interpreted as either $(\mathbf{u} + \mathbf{v}) + \mathbf{w}$ or $\mathbf{u} + (\mathbf{v} + \mathbf{w})$.

5.2 Main Examples

Important examples of vector spaces over $\mathbb{R}$ are the following.

1. $\mathbb{R}^n$ with addition and scalar multiplication of vectors: we have seen in Section 1.3 that these operations satisfy all the vector space axioms.

2. Subspaces of $\mathbb{R}^n$: when you take a subspace S of $\mathbb{R}^n$ it is closed under addition and scalar multiplication of vectors. In addition, we have that $\mathbf{0} \in S$, and due to closure of scalar multiplication we also have that $\mathbf{u} \in S$ implies $-\mathbf{u} = (-1)\mathbf{u} \in S$. All the other axioms (associativity, commutativity, etc.) follow because the operations are the same as in $\mathbb{R}^n$, and thus S 'inherits' these properties.

3. $\mathbb{R}[t] := \{p(t) : p \text{ is a polynomial in the variable } t\}$ with the usual addition and scalar multiplication of polynomials. For instance,

$$(2t+3t^2)+(5+6t^2+7t^3) = 5+2t+9t^2+7t^3, (-3)(2-t+t^4) = -6+3t-3t^4.$$

Of course, you have also learned how to multiply two polynomials, but that operation is not part of the vector space structure.

4. $\mathbb{R}_n[t] := \{p(t) \in \mathbb{R}[t] : \deg p(t) \le n\}$, with the addition and scalar multiplication of polynomials.

   The **degree** of a polynomial $p(t)$ (notation: $\deg p(t)$) is the highest power of t that appears in the polynomial. For instance, if $p(t) = 1+5t-7t^3$, then $\deg p(t) = 3$. If $p(t)$ is a constant nonzero polynomial, such as $p(t) \equiv -2$, then its degree is 0. (We use the notation $\equiv$ to emphasize that the equality holds for all t.) The constant zero polynomial, $p(t) \equiv 0$ (also denoted by $\mathbf{0}(t)$), has, by convention, degree $-\infty^*$.

5. $\mathbb{R}^{m \times n}$ with the usual addition and scalar multiplication of matrices. We have seen in Proposition 3.1.1 that $\mathbb{R}^{m \times n}$ satisfies all the vector space axioms.

6. $\text{Upper}_n = \{U \in \mathbb{R}^{n \times n} : U \text{ is upper triangular}\}$ with the addition and scalar multiplication of matrices. It is clear that Upper_n is closed under addition and scalar multiplication, and that $0 \in \text{Upper}_n$. Also, if $A \in \text{Upper}_n$, then $-A \in \text{Upper}_n$. The other vector space axioms follow as the axioms are the same as in $\mathbb{R}^{n \times n}$, and thus Upper_n 'inherits' these properties.

---

*While we are not multiplying polynomials in the vector space context, this convention is useful for the rule $\deg p(t)q(t) = \deg p(t) + \deg q(t)$.

We actually have that $\mathbb{R}_n[t]$ is a subspace of $\mathbb{R}[t]$, and that Upper_n is a subspace of $\mathbb{R}^{n\times n}$. The general definition is as follows.

Given a vector space V over $\mathbb{F}$, and $W \subseteq V$. If W with the operations as defined on V, is itself a vector space, we call W a **subspace** of V.

Theorem 5.2.1. *Given a vector space V over $\mathbb{F}$, and $W \subseteq V$, then W is a subspace of V if and only if*

(i) $\mathbf{0} \in W$.

(ii) W *is closed under addition: for all* $\mathbf{w}, \mathbf{y} \in W$, *we have* $\mathbf{w} + \mathbf{y} \in W$.

(iii) W *is closed under scalar multiplication: for all* $c \in \mathbb{F}$ *and* $\mathbf{w} \in W$, *we have that* $c\mathbf{w} \in W$.

Equivalently, W is a subspace of V if and only if

(i') $W \neq \emptyset$.

(ii') W *is closed under addition and scalar multiplication: for all* $c, d \in \mathbb{F}$ *and* $\mathbf{w}, \mathbf{y} \in W$, *we have that* $c\mathbf{w} + d\mathbf{y} \in W$.

Proof. If W is a vector space, then (i), (ii) and (iii) are clearly satisfied.

For the converse, we need to check that when W satisfies (i), (ii) and (iii), it satisfies all ten axioms in the definition of a vector space. Clearly properties (i), (ii) and (iii) above take care of axioms 1, 4 and 6 in the definition of a vector space. Axiom 5 follows from (iii) in combination with Lemma 5.1.2(ii). The other properties (associativity, commutativity, distributivity, unit multiplication) are satisfied as they hold for all elements of V, and thus also for elements of W.

The arguments showing the equivalence of (i), (ii) and (iii) with (i') and (ii') are the same as in Section 2.1. □

Similar as in Section 2.1 we have the intersection and sum of subspaces. Given two subspaces U and W of a vector space V, we let

$$U + W := \{\mathbf{v} \in V \;:\; \text{there exist } \mathbf{u} \in U \text{ and } \mathbf{w} \in W \text{ so that } \mathbf{v} = \mathbf{u} + \mathbf{w}\},$$

$$U \cap W := \{\mathbf{v} \in V \;:\; \mathbf{v} \in U \text{ and } \mathbf{v} \in W\}.$$

Proposition 5.2.2. *Given two subspaces U and W of a vector space V over $\mathbb{F}$, then $U + W$ and $U \cap W$ are also subspaces of V.*

The proof is the same as the proof of Proposition 2.1.7.

As before, when $U \cap W = \{\mathbf{0}\}$, then we refer to $U + W$ as a **direct sum** of U and W, and write $U \dot{+} W$.

5.3 Linear Independence, Span, and Basis

Let V be a vector space over $\mathbb{F}$. A set of vectors $\{\mathbf{v}_1, \ldots, \mathbf{v}_p\}$ in V is said to be **linearly independent** if the vector equation

$$c_1 \mathbf{v}_1 + c_2 \mathbf{v}_2 + \cdots + c_p \mathbf{v}_p = \mathbf{0}, \tag{5.1}$$

with $c_1, \ldots, c_p \in \mathbb{F}$, only has the solution $c_1 = 0, \ldots, c_p = 0$ (the **trivial solution**). The set $\{\mathbf{v}_1, \ldots, \mathbf{v}_p\}$ is said to be **linearly dependent** if (5.1) has a solution where not all of $c_1, \ldots, c_p$ are zero (a **nontrivial solution**). In such a case, (5.1) with at least one c_i nonzero gives a **linear dependence relation** among $\{\mathbf{v}_1, \ldots, \mathbf{v}_p\}$. An arbitrary set $S \subseteq V$ is said to be linearly independent if every finite subset of S is linearly independent. The set S is linearly dependent, if it is not linearly independent.

Example 5.3.1. Let $V = \mathbb{R}^{2 \times 2}$. Let us check whether

$$S = \left\{ \begin{bmatrix} 1 & 0 \\ 2 & 1 \end{bmatrix}, \begin{bmatrix} 1 & 1 \\ 1 & 1 \end{bmatrix}, \begin{bmatrix} 0 & 2 \\ 1 & 1 \end{bmatrix} \right\}$$

is linearly independent or not. Consider the equation

$$c_1 \begin{bmatrix} 1 & 0 \\ 2 & 1 \end{bmatrix} + c_2 \begin{bmatrix} 1 & 1 \\ 1 & 1 \end{bmatrix} + c_3 \begin{bmatrix} 0 & 2 \\ 1 & 1 \end{bmatrix} = \begin{bmatrix} 0 & 0 \\ 0 & 0 \end{bmatrix}.$$

Rewriting, we get

$$\begin{bmatrix} 1 & 1 & 0 \\ 0 & 1 & 2 \\ 2 & 1 & 1 \\ 1 & 1 & 1 \end{bmatrix} \begin{bmatrix} c_1 \\ c_2 \\ c_3 \end{bmatrix} = \begin{bmatrix} 0 \\ 0 \\ 0 \\ 0 \end{bmatrix}. \tag{5.2}$$

Bringing this 4×3 matrix in row echelon form gives

$$\begin{bmatrix} \boxed{1} & 1 & 0 \\ 0 & 1 & 2 \\ 2 & 1 & 1 \\ 1 & 1 & 1 \end{bmatrix} \rightarrow \begin{bmatrix} \boxed{1} & 1 & 0 \\ 0 & \boxed{1} & 2 \\ 0 & -1 & 1 \\ 0 & 0 & 1 \end{bmatrix} \rightarrow \begin{bmatrix} \boxed{1} & 1 & 0 \\ 0 & \boxed{1} & 2 \\ 0 & 0 & \boxed{3} \\ 0 & 0 & 1 \end{bmatrix} \rightarrow \begin{bmatrix} \boxed{1} & 1 & 0 \\ 0 & \boxed{1} & 2 \\ 0 & 0 & \boxed{3} \\ 0 & 0 & 0 \end{bmatrix}.$$

As there are pivots in all columns, the system (5.2) only has the trivial solution $c_1 = c_2 = c_3 = 0$. Thus S is linearly independent.

Next, consider

$$\hat{S} = \left\{ \begin{bmatrix} 1 & 0 \\ 2 & 1 \end{bmatrix}, \begin{bmatrix} 1 & 1 \\ 1 & 1 \end{bmatrix}, \begin{bmatrix} 2 & 1 \\ 3 & 2 \end{bmatrix} \right\}.$$

Following the same reasoning as above we arrive at the system

$$\begin{bmatrix} 1 & 1 & 2 \\ 0 & 1 & 1 \\ 2 & 1 & 3 \\ 1 & 1 & 2 \end{bmatrix} \begin{bmatrix} c_1 \\ c_2 \\ c_3 \end{bmatrix} = \begin{bmatrix} 0 \\ 0 \\ 0 \\ 0 \end{bmatrix}, \tag{5.3}$$

which after row reduction leads to

$$\begin{bmatrix} \boxed{1} & 0 & 1 \\ 0 & \boxed{1} & 1 \\ 0 & 0 & 0 \\ 0 & 0 & 0 \end{bmatrix} \begin{bmatrix} c_1 \\ c_2 \\ c_3 \end{bmatrix} = \begin{bmatrix} 0 \\ 0 \\ 0 \\ 0 \end{bmatrix}. \tag{5.4}$$

So, c_3 is a free variable. Letting $c_3 = 1$, we get $c_1 = -c_3 = -1$ and $c_2 = -c_3 = -1$, so we find the linear dependence relation

$$-\begin{bmatrix} 1 & 0 \\ 2 & 1 \end{bmatrix} - \begin{bmatrix} 1 & 1 \\ 1 & 1 \end{bmatrix} + \begin{bmatrix} 2 & 1 \\ 3 & 2 \end{bmatrix} = \begin{bmatrix} 0 & 0 \\ 0 & 0 \end{bmatrix},$$

and thus $\hat{S}$ is linearly dependent.   □

Example 5.3.2. Let $V = \mathbb{R}[t]$ (which has the neutral element $\mathbf{0}(t)$).
We claim that $\{1, t, t^2, t^3\}$ is linearly independent.

(Here '1' stands for the constant polynomial $p(t) \equiv 1$. Of course, 1 could also stand for the real number 1. If there might be confusion which one we are talking about, we could write the constant polynomial 1 as $\mathbf{1}(t)$. The above statement only makes sense if $1 \in V$, thus we must be referring to the constant polynomial.)

Let $c_0, c_1, c_2, c_3 \in \mathbb{R}$ be so that

$$c_0 1 + c_1 t + c_2 t^2 + c_3 t^3 = \mathbf{0}(t).$$

Let us take particular values of t, so that we get a traditional system of linear equations. We take $t = 0, 1, -1$ and 2. Then we get

$$\begin{cases} c_0 & = & 0, \\ c_0 + c_1 + c_2 + c_3 & = & 0, \\ c_0 - c_1 + c_2 - c_3 & = & 0, \\ c_0 + 2c_1 + 4c_2 + 8c_3 & = & 0. \end{cases}$$

Solving this system, we find that the only solution is $c_0 = c_1 = c_2 = c_3 = 0$, and thus linear independence follows.

In fact, for any $n \in \mathbb{N}$ we have that $\{1, t, \ldots, t^n\}$ is linearly independent. To prove this, one can use Vandermonde matrices and their connection to interpolation as explained in Exercise 4.5.6. You can also rely on knowledge from Calculus: if you have a polynomial of degree $\leq n$ with $n + 1$ (or more) roots, all its coefficients must be 0. $\qquad\square$

Given a set $\emptyset \neq S \subseteq V$ we define

$$\text{Span } S := \{c_1 \mathbf{v}_1 + \cdots + c_p \mathbf{v}_p \; : \; p \in \mathbb{N}, c_1, \ldots, c_p \in \mathbb{F}, \mathbf{v}_1, \ldots, \mathbf{v}_p \in S\}.$$

Thus, Span S consists of all linear combinations of a finite set of vectors in S. As before, it is straightforward to check that Span S is a subspace of V. Indeed, $\mathbf{0} \in$ Span S as one can choose $p = 1$, $c_1 = 0$, and any $\mathbf{v}_1 \in S$, to get that $\mathbf{0} = 0\mathbf{v}_1 \in$ Span S. Next, the sum of two linear combinations of vectors in S is again a linear combination of vectors of S. Finally, for $c \in \mathbb{F}$ we have that $c \sum_{j=1}^{p} c_j \mathbf{v}_j = \sum_{j=1}^{p} (cc_j)\mathbf{v}_j$ is again a linear combination of elements in S. Thus by Theorem 5.2.1 we have that Span S is a subspace of V.

Example 5.3.3. Let $V = \mathbb{R}[t]$, the vector space of polynomials in t. We claim that

$$\text{Span } \{1, t, t^2, t^3\} = \mathbb{R}_3[t].$$

We have that $1, t, t^2, t^3 \in \mathbb{R}_3[t]$, thus Span $\{1, t, t^2, t^3\} \subseteq \mathbb{R}_3[t]$ (since $\mathbb{R}_3[t]$ is closed under addition and scalar multiplication). Conversely, if $p(t) \in \mathbb{R}_3[t]$, then it has the form $p(t) = c_0 1 + c_1 t + c_2 t^2 + c_3 t^3$, and thus $p(t) \in$ Span $\{1, t, t^2, t^3\}$. This gives that $\mathbb{R}_3[t] \subseteq$ Span $\{1, t, t^2, t^3\}$. Combined, we thus find Span $\{1, t, t^2, t^3\} = \mathbb{R}_3[t]$.[†] $\qquad\square$

Often the same subspace can be represented in different ways. In the following example we describe W via a condition (polynomials with a root at 1) and as a span.

Example 5.3.4. Let $V = \mathbb{R}_3[t]$ and $W = \{p(t) \in V \; : \; p(1) = 0\}$. We claim that

$$\text{Span } \{t - 1, t^2 - 1, t^3 - 1\} = W. \tag{5.5}$$

Let us first make sure that W is a subspace. Clearly $\mathbf{0} \in W$. Also, if $p(t)$ and $q(t)$ are of degree ≤ 3 with $p(1) = 0 = q(1)$ and $c, d \in \mathbb{R}$, then the polynomial $cp(t) + dq(t)$ also has degree ≤ 3 and the property that its value at $t = 1$ equals 0. Consequently, $cp(t) + dq(t) \in W$.

Notice that each polynomial $t - 1$, $t^2 - 1$, and $t^3 - 1$ has a root at 1, and that they are of degree ≤ 3. Thus $\subseteq$ in (5.5) holds.

[†] It is a common technique to prove the equality of sets $X = Y$ by showing that they are subsets of one another; that is, $X \subseteq Y$ and $Y \subseteq X$.

To prove the converse inclusion $\supseteq$ in (5.5), let $p(t) = p_0+p_1t+p_2t^2+p_3t^3$ be an arbitrary element of W. The condition $p(1) = 0$ gives that $p_0+p_1+p_2+p_3 = 0$. This system of a single equation has p_1, p_2, p_3 as free variables, and $p_0 = -p_1 - p_2 - p_3$. Consequently,

$$p(t) = -p_1 - p_2 - p_3 + p_1t + p_2t^2 + p_3t^3 = p_1(t-1) + p_2(t^2-1) + p_3(t^3-1).$$

This shows that $p(t) \in \text{Span}\left\{t-1, t^2-1, t^3-1\right\}$, and we are done. $\square$

Definition 5.3.5. Let W be a vector space. We say that $S \subset W$ is a **basis** for W if the following two conditions are both satisfied:

(i) Span $S = W$.

(ii) S is linearly independent.

As we have seen before, if S has a finite number of elements, then for any other basis of W it will have the same number of elements.

Theorem 5.3.6. Let $\mathcal{B} = \{v_1, \ldots, v_n\}$ and $\mathcal{C} = \{w_1, \ldots, w_m\}$ be bases for the vector space W. Then $n = m$.

The proof is the same as in Theorem 2.4.4. The **dimension** of a vector space is the number of elements in a basis of W. When no finite number of elements of W span W, we say dim $W = \infty$.

Example 5.3.7. Combining Examples 5.3.3 and 5.3.2, we see that $\{1, t, t^2, t^3\}$ is a basis for $\mathbb{R}_3[t]$, which yields that dim $\mathbb{R}_3[t] = 4$.

More general, $\{1, t, \ldots, t^n\}$ is a basis for $\mathbb{R}_n[t]$ (which we call the **standard basis** for $\mathbb{R}_n[t]$). Thus dim $\mathbb{R}_n[t] = n + 1$. $\square$

Example 5.3.8. Let $V = \mathbb{R}^{2\times2}$ and

$$S = \left\{E_{11} = \begin{bmatrix} 1 & 0 \\ 0 & 0 \end{bmatrix}, E_{12} = \begin{bmatrix} 0 & 1 \\ 0 & 0 \end{bmatrix}, E_{21} = \begin{bmatrix} 0 & 0 \\ 1 & 0 \end{bmatrix}, E_{22} = \begin{bmatrix} 0 & 0 \\ 0 & 1 \end{bmatrix}\right\}.$$

Then S is a basis for V.

To check linear independence, suppose that

$$c_1E_{11} + c_2E_{12} + c_3E_{21} + c_4E_{22} = \begin{bmatrix} 0 & 0 \\ 0 & 0 \end{bmatrix}.$$

Thus

$$\begin{bmatrix} c_1 & c_2 \\ c_3 & c_4 \end{bmatrix} = \begin{bmatrix} 0 & 0 \\ 0 & 0 \end{bmatrix},$$

which yields $c_1 = c_2 = c_3 = c_4 = 0$. This gives linear independence.

Next, if we take an arbitrary element $A \in V$. Then

$$A = \begin{bmatrix} a & b \\ c & d \end{bmatrix} = aE_{11} + bE_{12} + cE_{21} + dE_{22} \in \text{Span} S.$$

Because both the linear independence and the spanning properties are satisfied, S is a basis for V. In particular, dim $\mathbb{R}^{2\times2} = 4$.   □

More general, if we let E_{ij} denote the $m \times n$ matrix which has a 1 in position (i,j) and zeros everywhere else, then

$$\{E_{ij} : 1 \leq i \leq m, 1 \leq j \leq n\}$$

is a basis for $\mathbb{R}^{m\times n}$ (the **standard basis** for $\mathbb{R}^{m\times n}$). This implies that dim $\mathbb{R}^{m\times n} = mn$.

Example 5.3.9. Let $W = \left\{ p(t) \in \mathbb{R}_2[t] : \int_{-1}^{1} p(t)dt = 0 \right\}$. Show that W is a subspace of $\mathbb{R}_2[t]$ and find a basis for W.

Clearly, the zero polynomial $\mathbf{0}(t)$ belongs to W as $\int_{-1}^{1} \mathbf{0}(t)dt = \int_{-1}^{1} 0 dt = 0$. Next, when $p(t), q(t) \in W$ and $c, d \in \mathbb{R}$, then

$$\int_{-1}^{1} (cp(t) + dq(t))dt = c \int_{-1}^{1} p(t)dt + d \int_{-1}^{1} q(t)dt = c\,0 + d\,0 = 0,$$

so $cp(t) + dq(t) \in W$. Thus by Theorem 5.2.1, W is a subspace of $\mathbb{R}_2[t]$.

To find a basis, let us take an arbitrary element $p(t) = p_0 + p_1 t + p_2 t^2 \in W$, which means that

$$\int_{-1}^{1} p(t)dt = 2p_0 + \frac{2}{3}p_2 = 0.$$

This yields the linear system

$$\begin{bmatrix} 2 & 0 & \frac{2}{3} \end{bmatrix} \begin{bmatrix} p_0 \\ p_1 \\ p_2 \end{bmatrix} = 0.$$

The coefficient matrix only has a pivot in column 1, so we let p_1 and p_2 be the free variables (as they correspond to the variables corresponding to the 2nd and 3rd column) and observe that $p_0 = -\frac{1}{3}p_2$. Expressing $p(t)$ in the free variables we get

$$p(t) = -\frac{1}{3}p_2 + p_1 t + p_2 t^2 = p_1 t + p_2 \left(t^2 - \frac{1}{3}\right).$$

Thus $p(t) \in \text{Span} \left\{t, t^2 - \frac{1}{3}\right\}$. As we started with an arbitrary $p(t) \in W$, we now proved that $W \subseteq \text{Span}\left\{t, t^2 - \frac{1}{3}\right\}$. Next, observe that $t \in W$ and $t^2 - \frac{1}{3} \in W$. Since W is a subspace of $\mathbb{R}_2[t]$, we consequently have $\text{Span}\left\{t, t^2 - \frac{1}{3}\right\} \subseteq W$. Due to both inclusions, we now obtain the equality $\text{Span}\left\{t, t^2 - \frac{1}{3}\right\} = W$.

Next, to check that $\left\{t, t^2 - \frac{1}{3}\right\}$ is linearly independent, let $c_1, c_2 \in \mathbb{R}$ be so that

$$c_1 t + c_2\left(t^2 - \frac{1}{3}\right) = \mathbf{0}(t).$$

Taking $t = 0$ and $t = 1$, we find that

$$\begin{cases} -\frac{1}{3}c_2 & = & 0, \\ c_1 + \frac{2}{3}c_2 & = & 0. \end{cases}$$

Thus $c_1 = c_2 = 0$, and linear independence follows.

Consequently, $\left\{t, t^2 - \frac{1}{3}\right\}$ is a basis for W. In particular, dim $W = 2$. $\qquad\square$

Example 5.3.10. Let $W = \{p(t) \in \mathbb{R}[t] \ : \ p(1) = 0\}$. We claim that W has as a basis

$$S = \left\{t - 1, t^2 - 1, t^3 - 1\right\}. \tag{5.6}$$

In Example 5.3.4 we showed that Span $S = W$, so it remains to show that S is linearly independent. For this, assume that $c_1, c_2, c_3 \in \mathbb{R}$ are such that

$$c_1(t - 1) + c_2(t^2 - 1) + c_3(t^3 - 1) = \mathbf{0}(t).$$

Letting $t = 0, -1$ and 2, we find

$$\begin{cases} -c_1 - c_2 - c_3 & = & 0, \\ -2c_1 \quad\quad - 2c_3 & = & 0, \\ c_1 + 3c_2 + 7c_3 & = & 0. \end{cases}$$

This yields $c_1 = c_2 = c_3 = 0$, showing linear independence.

Thus S is a basis for W, and dim $W = 3$ follows. $\qquad\square$

Example 5.3.11. Let $W = \text{Upper}_3$. It is not hard to see that W has as a basis

$$\left\{ \begin{bmatrix} 1 & 0 & 0 \\ 0 & 0 & 0 \\ 0 & 0 & 0 \end{bmatrix}, \begin{bmatrix} 0 & 1 & 0 \\ 0 & 0 & 0 \\ 0 & 0 & 0 \end{bmatrix}, \begin{bmatrix} 0 & 0 & 1 \\ 0 & 0 & 0 \\ 0 & 0 & 0 \end{bmatrix}, \begin{bmatrix} 0 & 0 & 0 \\ 0 & 1 & 0 \\ 0 & 0 & 0 \end{bmatrix}, \begin{bmatrix} 0 & 0 & 0 \\ 0 & 0 & 1 \\ 0 & 0 & 0 \end{bmatrix}, \begin{bmatrix} 0 & 0 & 0 \\ 0 & 0 & 0 \\ 0 & 0 & 1 \end{bmatrix} \right\}.$$

Thus dim $\text{Upper}_3 = 6$.

In general $\{E_{ij} : 1 \le i \le j \le n\}$ is a basis for Upper_n, yielding that

$$\dim \text{Upper}_n = \sum_{i=1}^{n} i = \frac{n(n+1)}{2}.$$

$\qquad\square$

Example 5.3.12. Let $W = \left\{ \begin{bmatrix} a & b & c \\ b & a & b \\ c & b & a \end{bmatrix} : a, b, c \in \mathbb{R} \right\}$. Here

$$S = \left\{ \begin{bmatrix} 1 & 0 & 0 \\ 0 & 1 & 0 \\ 0 & 0 & 1 \end{bmatrix}, \begin{bmatrix} 0 & 1 & 0 \\ 1 & 0 & 1 \\ 0 & 1 & 0 \end{bmatrix}, \begin{bmatrix} 0 & 0 & 1 \\ 0 & 0 & 0 \\ 1 & 0 & 0 \end{bmatrix} \right\}$$

is a basis. Indeed, if $A \in W$, then

$$A = a \begin{bmatrix} 1 & 0 & 0 \\ 0 & 1 & 0 \\ 0 & 0 & 1 \end{bmatrix} + b \begin{bmatrix} 0 & 1 & 0 \\ 1 & 0 & 1 \\ 0 & 1 & 0 \end{bmatrix} + c \begin{bmatrix} 0 & 0 & 1 \\ 0 & 0 & 0 \\ 1 & 0 & 0 \end{bmatrix} \in \text{Span } S.$$

In addition, if

$$a \begin{bmatrix} 1 & 0 & 0 \\ 0 & 1 & 0 \\ 0 & 0 & 1 \end{bmatrix} + b \begin{bmatrix} 0 & 1 & 0 \\ 1 & 0 & 1 \\ 0 & 1 & 0 \end{bmatrix} + c \begin{bmatrix} 0 & 0 & 1 \\ 0 & 0 & 0 \\ 1 & 0 & 0 \end{bmatrix} = \begin{bmatrix} 0 & 0 & 0 \\ 0 & 0 & 0 \\ 0 & 0 & 0 \end{bmatrix},$$

then $a = b = c = 0$. Thus S is linearly independent.

This proves that S is a basis for W, and consequently $\dim W = 3$. □

Example 5.3.13. Let $W = \left\{ \begin{bmatrix} a & b \\ c & d \end{bmatrix} : a + b + c + d = 0 \right\}$.

Notice that $a = -b - c - d$, and thus an element in W is of the form

$$\begin{bmatrix} -b - c - d & b \\ c & d \end{bmatrix} = b \begin{bmatrix} -1 & 1 \\ 0 & 0 \end{bmatrix} + c \begin{bmatrix} -1 & 0 \\ 1 & 0 \end{bmatrix} + d \begin{bmatrix} -1 & 0 \\ 0 & 1 \end{bmatrix}.$$

It is not hard to see that

$$\left\{ \begin{bmatrix} -1 & 1 \\ 0 & 0 \end{bmatrix}, \begin{bmatrix} -1 & 0 \\ 1 & 0 \end{bmatrix}, \begin{bmatrix} -1 & 0 \\ 0 & 1 \end{bmatrix} \right\}$$

is a basis for W. Thus $\dim W = 3$. □

For the dimension of a sum of subspaces we have the following general rule.

Theorem 5.3.14. *For finite-dimensional subspaces U and W of V we have that*

$$\dim(U + W) = \dim U + \dim W - \dim(U \cap W).$$

In particular, for a direct sum we have $\dim(U \dotplus W) = \dim U + \dim W$.

We will make use of the following results.

Proposition 5.3.15. *Let* $\mathcal{B} = \{\mathbf{v}_1, \ldots, \mathbf{v}_n\}$ *be a basis for the vector space* V, *and let* $\mathcal{C} = \{\mathbf{w}_1, \ldots, \mathbf{w}_m\}$ *be a set of vectors in* V *with* $m > n$. *Then* $\mathcal{C}$ *is linearly dependent.*

Proof. The proof is the same as the proof of Proposition 2.4.3. $\square$

Proposition 5.3.16. *Let* V *be a vector space of dimension* n, *and let* $\{\mathbf{v}_1, \ldots, \mathbf{v}_k\} \subset V$ *be linearly independent. Then there exist vectors* $\mathbf{v}_{k+1}, \ldots \mathbf{v}_n \in V$ *so that* $\{\mathbf{v}_1, \ldots, \mathbf{v}_n\}$ *is a basis for* V.

Proof. By Proposition 5.3.15 we must have that $k \leq n$. If $k < n$, then $\{\mathbf{v}_1, \ldots, \mathbf{v}_k\}$ is not a basis (as dim $V = n$), and thus there must exists a vector $\mathbf{v}_{k+1} \in V$ so that $\mathbf{v}_{k+1} \notin \text{Span}\{\mathbf{v}_1, \ldots, \mathbf{v}_k\}$. But then $\{\mathbf{v}_1, \ldots, \mathbf{v}_k, \mathbf{v}_{k+1}\}$ is linearly independent (see Exercise 5.5.17(a)). If $k+1 < n$, we can repeat the reasoning in the last two lines, and find a vector $\mathbf{v}_{k+2}$ so that $\{\mathbf{v}_1, \ldots, \mathbf{v}_{k+1}, \mathbf{v}_{k+2}\}$ is linearly independent. We can repeat this process until we have found a linearly independent set $\mathcal{B} = \{\mathbf{v}_1, \ldots, \mathbf{v}_n\}$. We now claim that $\mathcal{B}$ is a basis. Since $\mathcal{B}$ is linearly independent, it suffices to show that $\mathcal{B}$ spans V. If not, then we can find a vector $\mathbf{v}_{n+1} \in V$ so that $\mathbf{v}_{n+1} \notin \text{Span}\mathcal{B}$. As before, this gives us a linearly independent set $\{\mathbf{v}_1, \ldots, \mathbf{v}_n, \mathbf{v}_{n+1}\}$ in V. But this contradicts dim $V = n$. Thus we have that $\mathcal{B}$ spans V. $\square$

Proof of Theorem 5.3.14. Let $\{\mathbf{v}_1, \ldots, \mathbf{v}_p\}$ be a basis for $U \cap W$. Next apply Proposition 5.3.16 and find $\mathbf{u}_1, \ldots, \mathbf{u}_k$ so that $\{\mathbf{v}_1, \ldots, \mathbf{v}_p, \mathbf{u}_1, \ldots, \mathbf{u}_k\}$ is a basis for U. Similarly, find $\mathbf{w}_1, \ldots, \mathbf{w}_l$ so that $\{\mathbf{v}_1, \ldots, \mathbf{v}_p, \mathbf{w}_1, \ldots, \mathbf{w}_l\}$ is a basis for W. We next show that $\{\mathbf{v}_1, \ldots, \mathbf{v}_p, \mathbf{u}_1, \ldots, \mathbf{u}_k, \mathbf{w}_1, \ldots, \mathbf{w}_l\}$ is a basis for $U + W$.

First to show the spanning property, let $\mathbf{v}$ be in $U + W$. Due to the definition of $U + W$, there exists a $\mathbf{u} \in U$ and a $\mathbf{w} \in W$ so that $\mathbf{v} = \mathbf{u} + \mathbf{w}$. As $\mathbf{u} \in U$, there exists a_i and b_i so that

$$\mathbf{u} = \sum_{i=1}^{p} a_i \mathbf{v}_i + \sum_{i=1}^{k} b_i \mathbf{u}_i.$$

As $\mathbf{w} \in W$, there exists c_i and d_i so that

$$\mathbf{w} = \sum_{i=1}^{p} c_i \mathbf{v}_i + \sum_{i=1}^{l} d_i \mathbf{w}_i.$$

Then $\mathbf{v} = \mathbf{u} + \mathbf{w} = \sum_{i=1}^{p}(a_i + c_i)\mathbf{v}_i + \sum_{i=1}^{k} b_i \mathbf{u}_i + \sum_{i=1}^{l} d_i \mathbf{w}_i$, and thus $\{\mathbf{v}_1, \ldots, \mathbf{v}_p, \mathbf{u}_1, \ldots, \mathbf{u}_k, \mathbf{w}_1, \ldots, \mathbf{w}_l\}$ spans $U + W$.

Next, to show linear independence, suppose that

$$\sum_{i=1}^{p} a_i \mathbf{v}_i + \sum_{i=1}^{k} b_i \mathbf{u}_i + \sum_{i=1}^{l} c_i \mathbf{w}_i = \mathbf{0}.$$

Then

$$\sum_{i=1}^{p} a_i \mathbf{v}_i + \sum_{i=1}^{k} b_i \mathbf{u}_i = -\sum_{i=1}^{l} c_i \mathbf{w}_i \in U \cap W.$$

As $\{\mathbf{v}_1, \ldots, \mathbf{v}_p\}$ is a basis for $U \cap W$, there exist d_i so that

$$-\sum_{i=1}^{l} c_i \mathbf{w}_i = \sum_{i=1}^{p} d_i \mathbf{v}_i.$$

Then $\sum_{i=1}^{p} d_i \mathbf{v}_i + \sum_{i=1}^{l} c_i \mathbf{w}_i = \mathbf{0}$. As $\{\mathbf{v}_1, \ldots, \mathbf{v}_p, \mathbf{w}_1, \ldots, \mathbf{w}_l\}$ is linearly independent, we get that $d_1 = \cdots = d_p = c_1 = \cdots = c_l = 0$. But then we get that $\sum_{i=1}^{p} a_i \mathbf{v}_i + \sum_{i=1}^{k} b_i \mathbf{u}_i = \mathbf{0}$. Using now that $\{\mathbf{v}_1, \ldots, \mathbf{v}_p, \mathbf{u}_1, \ldots, \mathbf{u}_k\}$ is linearly independent, we get $a_1 = \cdots = a_p = b_1 = \cdots = b_k = 0$. This shows that $\{\mathbf{v}_1, \ldots, \mathbf{v}_p, \mathbf{u}_1, \ldots, \mathbf{u}_k, \mathbf{w}_1, \ldots, \mathbf{w}_l\}$ is linearly independent, proving that it is a basis for $U + W$.

Thus $\dim U + W = p + k + l = (p + k) + (p + l) - p = \dim U + \dim W - \dim(U \cap W)$.

When it is a direct sum $\dim(U \cap W) = 0$, so $\dim(U \dot{+} W) = \dim U + \dim W$ immediately follows. $\qquad \square$

5.4 Coordinate Systems

Coordinate systems allow one to view an n dimensional vector space over $\mathbb{R}$ as a copy of the vector space $\mathbb{R}^n$. It requires choosing a basis and representing all vectors in the vector space relative to this chosen basis. The principles and the proofs are the same as we have seen in Section.2.5, and therefore we focus in this section on new examples. The following result summarizes the coordinate system concept.

Theorem 5.4.1. *Let* $\mathcal{B} = \{\mathbf{v}_1, \ldots, \mathbf{v}_n\}$ *be a basis for a vector space* V *over* $\mathbb{F}$. *Then for each* $\mathbf{v} \in V$ *there exists unique* $c_1, \ldots, c_n \in \mathbb{F}$ *so that*

$$\mathbf{v} = c_1 \mathbf{v}_1 + \cdots + c_n \mathbf{v}_n. \tag{5.7}$$

In this case, we write

$$[\mathbf{v}]_\mathcal{B} = \begin{bmatrix} c_1 \\ \vdots \\ c_n \end{bmatrix} \in \mathbb{F}^n,$$

and refer to $c_1, \ldots, c_n$ as the **coordinates** *of $\mathbf{v}$ relative to the basis $\mathcal{B}$. The coordinate vectors satisfy*

$$[\alpha \mathbf{v} + \beta \mathbf{w}]_\mathcal{B} = \alpha [\mathbf{v}]_\mathcal{B} + \beta [\mathbf{w}]_\mathcal{B}, \text{ where } \mathbf{v}, \mathbf{w} \in V, \alpha, \beta \in \mathbb{F}.$$

Example 5.4.2. Let $V = \mathbb{R}_3[t]$ and $\mathcal{B} = \left\{ 1, t - 1, t^2 - 2t + 1, t^3 - 3t^2 + 3t - 1 \right\}$. Find $[t^3 + t^2 + t + 1]_\mathcal{B}$.

We need to find $c_1, c_2, c_3, c_4 \in \mathbb{R}$ so that

$$c_1 1 + c_2(t - 1) + c_3(t^2 - 2t + 1) + c_4(t^3 - 3t^2 + 3t - 1) = t^3 + t^2 + t + 1.$$

Equating the coefficients of $1, t, t^2, t^3$, setting up the augmented matrix, and row reducing gives

$$\begin{bmatrix} 1 & -1 & 1 & -1 & | & 1 \\ 0 & 1 & -2 & 3 & | & 1 \\ 0 & 0 & 1 & -3 & | & 1 \\ 0 & 0 & 0 & 1 & | & 1 \end{bmatrix} \rightarrow \begin{bmatrix} 1 & -1 & 1 & 0 & | & 2 \\ 0 & 1 & -2 & 0 & | & -2 \\ 0 & 0 & 1 & 0 & | & 4 \\ 0 & 0 & 0 & 1 & | & 1 \end{bmatrix} \rightarrow$$

$$\begin{bmatrix} 1 & -1 & 0 & 0 & | & -2 \\ 0 & 1 & 0 & 0 & | & 6 \\ 0 & 0 & 1 & 0 & | & 4 \\ 0 & 0 & 0 & 1 & | & 1 \end{bmatrix} \rightarrow \begin{bmatrix} 1 & 0 & 0 & 0 & | & 4 \\ 0 & 1 & 0 & 0 & | & 6 \\ 0 & 0 & 1 & 0 & | & 4 \\ 0 & 0 & 0 & 1 & | & 1 \end{bmatrix}.$$

Thus we find $[t^3 + t^2 + t + 1]_\mathcal{B} = \begin{bmatrix} 4 \\ 6 \\ 4 \\ 1 \end{bmatrix}$. $\qquad \square$

Example 5.4.3. As in Example 5.3.12, let $W = \left\{ \begin{bmatrix} a & b & c \\ b & a & b \\ c & b & a \end{bmatrix} : a, b, c \in \mathbb{R} \right\}$

with basis

$$\mathcal{B} = \left\{ \begin{bmatrix} 1 & 0 & 0 \\ 0 & 1 & 0 \\ 0 & 0 & 1 \end{bmatrix}, \begin{bmatrix} 0 & 1 & 0 \\ 1 & 0 & 1 \\ 0 & 1 & 0 \end{bmatrix}, \begin{bmatrix} 0 & 0 & 1 \\ 0 & 0 & 0 \\ 1 & 0 & 0 \end{bmatrix} \right\}.$$

Since

$$\begin{bmatrix} a & b & c \\ b & a & b \\ c & b & a \end{bmatrix} = a \begin{bmatrix} 1 & 0 & 0 \\ 0 & 1 & 0 \\ 0 & 0 & 1 \end{bmatrix} + b \begin{bmatrix} 0 & 1 & 0 \\ 1 & 0 & 1 \\ 0 & 1 & 0 \end{bmatrix} + c \begin{bmatrix} 0 & 0 & 1 \\ 0 & 0 & 0 \\ 1 & 0 & 0 \end{bmatrix},$$

we find

$$
[\begin{bmatrix} a & b & c \\ b & a & b \\ c & b & a \end{bmatrix}]_{\mathcal{B}} = \begin{bmatrix} a \\ b \\ c \end{bmatrix}.
$$

□

Example 5.4.4. As in Example 5.3.10 let $W = \{p(t) \in \mathbb{R}[t] \ : \ p(1) = 0\}$ with basis

$$
\mathcal{B} = \{t - 1, t^2 - 1, t^3 - 1\}. \tag{5.8}
$$

Find $[t^3 - 3t^2 + 3t - 1]_{\mathcal{B}}$.

First observe that $t^3 - 3t^2 + 3t - 1 \in W$ since its degree is ≤ 3 and has 1 as a root. Thus the question makes sense.

We need to find c_1, c_2, c_3 so that

$$
t^3 - 3t^2 + 3t - 1 = c_1(t - 1) + c_2(t^2 - 1) + c_3(t^3 - 1).
$$

Equating the coefficients of $1, t, t^2, t^3$, setting up the augmented matrix, and row reducing gives

$$
\begin{bmatrix} -1 & -1 & -1 & | & -1 \\ 1 & 0 & 0 & | & 3 \\ 0 & 1 & 0 & | & -3 \\ 0 & 0 & 1 & | & 1 \end{bmatrix} \rightarrow \begin{bmatrix} 1 & 0 & 0 & | & 3 \\ 0 & 1 & 0 & | & -3 \\ 0 & 0 & 1 & | & 1 \\ 0 & 0 & 0 & | & 0 \end{bmatrix}.
$$

Thus we find $[t^3 - 3t^2 + 3t - 1]_{\mathcal{B}} = \begin{bmatrix} 3 \\ -3 \\ 1 \end{bmatrix}$.

□

Example 5.4.5. Let $W = \mathbb{R}^{2\times 2}$ with basis

$$
\mathcal{B} = \left\{ \begin{bmatrix} 2 & 1 \\ 1 & 1 \end{bmatrix}, \begin{bmatrix} 1 & 2 \\ 1 & 1 \end{bmatrix}, \begin{bmatrix} 1 & 1 \\ 2 & 1 \end{bmatrix}, \begin{bmatrix} 1 & 1 \\ 1 & 2 \end{bmatrix} \right\}.
$$

Find $[\begin{bmatrix} 1 & 1 \\ 1 & 1 \end{bmatrix}]_{\mathcal{B}}, [\begin{bmatrix} 2 & 1 \\ 1 & 1 \end{bmatrix}]_{\mathcal{B}}, [\begin{bmatrix} 1 & 0 \\ 0 & 0 \end{bmatrix}]_{\mathcal{B}}$.

We need to find c_1, c_2, c_3, c_4 so that

$$
\begin{bmatrix} 1 & 1 \\ 1 & 1 \end{bmatrix} = c_1 \begin{bmatrix} 2 & 1 \\ 1 & 1 \end{bmatrix} + c_2 \begin{bmatrix} 1 & 2 \\ 1 & 1 \end{bmatrix} + c_3 \begin{bmatrix} 1 & 1 \\ 2 & 1 \end{bmatrix} + c_4 \begin{bmatrix} 1 & 1 \\ 1 & 2 \end{bmatrix}
$$

$$
= \begin{bmatrix} 2c_1 + c_2 + c_3 + c_4 & c_1 + 2c_2 + c_3 + c_4 \\ c_1 + c_2 + 2c_3 + c_4 & c_1 + c_2 + c_3 + 2c_4 \end{bmatrix}.
$$

Thus we get the system

$$\begin{cases} 2c_1 + c_2 + c_3 + c_4 = 1, \\ c_1 + 2c_2 + c_3 + c_4 = 1, \\ c_1 + c_2 + 2c_3 + c_4 = 1, \\ c_1 + c_2 + c_3 + 2c_4 = 1. \end{cases}$$

Solving the system, we get $c_1 = c_2 = c_3 = c_4 = \frac{1}{5}$. Thus

$$\left[\begin{bmatrix} 1 & 1 \\ 1 & 1 \end{bmatrix}\right]_{\mathcal{B}} = \begin{bmatrix} \frac{1}{5} \\ \frac{1}{5} \\ \frac{1}{5} \\ \frac{1}{5} \end{bmatrix}.$$

Since $\begin{bmatrix} 2 & 1 \\ 1 & 1 \end{bmatrix}$ is the first basis element, we simply get

$$\left[\begin{bmatrix} 2 & 1 \\ 1 & 1 \end{bmatrix}\right]_{\mathcal{B}} = \begin{bmatrix} 1 \\ 0 \\ 0 \\ 0 \end{bmatrix}.$$

Finally, since $\begin{bmatrix} 1 & 0 \\ 0 & 0 \end{bmatrix} = \begin{bmatrix} 2 & 1 \\ 1 & 1 \end{bmatrix} - \begin{bmatrix} 1 & 1 \\ 1 & 1 \end{bmatrix}$, we get

$$\left[\begin{bmatrix} 1 & 0 \\ 0 & 0 \end{bmatrix}\right]_{\mathcal{B}} = \left[\begin{bmatrix} 2 & 1 \\ 1 & 1 \end{bmatrix}\right]_{\mathcal{B}} - \left[\begin{bmatrix} 1 & 1 \\ 1 & 1 \end{bmatrix}\right]_{\mathcal{B}} = \begin{bmatrix} 1 \\ 0 \\ 0 \\ 0 \end{bmatrix} - \begin{bmatrix} \frac{1}{5} \\ \frac{1}{5} \\ \frac{1}{5} \\ \frac{1}{5} \end{bmatrix} = \begin{bmatrix} \frac{4}{5} \\ -\frac{1}{5} \\ -\frac{1}{5} \\ -\frac{1}{5} \end{bmatrix}.$$

Of course we also could have set up a system of equations to compute the coordinates of $\begin{bmatrix} 1 & 0 \\ 0 & 0 \end{bmatrix}$, but we were able to make use of previous calculations here. □

It is important to remember that the number of entries in the coordinate vector is exactly the number of elements in the basis.

5.5   Exercises

Exercise 5.5.1. Use the vector space axioms to prove that in a vector space it holds that

$$\mathbf{u} + \mathbf{w} = \mathbf{v} + \mathbf{w} \quad \text{implies} \quad \mathbf{u} = \mathbf{v}.$$

Exercise 5.5.2. Use the vector space axioms to prove that in a vector space it holds that

$$\mathbf{u} + \mathbf{u} = \mathbf{u} \quad \text{implies} \quad \mathbf{u} = \mathbf{0}.$$

Exercise 5.5.3. Use the vector space axioms to prove that in a vector space it holds that

$$\mathbf{u} + \mathbf{u} = 2\mathbf{u}.$$

Exercise 5.5.4. Let V be a vector space over $\mathbb{R}$. Let $\mathbf{v} \in V$ and $c \in \mathbb{R}$. Show that $c\mathbf{v} = \mathbf{0}$ if and only if $c = 0$ or $\mathbf{v} = \mathbf{0}$.

Exercise 5.5.5. Let $V = \mathbb{R}$ and define the following 'addition'

$$x \oplus y = \max\{x, y\}, \qquad x, y \in V,$$

and 'scalar multiplication'

$$c \odot x = cx, \qquad x \in V, c \in \mathbb{R}.$$

Determine which of the vector space axioms are satisfied and which ones fail, for this definition of 'addition' and 'scalar multiplication'.

Exercise 5.5.6. Let $V = \{(x, y) : x, y \in \mathbb{R}\}$ and define the following 'addition'

$$(x, y) \oplus (v, w) = (0, y + w), \qquad (x, y), (v, w) \in V,$$

and 'scalar multiplication'

$$c \odot (x, y) = (cx, cy), \qquad (x, y) \in V, c \in \mathbb{R}.$$

Determine which of the vector space axioms are satisfied and which ones fail, for this definition of 'addition' and 'scalar multiplication'.

Exercise 5.5.7. Let $V = \mathbb{R}_+ = \{x \in \mathbb{R} : x > 0\}$, and define the following 'addition'

$$x \oplus y = xy, \qquad x, y \in V,$$

and 'scalar multiplication'

$$c \odot x = x^c, \qquad x \in V, c \in \mathbb{R}.$$

Show that V with these two operations forms a vector space over $\mathbb{R}$. (Note that the neutral element of 'addition' is 1.)

Exercise 5.5.8. Let $V = \mathbb{C} = \{a + bi : a, b \in \mathbb{R}\}$, and define the usual addition

$$(a + bi) + (\alpha + \beta i) = (a + \alpha) + (b + \beta)i, \qquad a + bi, \alpha + \beta i \in V,$$

and scalar multiplication

$$c(a + bi) = (ca) + (cb)i, a + bi \in V, c \in \mathbb{R}.$$

Show that V with these two operations forms a vector space over $\mathbb{R}$.

Exercise 5.5.9. Let $V = \{f \mid f : (0, \infty) \to \mathbb{R} \text{ is a continuous function}\}$, and define addition and scalar multiplication by

$$(f + g)(t) = f(t) + g(t), \quad (cf)(t) = cf(t), \quad f, g \in V, c \in \mathbb{R}.$$

Show that V is a vector space over $\mathbb{R}$.

Exercise 5.5.10. For the following choices of V and W, determine whether W is a subspace of V.

(a) $V = \mathbb{R}_2[t]$ and

$$W = \{p(t) \in V \ : p(1) + p(2) + p(3) = 0\}.$$

(b) $V = \mathbb{R}^{2 \times 2}$ and

$$W = \left\{ \begin{bmatrix} a & b \\ c & d \end{bmatrix} \in \mathbb{R}^{2 \times 2} : a = d \right\}.$$

(c) $V = \mathbb{R}_2[t]$ and

$$W = \{p(t) \in V \ : p(1) = p(2)p(3)\}.$$

(d) $V = \mathbb{R}_2[t]$ and

$$W = \{p(t) \in V \ : p(1) = p(2) \text{ and } p(3) = 0\}.$$

(e) Let $V = \mathbb{R}^3$,

$$W = \left\{ \begin{bmatrix} x_1 \\ x_2 \\ x_3 \end{bmatrix} : x_1, x_2, x_3 \in \mathbb{R}, x_1 - 2x_2 + x_3^2 = 0 \right\}.$$

(f) $V = \mathbb{R}^{3 \times 3}$,

$$W = \left\{ \begin{bmatrix} a & b & c \\ 0 & a & b \\ 0 & 0 & a \end{bmatrix} : a, b, c \in \mathbb{R} \right\}.$$

(g) $V = \mathbb{R}_2[t]$ and

$$W = \{p(t) \in V \ : \ p(2) = 0\}.$$

(h) $V = \mathbb{R}_1[t]$ and

$$W = \left\{ p(t) \in V \ : \int_0^1 p(t)e^t dt = 0 \right\}.$$

(Remember that $te^t - e^t$ is an anti-derivative of te^t.)

Exercise 5.5.11. For the following vector spaces V over $\mathbb{R}$ and vectors, determine whether the vectors are linearly independent or linearly dependent. In the latter case, please provide a linear dependence relation between the vectors.

(a) Let $V = \mathbb{R}_3[t]$ and consider the vectors

$$1 + t\ , 1 - t + t^3,\ 2 - 2t^3.$$

(b) Let $V = \mathbb{R}^{1 \times 4}$, and consider the vectors

$$\begin{bmatrix} 1 & 1 & 1 & 2 \end{bmatrix}, \begin{bmatrix} 1 & 2 & 4 & 1 \end{bmatrix}, \begin{bmatrix} 0 & 1 & 3 & 4 \end{bmatrix}.$$

(c) Let $V = \mathbb{R}_2[t]$, and consider the vectors

$$1 + 2t - t^2,\ 2 - t - t^2,\ -1 + 8t - t^2.$$

(d) Let $V = \mathbb{R}^4$ and consider the vectors

$$\begin{bmatrix} 4 \\ 0 \\ 2 \\ 3 \end{bmatrix}, \begin{bmatrix} 2 \\ 1 \\ 0 \\ 3 \end{bmatrix}, \begin{bmatrix} 1 \\ 2 \\ 1 \\ 0 \end{bmatrix}.$$

(e) Let $V = \mathbb{R}^{2 \times 2}$, and consider the vectors

$$\begin{bmatrix} 0 & 1 \\ -1 & 0 \end{bmatrix}, \begin{bmatrix} 1 & 1 \\ 1 & 0 \end{bmatrix}, \begin{bmatrix} -1 & 1 \\ -1 & 0 \end{bmatrix}.$$

(f) Let $V = \mathbb{R}^{3 \times 2}$, and consider the vectors

$$\begin{bmatrix} 3 & 4 \\ 1 & 0 \\ 1 & 0 \end{bmatrix}, \begin{bmatrix} 1 & 1 \\ 4 & 2 \\ 1 & 2 \end{bmatrix}, \begin{bmatrix} 1 & 2 \\ 3 & 1 \\ 1 & 2 \end{bmatrix}.$$

(g) Let $V = \mathbb{R}_2[t]$, and consider the vectors

$$2t + 5,\, 3t^2 + 1.$$

(h) Let $V = \{f \mid f : (0, \infty) \to \mathbb{R}$ is a continuous function$\}$, and consider the vectors

$$t, t^2, \frac{1}{t}.$$

Exercise 5.5.12. Do the following sets of vectors form a basis for $\mathbb{R}_2[t]$?

(a) $\{3 + 2t, 1 + 3t^2, -4t + 8t^2\}$.

(b) $\{1 + 2t, t - t^2, t + t^2\}$.

Exercise 5.5.13. Determine a basis for the subspaces W in Exercise 5.5.10.

Exercise 5.5.14. This exercise generalizes Example 5.3.4. Let $a \in \mathbb{R}$, $V = \mathbb{R}_3[t]$ and $W = \{p(t) \in V : p(a) = 0\}$. Show that W is a subspace, and that $\{t - a, t^2 - a^2, t^3 - a^3\}$ is a basis for W.

Exercise 5.5.15. For V as in Exercise 5.5.7 show that for $1 \neq x \in \mathbb{R}_+$ we have that $\{x\}$ is a basis for V.

Exercise 5.5.16. For V as in Exercise 5.5.8 find a basis for V. Conclude that $\dim_{\mathbb{R}} V = 2$. (We write $\dim_{\mathbb{R}}$ to emphasize that V is defined as a vector space over $\mathbb{R}$.)

Exercise 5.5.17.

(a) Show that if the set $\{v_1, \ldots, v_k\}$ is linearly independent, and v_{k+1} is not in $\mathrm{Span}\{v_1, \ldots, v_k\}$, then the set $\{v_1, \ldots, v_k, v_{k+1}\}$ is linearly independent.

(b) Let W be a subspace of an n-dimensional vector space V, and let $\{v_1, \ldots, v_p\}$ be a basis for W. Show that there exist vectors $v_{p+1}, \ldots, v_n \in V$ so that $\{v_1, \ldots, v_p, v_{p+1}, \ldots, v_n\}$ is a basis for V.

(Hint: Once $v_1, \ldots, v_k$ are found and $k < n$, observe that one can choose $v_{k+1} \in V \setminus (\mathrm{Span}\{v_1, \ldots, v_k\})$. Argue that this process stops when $k = n$, and that at that point a basis for V is found.)

Exercise 5.5.18. For the following choices of subspaces U and W in V, find bases for $U + W$ and $U \cap W$.

(a) $V = \mathbb{R}_5[t]$, $U = \mathrm{Span}\{t + 1, t^2 - 1\}$, $W = \{p(t) : p(2) = 0\}$.

(b) $V = \mathbb{R}^4$,

$$U = \mathrm{Span}\left\{\begin{bmatrix}3\\0\\2\\1\end{bmatrix}, \begin{bmatrix}2\\1\\0\\0\end{bmatrix}\right\}, \quad W = \mathrm{Span}\left\{\begin{bmatrix}1\\2\\1\\0\end{bmatrix}, \begin{bmatrix}4\\4\\1\\1\end{bmatrix}\right\}.$$

Exercise 5.5.19. Let $\{v_1, v_2, v_3, v_4, v_5\}$ be linearly independent vectors in a vector space V. Determine whether the following sets are linearly dependent or linearly independent.

(a) $\{\mathbf{v}_1 + \mathbf{v}_2 + \mathbf{v}_3 + \mathbf{v}_4, \mathbf{v}_1 - \mathbf{v}_2 + \mathbf{v}_3 - \mathbf{v}_4, \mathbf{v}_1 - \mathbf{v}_2 - \mathbf{v}_3 - \mathbf{v}_4\}$.

(b) $\{\mathbf{v}_1 + \mathbf{v}_2, \mathbf{v}_2 + \mathbf{v}_3, \mathbf{v}_3 + \mathbf{v}_4, \mathbf{v}_4 + \mathbf{v}_5, \mathbf{v}_5 + \mathbf{v}_2\}$.

(c) $\{\mathbf{v}_1 + \mathbf{v}_3, \mathbf{v}_4 - \mathbf{v}_2, \mathbf{v}_5 + \mathbf{v}_1, \mathbf{v}_4 - \mathbf{v}_2, \mathbf{v}_5 + \mathbf{v}_3, \mathbf{v}_1 + \mathbf{v}_2\}$.

Exercise 5.5.20. For the following choices of vector spaces V over $\mathbb{R}$, bases $\mathcal{B}$ and vectors $\mathbf{v}$, determine $[\mathbf{v}]_\mathcal{B}$.

(a) $V = \mathbb{R}^4$,

$$\mathcal{B} = \left\{ \begin{bmatrix} 3 \\ 0 \\ 2 \\ 1 \end{bmatrix}, \begin{bmatrix} 2 \\ 1 \\ 0 \\ 0 \end{bmatrix}, \begin{bmatrix} 1 \\ 2 \\ 1 \\ 0 \end{bmatrix}, \begin{bmatrix} 0 \\ 2 \\ 1 \\ 0 \end{bmatrix} \right\}, \quad \mathbf{v} = \begin{bmatrix} 1 \\ 3 \\ 2 \\ 2 \end{bmatrix}.$$

(b) $V = \mathbb{R}^{2 \times 2}$,

$$\mathcal{B} = \left\{ \begin{bmatrix} 0 & 1 \\ -1 & -1 \end{bmatrix}, \begin{bmatrix} 1 & 1 \\ 1 & -1 \end{bmatrix}, \begin{bmatrix} 1 & 0 \\ -1 & -1 \end{bmatrix}, \begin{bmatrix} 1 & 1 \\ -1 & -1 \end{bmatrix} \right\}, \mathbf{v} = \begin{bmatrix} -2 & 3 \\ -5 & 10 \end{bmatrix}.$$

(c) $V = \mathbb{R}^{2 \times 2}$,

$$\mathcal{B} = \{E_{11}, E_{12}, E_{21}, E_{22}\}, \mathbf{v} = \begin{bmatrix} -1 & 5 \\ -7 & 1 \end{bmatrix}.$$

(d) $V = \mathrm{Span}\mathcal{B}$,

$$\mathcal{B} = \left\{ \begin{bmatrix} 3 & 4 \\ 1 & 0 \\ 1 & 0 \end{bmatrix}, \begin{bmatrix} 1 & 1 \\ 4 & 2 \\ 1 & 2 \end{bmatrix}, \begin{bmatrix} 1 & 2 \\ 3 & 3 \\ 3 & 0 \end{bmatrix} \right\}, \mathbf{v} = \begin{bmatrix} 5 & 7 \\ 8 & 5 \\ 5 & 2 \end{bmatrix}.$$

(e) $V = \mathbb{R}_3[t]$, $\mathcal{B} = \{1 + t, t + t^2, 2 - 3t^2, t^3\}$, $\mathbf{v} = 2 - t - t^3$.

(f) $V = \mathrm{Span}\mathcal{B}$,

$$\mathcal{B} = \left\{ \begin{bmatrix} 1 & 0 \\ 0 & 0 \end{bmatrix}, \begin{bmatrix} 1 & 0 \\ 0 & 1 \end{bmatrix}, \begin{bmatrix} 1 & 1 \\ 0 & 0 \end{bmatrix}, \begin{bmatrix} 1 & 0 \\ 1 & 0 \end{bmatrix} \right\}, \mathbf{v} = \begin{bmatrix} 1 & 1 \\ 1 & 1 \end{bmatrix}.$$

(g) $V = \mathbb{R}_2[t]$, $\mathcal{B} = \{1 - t^2, t - 3t^2, 3 + 2t^2\}$, $\mathbf{v} = -9 + 4t - 18t^2$.

(h) $V = \mathrm{Span}\mathcal{B}$, $\mathcal{B} = \{t, t^2, \frac{1}{t}\}$, $\mathbf{v} = \frac{t^3 + 3t^2 + 5}{t}$.

Exercise 5.5.21. True or False? Justify each answer.

(i) The set $H = \{A \in \mathbb{R}^{3 \times 3} : A = -A^T\}$ is a subspace of $\mathbb{R}^{3 \times 3}$.

(ii) If $\dim V = p$, then the vectors $\{\mathbf{v}_1, \ldots, \mathbf{v}_{p-1}\}$ do not span V.

(iii) The set $H = \left\{ \begin{bmatrix} x_1 \\ x_2 \end{bmatrix} : x_1 + x_2 = 0 \right\}$ is a subspace of $\mathbb{R}^2$.

(iv) The set $H = \{p \in \mathbb{R}_2[t] : p(0) = 5p(1)\}$ is a subspace of $\mathbb{R}_2[t]$.

(v) The set $H = \{p \in \mathbb{R}_2[t] : p(0) = p(1)^2\}$ is a subspace of $\mathbb{R}_2[t]$.

Exercise 5.5.22. When looking for a function that interpolates measured data, one can use Linear Algebra if one has some of idea of the types of functions one is looking for. For instance, if one expects there to be exponential growth but does not know at what rate(s), one can look at the vector space

$$W = \text{Span}\{e^t, e^{2t}, \ldots, e^{Nt}\}.$$

If one expects the data to come from an oscillating source, one can look at the vector space

$$W = \text{Span}\{\cos(t), \cos(2t), \ldots, \cos(Nt)\}.$$

This is a very useful vector space when one wants to analyze sound signals; the number k in $\cos(kx)$ is referred to as the **frequency**.

Let us do a simple example. Suppose we have following data.

t	1	2	3	4	5	6	7
$f(t)$	2.0481	−0.1427	−2.0855	−2.5789	−1.5237	0.1602	1.3551

Let us assume that the function we are trying to find is an element of

$$W = \text{Span}\{\cos(t), \cos(2t), \ldots, \cos(7t)\}.$$

Then we have that $f(t) = \sum_{k=1}^{7} c_k \cos(kt)$. To determine $f(t)$, we just need to solve for the coefficients $c_1, \ldots, c_7$. Plugging in $t = 1, \ldots, 7$, we obtain the following system of equations:

$$\begin{bmatrix} \cos(1) & \cos(2) & \cdots & \cos(7) \\ \cos(2) & \cos(4) & \cdots & \cos(14) \\ \vdots & \vdots & & \vdots \\ \cos(7) & \cos(14) & \cdots & \cos(49) \end{bmatrix} \begin{bmatrix} c_1 \\ c_2 \\ \vdots \\ c_7 \end{bmatrix} = \begin{bmatrix} 2.0481 \\ -0.1427 \\ \vdots \\ 1.3551 \end{bmatrix}.$$

Solving this system, we find $c_1 = 1$, $c_7 = 2$, and $c_2 = \cdots = c_6 = 0$, giving the solution

$$f(t) = \cos(t) + 2\cos(7t).$$

Thus this signal is comprised of a low frequency signal ($\cos(t)$) and a high frequency signal of double the strength ($2\cos(7t)$). It could be the combination of two voices, for instance. This technique is also used to analyze chemical compounds (using spectroscopy), where the different frequencies indicate the presence of certain components within the compound. The matrix above is closely related to the discrete cosine transform. Indeed, with a change of variables, one obtains the matrix of the (second) **discrete cosine transform** (**DCT**) whose (k,j)th entry is $\cos(\frac{(k-\frac{1}{2})(j-1)\pi}{N})$. The DCT is used in many applications as a search will quickly reveal.

- Suppose we have data

t	$\frac{\pi}{4}$	$\frac{\pi}{2}$	$\frac{3\pi}{4}$
$f(t)$	$-\frac{1}{2}\sqrt{2}$	-1	$\frac{1}{2}\sqrt{2}$

Find $f(t) \in \text{Span}\{\cos(t), \cos(2t), \cos(3t)\}$ fitting the given data.

6

Linear Transformations

CONTENTS

6.1   Definition of a Linear Transformation

Linear transformations (or linear maps) are those functions between vector spaces that respect the vector space structure of addition and scalar multiplication. The definition is as follows.

Definition 6.1.1. Let V and W be vector spaces over $\mathbb{F}$. A function $T : V \to W$ is called **linear** if

(i) $T(\mathbf{u} + \mathbf{v}) = T(\mathbf{u}) + T(\mathbf{v})$ for all $\mathbf{u}, \mathbf{v} \in V$, and

(ii) $T(c\mathbf{u}) = cT(\mathbf{u})$ for all $c \in \mathbb{F}$ and all $\mathbf{u} \in V$.

In this case, we say that T is a **linear transformation** or a **linear map**.

When T is linear, we must have that $T(\mathbf{0}) = \mathbf{0}$. Indeed, by using (ii) we have $T(\mathbf{0}) = T(0 \cdot \mathbf{0}) = 0T(\mathbf{0}) = \mathbf{0}$, where in the first and last step we used Lemma 5.1.2.

We can also combine (i) and (ii) in one statement as the next lemma shows. The proof is straightforward and left out.

Lemma 6.1.2. *The map $T : V \to W$ is linear if and only if*

$$T(c\mathbf{u} + d\mathbf{v}) = cT(\mathbf{u}) + dT(\mathbf{v}) \quad \textit{for all } \mathbf{u}, \mathbf{v} \in V \textit{ and all } c, d \in \mathbb{F}.$$

Example 6.1.3. Let $T : \mathbb{R}^2 \to \mathbb{R}^3$ be defined by

$$T \begin{bmatrix} x_1 \\ x_2 \end{bmatrix} = \begin{bmatrix} 2x_1 + x_2 \\ x_1 + x_2 \\ x_2 \end{bmatrix}.$$

Then

$$T(\begin{bmatrix} x_1 \\ x_2 \end{bmatrix} + \begin{bmatrix} y_1 \\ y_2 \end{bmatrix}) = T \begin{bmatrix} x_1 + y_1 \\ x_2 + y_2 \end{bmatrix} = \begin{bmatrix} 2(x_1 + y_1) + x_2 + y_2 \\ x_1 + y_1 + x_2 + y_2 \\ x_2 + y_2 \end{bmatrix} =$$

$$\begin{bmatrix} 2x_1 + x_2 \\ x_1 + x_2 \\ x_2 \end{bmatrix} + \begin{bmatrix} 2y_1 + y_2 \\ y_1 + y_2 \\ y_2 \end{bmatrix} = T \begin{bmatrix} x_1 \\ x_2 \end{bmatrix} + T \begin{bmatrix} y_1 \\ y_2 \end{bmatrix},$$

and

$$T(c \begin{bmatrix} x_1 \\ x_2 \end{bmatrix}) = T \begin{bmatrix} cx_1 \\ cx_2 \end{bmatrix} = \begin{bmatrix} 2cx_1 + cx_2 \\ cx_1 + cx_2 \\ cx_2 \end{bmatrix} = c \begin{bmatrix} 2x_1 + x_2 \\ x_1 + x_2 \\ x_2 \end{bmatrix} = cT \begin{bmatrix} x_1 \\ x_2 \end{bmatrix}.$$

Thus T is linear.   □

Example 6.1.4. Let $T : \mathbb{R}^3 \to \mathbb{R}^2$ be defined by

$$T \begin{bmatrix} x_1 \\ x_2 \\ x_3 \end{bmatrix} = \begin{bmatrix} x_1 x_2 \\ x_1 + x_2 + x_3 \end{bmatrix}.$$

Let us start by checking condition (i):

$$T \begin{bmatrix} x_1 \\ x_2 \\ x_3 \end{bmatrix} + T \begin{bmatrix} y_1 \\ y_2 \\ y_3 \end{bmatrix} = \begin{bmatrix} x_1 x_2 + y_1 y_2 \\ x_1 + x_2 + x_3 + y_1 + y_2 + y_3 \end{bmatrix} \qquad (6.1)$$

and

$$T(\begin{bmatrix} x_1 \\ x_2 \\ x_3 \end{bmatrix} + \begin{bmatrix} y_1 \\ y_2 \\ y_3 \end{bmatrix}) = \begin{bmatrix} (x_1 + y_1)(x_2 + y_2) \\ x_1 + x_2 + x_3 + y_1 + y_2 + y_3 \end{bmatrix}. \qquad (6.2)$$

Notice that the first components in the right hand sides of (6.1) and (6.2) differ (by a term $x_1 y_2 + y_1 x_2$), so T does not seem to be linear. Let us find a counterexample (and make sure $x_1 y_2 + y_1 x_2 \neq 0$). Take for instance

$$\mathbf{x} = \mathbf{y} = \begin{bmatrix} 1 \\ 1 \\ 1 \end{bmatrix}.$$

Then

$$T(\mathbf{x} + \mathbf{y}) = T \begin{bmatrix} 2 \\ 2 \\ 2 \end{bmatrix} = \begin{bmatrix} 4 \\ 6 \end{bmatrix} \neq \begin{bmatrix} 1 \\ 3 \end{bmatrix} + \begin{bmatrix} 1 \\ 3 \end{bmatrix} = T(\mathbf{x}) + T(\mathbf{y}).$$

Thus T is not linear.   □

Notice that in order to show that a function is not linear, one only needs to provide one example where the above rule (i) or (ii) is not satisfied.

Here are some more examples and non-examples:

- $T : \mathbb{R}^2 \to \mathbb{R}^3$ defined via $T(\begin{bmatrix} x_1 \\ x_2 \end{bmatrix}) = \begin{bmatrix} \sin(x_1) \\ 5x_1 - x_2 \\ x_1 \end{bmatrix}$ is not linear. For instance,

  $2T(\begin{bmatrix} \frac{\pi}{2} \\ 0 \end{bmatrix}) \neq T(2 \begin{bmatrix} \frac{\pi}{2} \\ 0 \end{bmatrix})$.

- $T : \mathbb{R} \to \mathbb{R}^2$ defined via $T([x_1]) = \begin{bmatrix} 2x_1 \\ 3x_1 + 6 \end{bmatrix}$ is not linear. Indeed, observe that $T(\mathbf{0}) \neq \mathbf{0}$.

- $T : \mathbb{R}^3 \to \mathbb{R}^2$ defined via $T(\begin{bmatrix} x_1 \\ x_2 \\ x_3 \end{bmatrix}) = \begin{bmatrix} 2x_1 + 3x_2 \\ x_1 + 5x_2 - 7x_3 \end{bmatrix}$ is linear, as we

  show below.

The last linear map can be written as

$$T(\begin{bmatrix} x_1 \\ x_2 \\ x_3 \end{bmatrix}) = \begin{bmatrix} 2 & 3 & 0 \\ 1 & 5 & -7 \end{bmatrix} \begin{bmatrix} x_1 \\ x_2 \\ x_3 \end{bmatrix}.$$

Also the linear map in Example 6.1.3 can be written similarly in the form

$$T(\begin{bmatrix} x_1 \\ x_2 \end{bmatrix}) = \begin{bmatrix} 2 & 1 \\ 1 & 1 \\ 0 & 1 \end{bmatrix} \begin{bmatrix} x_1 \\ x_2 \end{bmatrix}.$$

In general linear maps from $\mathbb{R}^n$ to $\mathbb{R}^m$ are described by multiplication by a matrix.

Proposition 6.1.5. *Let* $T : \mathbb{R}^n \to \mathbb{R}^m$. *Then* T *is linear if and only if there exists a matrix* $A \in \mathbb{R}^{m \times n}$ *so that* $T(\mathbf{x}) = A\mathbf{x}$ *for all* $\mathbf{x} \in \mathbb{R}^n$. *This matrix is given by*

$$A = \begin{bmatrix} T(\mathbf{e}_1) & T(\mathbf{e}_2) & \cdots & T(\mathbf{e}_n) \end{bmatrix}, \tag{6.3}$$

where $\{\mathbf{e}_1, \ldots, \mathbf{e}_n\}$ *is the standard basis of* $\mathbb{R}^n$.

Proof. If $T(\mathbf{x}) = A\mathbf{x}$, then linearity of T follows directly from the rules on matrix vector multiplication: $A(\mathbf{x} + \mathbf{y}) = A\mathbf{x} + A\mathbf{y}$ and $A(c\mathbf{x}) = cA\mathbf{x}$.

Conversely, suppose that T is linear. If we let $\mathbf{x} = (x_i)_{i=1}^n$ then $\mathbf{x} = \sum_{i=1}^n x_i \mathbf{e}_i$. Since T is linear we then get that

$$T(\mathbf{x}) = T(\sum_{i=1}^n x_i \mathbf{e}_i) = \sum_{i=1}^n x_i T(\mathbf{e}_i) = \begin{bmatrix} T(\mathbf{e}_1) & \cdots & T(\mathbf{e}_n) \end{bmatrix} \begin{bmatrix} x_1 \\ \vdots \\ x_n \end{bmatrix} = A\mathbf{x},$$

where A is given in (6.3).   □

We call the matrix A in (6.3) the **standard matrix** of the linear map T.

Example 6.1.6. Let $T : \mathbb{R}^2 \to \mathbb{R}^2$ be the linear map that rotates a vector counter clockwise by an angle α. Determine the standard matrix of T.

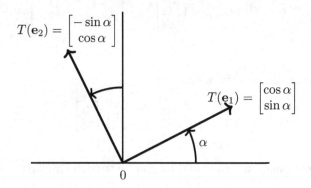

Figure 6.1: Rotation over an angle α.

From the figure we see that

$$T(\mathbf{x}) = \begin{bmatrix} T(\mathbf{e}_1) & T(\mathbf{e}_2) \end{bmatrix} \begin{bmatrix} x_1 \\ x_2 \end{bmatrix} = \begin{bmatrix} \cos \alpha & -\sin \alpha \\ \sin \alpha & \cos \alpha \end{bmatrix} \begin{bmatrix} x_1 \\ x_2 \end{bmatrix}.$$

□

Example 6.1.7. Let $T : \mathbb{R}^3 \to \mathbb{R}^3$ be the linear map that reflects a vector in the xy plane. Determine the standard matrix of T.

We have that $T(\mathbf{e}_1) = \mathbf{e}_1$, $T(\mathbf{e}_2) = \mathbf{e}_2$ and $T(\mathbf{e}_3) = -\mathbf{e}_3$. Thus we find

$$T(\mathbf{x}) = \begin{bmatrix} T(\mathbf{e}_1) & T(\mathbf{e}_2) & T(\mathbf{e}_3) \end{bmatrix} \begin{bmatrix} x_1 \\ x_2 \\ x_3 \end{bmatrix} = \begin{bmatrix} 1 & 0 & 0 \\ 0 & 1 & 0 \\ 0 & 0 & -1 \end{bmatrix} \begin{bmatrix} x_1 \\ x_2 \\ x_3 \end{bmatrix}.$$

□

Let us also give some examples on other vector spaces.

Example 6.1.8. Let $T : \mathbb{R}_2[t] \to \mathbb{R}^2$ be given by

$$T(p(t)) = \begin{bmatrix} p(5) \\ \int_0^2 p(t)dt \end{bmatrix}.$$

For instance, $T(t^2) = \begin{bmatrix} 25 \\ \frac{8}{3} \end{bmatrix}$. Is T linear?

We notice that if we take a sum of two polynomials, say of $p(t)$ and $q(t)$ and then plug in $t = 5$, the answer is simply $p(5) + q(5)$. Also $\int_0^2 p(t) + q(t)dt = \int_0^2 p(t)dt + \int_0^2 q(t)dt$. Both actions also work well with scalar multiplication, so it seems that T is linear. To prove this, we compute

$$T(c\ p(t) + d\ q(t)) = \begin{bmatrix} c\ p(5) + d\ q(5) \\ \int_0^2 c\ p(t) + d\ q(t)dt \end{bmatrix} = \begin{bmatrix} c\ p(5) \\ c\int_0^2 p(t)dt \end{bmatrix} + \begin{bmatrix} d\ q(5) \\ d\int_0^2 q(t)dt \end{bmatrix} =$$

$cT(p(t)) + dT(q(t))$, and thus T is linear. □

Example 6.1.9. Let $T : \mathbb{R}^{2\times 2} \to \mathbb{R}^{1\times 2}$ be given by

$$T(A) = T(\begin{bmatrix} a_{11} & a_{12} \\ a_{21} & a_{22} \end{bmatrix}) = \begin{bmatrix} a_{11} - a_{12} & a_{21} - a_{22} \end{bmatrix}.$$

Is T linear?

Let us compare $T(cA + dB)$ with $cT(A) + dT(B)$:

$$T(cA + dB) = T\begin{bmatrix} ca_{11} + db_{11} & ca_{12} + db_{12} \\ ca_{21} + db_{21} & ca_{22} + db_{22} \end{bmatrix} =$$

$$\begin{bmatrix} ca_{11} + db_{11} - (ca_{12} + db_{12}) & ca_{21} + db_{21} - (ca_{22} + db_{22}) \end{bmatrix} =$$

$$c\begin{bmatrix} a_{11} - a_{12} & a_{21} - a_{22} \end{bmatrix} + d\begin{bmatrix} b_{11} - b_{12} & b_{21} - b_{22} \end{bmatrix} = cT(A) + dT(B),$$

so T is linear. □

Example 6.1.10. Let $T : \mathbb{R}_2[t] \to \mathbb{R}^2$ be given by

$$T(p(t)) = \begin{bmatrix} p(1)p(2) \\ \int_1^3 p(t)dt \end{bmatrix}.$$

Is T linear?

Looking at the first component we have to compare $(p(1) + q(1))(p(2) + q(2))$ with $p(1)p(2) + q(1)q(2)$, which are in general not equal. So we suspect that T is not linear. Let us take $p(t) = t = q(t)$. Then we get that

$$T(p(t) + q(t)) = T(2t) = \begin{bmatrix} (2)(4) \\ \int_1^3 2tdt \end{bmatrix} = \begin{bmatrix} 8 \\ 8 \end{bmatrix},$$

while

$$T(p(t)) + T(q(t)) = \begin{bmatrix} (1)(2) \\ \int_1^3 tdt \end{bmatrix} + \begin{bmatrix} (1)(2) \\ \int_1^3 tdt \end{bmatrix} = \begin{bmatrix} 4 \\ 8 \end{bmatrix}.$$

Thus for this choice of $p(t)$ and $q(t)$ we have that $T(p(t) + q(t)) \neq T(p(t)) + T(q(t))$, and thus T is not linear. $\qquad\square$

Given two linear transformations $T : V \to W$ and $S : W \to X$, then the **composition** $S \circ T : V \to X$ of S and T is defined by

$$S \circ T(\mathbf{v}) = S(T(\mathbf{v})).$$

The composition of two linear transformations is again a linear transformation. We leave this as an exercise (see Exercise 6.5.4).

6.2   Range and Kernel of Linear Transformations

There are two subspaces associated with a linear transformation: (i) the range (which lies in the co-domain) and (ii) the kernel (which lies in the domain). These subspaces provide us with crucial information about the linear transformation. We start by discussing the range.

Definition 6.2.1. Let $T : V \to W$ be a linear map. The **range** of T is

$$\operatorname{Ran} T := \{\mathbf{w} \in W \ : \ \text{there exists a } \mathbf{v} \in V \text{ so that } T(\mathbf{v}) = \mathbf{w}\}.$$

Example 6.2.2. If $T : \mathbb{R}^n \to \mathbb{R}^m$ is given by $T(\mathbf{x}) = A\mathbf{x}$ with $A \in \mathbb{R}^{m \times n}$, then

$$\operatorname{Ran} T = \operatorname{Col} A.$$

Indeed, $T(\mathbf{x}) = \mathbf{b}$ if and only if $A\mathbf{x} = \mathbf{b}$. Consequently, $\mathbf{b} \in \operatorname{Ran} T$ if and only if $\mathbf{b} \in \operatorname{Col} A$. $\qquad\square$

Definition 6.2.3. We say that $T : V \to W$ is **onto** (or **surjective**) if $\operatorname{Ran} T = W$. Equivalently, T is onto if and only if for every $\mathbf{w} \in W$ there exists a $\mathbf{v} \in V$ so that $T(\mathbf{v}) = \mathbf{w}$.

Note that in Example 6.2.2 the linear map T, given by $T(\mathbf{x}) = A\mathbf{x}$, is onto if and only if every row of A has a pivot.

We know already that the column space of a matrix $A \in \mathbb{R}^{m \times n}$ is a subspace of $\mathbb{R}^m$, and a basis can be found by taking the pivot columns of the n columns of A. The following statement is a generalization of this.

Proposition 6.2.4. *Let* $T : V \to W$ *be a linear map. Then* Ran T *is a subspace of* W. *Moreover, if* $\{\mathbf{v}_1, \ldots, \mathbf{v}_p\}$ *is a basis for* V, *then* Ran $T =$ Span $\{T(\mathbf{v}_1), \ldots, T(\mathbf{v}_p)\}$. *In particular* dim Ran $T \leq$ dim V.

Proof. First observe that $T(\mathbf{0}) = \mathbf{0}$ gives that $\mathbf{0} \in$ Ran T. Next, let $\mathbf{w}$, $\hat{\mathbf{w}} \in$ Ran T and $c \in \mathbb{F}$. Then there exist $\mathbf{v}$, $\hat{\mathbf{v}} \in V$ so that $T(\mathbf{v}) = \mathbf{w}$ and $T(\hat{\mathbf{v}}) = \hat{\mathbf{w}}$. Then $\mathbf{w} + \hat{\mathbf{w}} = T(\mathbf{v} + \hat{\mathbf{v}}) \in$ Ran T and $c\mathbf{w} = T(c\mathbf{v}) \in$ Ran T. Thus, by Theorem 5.2.1, Ran T is a subspace of W.

Clearly, $T(\mathbf{v}_1), \ldots, T(\mathbf{v}_p) \in$ Ran T, and since Ran T is a subspace we have that Span $\{T(\mathbf{v}_1), \ldots, T(\mathbf{v}_p)\} \subseteq$ Ran T. For the converse inclusion, let $\mathbf{w} \in$ Ran T. Then there exists a $\mathbf{v} \in V$ so that $T(\mathbf{v}) = \mathbf{w}$. As $\{\mathbf{v}_1, \ldots, \mathbf{v}_p\}$ is a basis for V, there exist $c_1, \ldots, c_p \in \mathbb{R}$ so that $\mathbf{v} = c_1\mathbf{v}_1 + \cdots + c_p\mathbf{v}_p$. Then

$$\mathbf{w} = T(\mathbf{v}) = T(\sum_{j=1}^{p} c_j\mathbf{v}_j) = \sum_{j=1}^{p} c_j T(\mathbf{v}_j) \in \text{Span}\,\{T(\mathbf{v}_1), \ldots, T(\mathbf{v}_p)\}.$$

Thus Ran $T \subseteq$ Span $\{T(\mathbf{v}_1), \ldots, T(\mathbf{v}_p)\}$. We have shown both inclusions, and consequently Ran $T =$ Span $\{T(\mathbf{v}_1), \ldots, T(\mathbf{v}_p)\}$ follows. $\quad\square$

Definition 6.2.5. The **kernel** of T is

$$\text{Ker}\, T := \{\mathbf{v} \in V \,:\, T(\mathbf{v}) = \mathbf{0}\}.$$

Example 6.2.6. If $T : \mathbb{R}^n \to \mathbb{R}^m$ is given by $T(\mathbf{x}) = A\mathbf{x}$ with $A \in \mathbb{R}^{m \times n}$, then

$$\text{Ker}\, T = \text{Nul}\, A.$$

Indeed, $T(\mathbf{x}) = \mathbf{0}$ if and only if $A\mathbf{x} = \mathbf{0}$, which happens if and only if $\mathbf{x} \in$ Nul A. $\quad\square$

We know already that the null space of a matrix $A \in \mathbb{R}^{m \times n}$ is a subspace of $\mathbb{R}^n$. The following statement is a generalization of this.

Proposition 6.2.7. *Let* $T : V \to W$ *be a linear map. Then* Ker T *is subspace of* V.

Proof. First observe that $T(\mathbf{0}) = \mathbf{0}$ gives that $\mathbf{0} \in$ Ker T. Next, let $\mathbf{v}$, $\hat{\mathbf{v}} \in$ Ker T and $c \in \mathbb{F}$. Then $T(\mathbf{v} + \hat{\mathbf{v}}) = T(\mathbf{v}) + T(\hat{\mathbf{v}}) = \mathbf{0} + \mathbf{0} = \mathbf{0}$ and $T(c\mathbf{v}) = cT(\mathbf{v}) = c\mathbf{0} = \mathbf{0}$, so $\mathbf{v} + \hat{\mathbf{v}}, c\mathbf{v} \in$ Ker T. Thus, by Theorem 5.2.1, Ker T is a subspace of V. $\quad\square$

Definition 6.2.8. We say that $T : V \to W$ is **one-to-one** (or **injective**) if $T(\mathbf{v}) = T(\mathbf{w})$ only holds when $\mathbf{v} = \mathbf{w}$. Equivalently, T is one-to-one if $\mathbf{u} \neq \mathbf{v}$ implies $T(\mathbf{u}) \neq T(\mathbf{v})$.

Note that in Example 6.2.6 the linear map T, given by $T(\mathbf{x}) = A\mathbf{x}$, is one-to-one if and only if all columns of A are pivot columns.

We have the following way to check wether a linear map is injective.

Lemma 6.2.9. *The linear map T is one-to-one if and only if* $\operatorname{Ker} T = \{\mathbf{0}\}$.

Proof. Suppose that T is one-to-one, and $\mathbf{v} \in \operatorname{Ker} T$. Then $T(\mathbf{v}) = \mathbf{0} = T(\mathbf{0})$, where in the last step we used that T is linear. Since T is one-to-one, $T(\mathbf{v}) = T(\mathbf{0})$ implies that $\mathbf{v} = \mathbf{0}$. Thus $\operatorname{Ker} T = \{\mathbf{0}\}$.

Next, suppose that $\operatorname{Ker} T = \{\mathbf{0}\}$, and let $T(\mathbf{v}) = T(\mathbf{w})$. Then, using linearity we get $\mathbf{0} = T(\mathbf{v}) - T(\mathbf{w}) = T(\mathbf{v} - \mathbf{w})$, implying that $\mathbf{v} - \mathbf{w} \in \operatorname{Ker} T = \{\mathbf{0}\}$, and thus $\mathbf{v} - \mathbf{w} = \mathbf{0}$. Thus $\mathbf{v} = \mathbf{w}$, and we can conclude that T is one-to-one.

$\square$

Example 6.2.10. Let $V = \mathbb{R}_3[t]$, $W = \mathbb{R}^2$, and

$$T(p(t)) = \begin{bmatrix} p(1) \\ \int_0^2 p(t)dt \end{bmatrix}.$$

Determine bases for $\operatorname{Ker} T$ and $\operatorname{Ran} T$.

We start with the kernel. Let $p(t) = a + bt + ct^2 + dt^3 \in \operatorname{Ker} T$, then

$$\mathbf{0} = T(p(t)) = \begin{bmatrix} a+b+c+d \\ 2a + 2b + \frac{8}{3}c + 4d \end{bmatrix} = \begin{bmatrix} 1 & 1 & 1 & 1 \\ 2 & 2 & \frac{8}{3} & 4 \end{bmatrix} \begin{bmatrix} a \\ b \\ c \\ d \end{bmatrix}.$$

Row reducing

$$\begin{bmatrix} 1 & 1 & 1 & 1 & | & 0 \\ 2 & 2 & \frac{8}{3} & 4 & | & 0 \end{bmatrix} \to \begin{bmatrix} 1 & 1 & 1 & 1 & | & 0 \\ 0 & 0 & \frac{2}{3} & 2 & | & 0 \end{bmatrix} \to \begin{bmatrix} 1 & 1 & 0 & -2 & | & 0 \\ 0 & 0 & 1 & 3 & | & 0 \end{bmatrix}, \quad (6.4)$$

gives that b and d are free variables and $a = -b + 2d$, $c = -3d$. Thus

$$p(t) = (-b + 2d) + bt + (-3d)t^2 + dt^3 = b(-1 + t) + d(2 - 3t^2 + t^3).$$

We get that $\operatorname{Ker} T = \operatorname{Span}\{-1 + t, 2 - 3t^2 + t^3\}$. In fact, since these two polynomials are linearly independent, they form a basis for $\operatorname{Ker} T$.

As $\{1, t, t^2, t^3\}$ is a basis for $\mathbb{R}_3[t]$, we get that

$$\text{Ran } T = \text{Span}\left\{T(1), T(t), T(t^2), T(t^3)\right\}$$

$$= \text{Span}\left\{\begin{bmatrix} 1 \\ 2 \end{bmatrix}, \begin{bmatrix} 1 \\ 2 \end{bmatrix}, \begin{bmatrix} 1 \\ \frac{8}{3} \end{bmatrix}, \begin{bmatrix} 1 \\ 4 \end{bmatrix}\right\} = \text{Span}\left\{\begin{bmatrix} 1 \\ 2 \end{bmatrix}, \begin{bmatrix} 1 \\ \frac{8}{3} \end{bmatrix}\right\}.$$

In the last step, we reduced the set of vectors to a basis for Ran T by just keeping the columns corresponding to pivot columns in (6.4).

Notice that

$$\dim \text{Ker } T + \dim \text{Ran } T = 2 + 2 = 4 = \dim \mathbb{R}_3[t].$$

As the next result shows, this is not a coincidence. $\qquad\square$

Theorem 6.2.11. *Let* $T : V \to W$ *be linear, and suppose that* $\dim V < \infty$. *Then*

$$\dim \text{Ker } T + \dim \text{Ran } T = \dim V. \tag{6.5}$$

Notice that this theorem generalizes the rule that for $A \in \mathbb{R}^{m \times n}$:

$$\dim \text{Nul } A + \dim \text{Col } A = n.$$

Proof. Let $\{v_1, \ldots, v_p\}$ be a basis for Ker $T(\subseteq V)$, and $\{w_1, \ldots, w_q\}$ a basis for Ran T (notice that by Proposition 6.2.4 it follows that Ran T is finite dimensional as V is finite dimensional). Let $x_1, \ldots, x_q \in V$ be so that $T(x_j) = w_j$, $j = 1, \ldots, q$. We claim that $\mathcal{B} = \{v_1, \ldots, v_p, x_1, \ldots, x_q\}$ is a basis for V, which then yields that $\dim V = p + q = \dim \text{Ker } T + \dim \text{Ran } T$.

Let $v \in V$. Then $T(v) \in$ Ran T, and thus there exist $b_1, \ldots, b_q$ so that $T(v) = \sum_{j=1}^{q} b_j w_j$. Then

$$T(v - \sum_{j=1}^{q} b_j x_j) = T(v) - \sum_{j=1}^{q} b_j w_j = 0.$$

Thus $v - \sum_{j=1}^{q} b_j x_j \in$ Ker T. Therefore, there exist $a_1, \ldots, a_p \in \mathbb{R}$ so that $v - \sum_{j=1}^{q} b_j x_j = \sum_{j=1}^{p} a_j v_j$. Consequently, $v = \sum_{j=1}^{p} a_j v_j + \sum_{j=1}^{q} b_j x_j \in$ Span $\mathcal{B}$. This proves that $V =$ Span $\mathcal{B}$.

It remains to show that $\mathcal{B}$ is linearly independent, so assume $\sum_{j=1}^{p} a_j v_j + \sum_{j=1}^{q} b_j x_j = 0$. Then

$$0 = T(\sum_{j=1}^{p} a_j v_j + \sum_{j=1}^{q} b_j x_j) = \sum_{j=1}^{p} a_j T(v_j) + \sum_{j=1}^{q} b_j T(x_j) = \sum_{j=1}^{q} b_j w_j,$$

where we use that $\mathbf{v}_j \in \operatorname{Ker} T$, $j = 1, \ldots, p$. As $\{\mathbf{w}_1, \ldots, \mathbf{w}_q\}$ is linearly independent, we now get that $b_1 = \cdots = b_q = 0$. But then we obtain that $\sum_{j=1}^{p} a_j \mathbf{v}_j = \mathbf{0}$, and as $\{\mathbf{v}_1, \ldots, \mathbf{v}_p\}$ is linearly independent, we get $a_1 = \cdots = a_p = 0$. Thus $\sum_{j=1}^{p} a_j \mathbf{v}_j + \sum_{j=1}^{q} b_j \mathbf{x}_j = \mathbf{0}$ implies $a_1 = \cdots = a_p = b_1 = \cdots = b_q = 0$, showing the linear independence of $\mathcal{B}$. $\qquad\square$

Definition 6.2.12. We say that T is **invertible** (or **bijective**) if T is both onto and one-to-one.

We let $id_V : V \to V$ denote the **identity mapping**; that is $id_V(\mathbf{v}) = \mathbf{v}$, $\mathbf{v} \in V$. When we do not need to emphasize the underlying space V, we just write id for id_V.

Proposition 6.2.13. *Let $T : V \to W$ be bijective. Then T has an inverse T^{-1}. That is, $T^{-1} : W \to V$ exists so that $T \circ T^{-1} = id_W$ and $T^{-1} \circ T = id_V$. Moreover, T^{-1} is linear. Conversely, if T has an inverse, then T is bijective.*

Proof. Let $\mathbf{w} \in W$. As T is onto, there exists a $\mathbf{v} \in V$ so that $T(\mathbf{v}) = \mathbf{w}$, and as T is one-to-one, this $\mathbf{v}$ is unique. Define $T^{-1}(\mathbf{w}) := \mathbf{v}$, making $T^{-1} : W \to V$ well-defined. It is straightforward to check that $T(T^{-1}(\mathbf{w})) = \mathbf{w}$ for all $\mathbf{w} \in W$, and $T^{-1}(T(\mathbf{v})) = \mathbf{v}$ for all $\mathbf{v} \in V$.

Next suppose $T^{-1}(\mathbf{w}) = \mathbf{v}$ and $T^{-1}(\hat{\mathbf{w}}) = \hat{\mathbf{v}}$. This means that $T(\mathbf{v}) = \mathbf{w}$ and $T(\hat{\mathbf{v}}) = \hat{\mathbf{w}}$. Thus $T(\mathbf{v} + \hat{\mathbf{v}}) = \mathbf{w} + \hat{\mathbf{w}}$. But then, by definition, $T^{-1}(\mathbf{w} + \hat{\mathbf{w}}) = \mathbf{v} + \hat{\mathbf{v}}$ and, consequently, $T^{-1}(\mathbf{w} + \hat{\mathbf{w}}) = T^{-1}(\mathbf{w}) + T^{-1}(\hat{\mathbf{w}})$. Similarly, one proves $T^{-1}(c\mathbf{w}) = cT^{-1}(\mathbf{w})$. Thus T^{-1} is linear.

Next, suppose that T has an inverse T^{-1}. Let $\mathbf{w} \in W$. Put $\mathbf{v} = T^{-1}(\mathbf{w})$. Then $T(\mathbf{v}) = \mathbf{w}$, and thus $\mathbf{w} \in \operatorname{Ran} T$. This shows that T is onto. Finally, suppose that $T(\mathbf{v}) = T(\hat{\mathbf{v}})$. Applying T^{-1} on both sides, gives $\mathbf{v} = T^{-1}(T(\mathbf{v})) = T^{-1}(T(\hat{\mathbf{v}})) = \hat{\mathbf{v}}$, showing that T is one-to-one. $\qquad\square$

Definition 6.2.14. A bijective linear map T is also called an **isomorphism**. We call two vector spaces V and W **isomorphic** if there exists an isomorphism $T : V \to W$.

The following example shows that $\mathbb{R}_{n-1}[t]$ and $\mathbb{R}^n$ are isomorphic.

Example 6.2.15. Let $T : \mathbb{R}_{n-1}[t] \to \mathbb{R}^n$ be defined by

$$T(a_0 + a_1 t + \cdots + a_{n-1} t^{n-1}) := \begin{bmatrix} a_0 \\ a_1 \\ \vdots \\ a_{n-1} \end{bmatrix}.$$

It is easy to see that T is an isomorphism. Indeed, T^{-1} is given by

$$T^{-1} \begin{bmatrix} a_0 \\ a_1 \\ \vdots \\ a_{n-1} \end{bmatrix} = a_0 + a_1 t + \cdots + a_{n-1} t^{n-1}.$$

Thus $\mathbb{R}_{n-1}[t]$ and $\mathbb{R}^n$ are isomorphic. $\qquad\square$

6.3   Matrix Representations of Linear Transformations

We have seen that any linear map from $\mathbb{R}^n$ to $\mathbb{R}^m$ can be represented by its standard matrix. This principle holds for a general linear map as well. What is needed is a choice of bases for the underlying spaces.

Theorem 6.3.1. *Given vector spaces V and W over $\mathbb{F}$, with bases $\mathcal{B} = \{\mathbf{v}_1, \ldots, \mathbf{v}_n\}$ and $\mathcal{C} = \{\mathbf{w}_1, \ldots, \mathbf{w}_m\}$, respectively. Let $T : V \to W$. Represent $T(\mathbf{v}_j)$ with respect to the basis $\mathcal{C}$:*

$$T(\mathbf{v}_j) = a_{1j}\mathbf{w}_1 + \cdots + a_{mj}\mathbf{w}_m \ \Leftrightarrow \ [T(\mathbf{v}_j)]_{\mathcal{C}} = \begin{bmatrix} a_{1j} \\ \vdots \\ a_{mj} \end{bmatrix}, j = 1, \ldots, n. \quad (6.6)$$

Introduce the matrix $[T]_{\mathcal{C}\leftarrow\mathcal{B}} = (a_{ij})_{i=1,j=1}^{m,\ n}$. Then we have that

$$T(\mathbf{v}) = \mathbf{w} \ \Leftrightarrow \ [\mathbf{w}]_{\mathcal{C}} = [T]_{\mathcal{C}\leftarrow\mathcal{B}}[\mathbf{v}]_{\mathcal{B}}. \quad (6.7)$$

Conversely, if $A = (a_{ij})_{i=1,j=1}^{m,\ n} \in \mathbb{F}^{m\times n}$ is given, then defining $T : V \to W$ via (6.6) and extending by linearity via $T(\sum_{j=1}^{n} c_j\mathbf{v}_j) := \sum_{j=1}^{n} c_j T(\mathbf{v}_j)$, yields a linear map $T : V \to W$ with matrix representation $[T]_{\mathcal{C}\leftarrow\mathcal{B}} = A$.

We can rewrite (6.6) as

$$[T]_{\mathcal{C}\leftarrow\mathcal{B}} = \begin{bmatrix} [T(\mathbf{v}_1)]_{\mathcal{C}} & \cdots & [T(\mathbf{v}_n)]_{\mathcal{C}} \end{bmatrix}. \quad (6.8)$$

Proof. The proof follows directly from the following observation. If

$$\mathbf{v} = c_1\mathbf{v}_1 + \cdots + c_n\mathbf{v}_n \ \Leftrightarrow \ [\mathbf{v}]_{\mathcal{B}} = \begin{bmatrix} c_1 \\ \vdots \\ c_n \end{bmatrix},$$

then

$$\mathbf{w} = T(\mathbf{v}) = \sum_{j=1}^{n} c_j T(\mathbf{v}_j) =$$

$$\sum_{j=1}^{n} c_j \left(\sum_{k=1}^{m} a_{kj} \mathbf{w}_k \right) \Leftrightarrow [\mathbf{w}]_{\mathcal{C}} = \begin{bmatrix} \sum_{j=1}^{n} a_{1j} c_j \\ \vdots \\ \sum_{j=1}^{n} a_{mj} c_j \end{bmatrix} = (a_{ij})_{i=1, j=1}^{m \quad n} \begin{bmatrix} c_1 \\ \vdots \\ c_n \end{bmatrix}.$$

□

Example 6.3.2. Let $V = \mathbb{R}^{2 \times 2}$ and $\mathcal{B}$ be the standard basis $\{E_{11}, E_{12}, E_{21}, E_{22}\}$. Define $T : V \to V$ via

$$T(A) = \begin{bmatrix} 1 & 2 \\ 3 & 4 \end{bmatrix} A \begin{bmatrix} 1 & 3 \\ 5 & 7 \end{bmatrix}.$$

Find the matrix representation $[T]_{\mathcal{B} \leftarrow \mathcal{B}}$.

Compute

$$T(E_{11}) = \begin{bmatrix} 1 & 3 \\ 3 & 9 \end{bmatrix} = E_{11} + 3E_{12} + 3E_{21} + 9E_{22},$$

$$T(E_{12}) = \begin{bmatrix} 5 & 7 \\ 15 & 21 \end{bmatrix} = 5E_{11} + 7E_{12} + 15E_{21} + 21E_{22},$$

$$T(E_{21}) = \begin{bmatrix} 2 & 6 \\ 4 & 12 \end{bmatrix} = 2E_{11} + 6E_{12} + 4E_{21} + 12E_{22},$$

$$T(E_{22}) = \begin{bmatrix} 10 & 14 \\ 20 & 28 \end{bmatrix} = 10E_{11} + 14E_{12} + 20E_{21} + 28E_{22}.$$

This gives that

$$[T]_{\mathcal{B} \leftarrow \mathcal{B}} = \begin{bmatrix} 1 & 5 & 2 & 10 \\ 3 & 7 & 6 & 14 \\ 3 & 15 & 4 & 20 \\ 9 & 21 & 12 & 28 \end{bmatrix}.$$

□

When $T(\mathbf{x}) = A\mathbf{x}$ and $S(\mathbf{y}) = B\mathbf{y}$ are represented by standard matrices A and B respectively, then we get that

$$(S \circ T)(\mathbf{x}) = S(T(\mathbf{x})) = S(A\mathbf{x}) = BA\mathbf{x}.$$

Thus, in this case BA is the standard matrix for $S \circ T$. The next result shows that such a rule holds for general linear maps, namely that the composition of linear maps corresponds to matrix multiplication of the matrix representations, when the bases match.

Theorem 6.3.3. *Let $T : V \to W$ and $S : W \to X$ be linear maps between finite-dimensional vector spaces over $\mathbb{F}$, and let $\mathcal{B}, \mathcal{C}$, and $\mathcal{D}$ be bases for V, W, and X, respectively. Then*

$$[S \circ T]_{\mathcal{D} \leftarrow \mathcal{B}} = [S]_{\mathcal{D} \leftarrow \mathcal{C}}[T]_{\mathcal{C} \leftarrow \mathcal{B}}. \tag{6.9}$$

Proof. Denoting

$$\mathcal{B} = \{\mathbf{v}_1, \ldots, \mathbf{v}_n\}, \mathcal{C} = \{\mathbf{w}_1, \ldots, \mathbf{w}_m\}, \mathcal{D} = \{\mathbf{x}_1, \ldots, \mathbf{x}_p\},$$

$$[S \circ T]_{\mathcal{D} \leftarrow \mathcal{B}} = (c_{ij})_{i=1,j=1}^{p \quad n}, \; [S]_{\mathcal{D} \leftarrow \mathcal{C}} = (b_{ij})_{i=1,j=1}^{p \quad m}, \; [T]_{\mathcal{C} \leftarrow \mathcal{B}} = (a_{ij})_{i=1,j=1}^{m \quad n}.$$

We thus have that

$$T(\mathbf{v}_j) = \sum_{i=1}^{m} a_{ij}\mathbf{w}_i, j = 1, \ldots, n, \quad S(\mathbf{w}_k) = \sum_{l=1}^{p} b_{lk}\mathbf{x}_l, k = 1, \ldots, m.$$

Then

$$(S \circ T)(\mathbf{v}_j) = S(T(\mathbf{v}_j)) = S(\sum_{i=1}^{m} a_{ij}\mathbf{w}_i) = \sum_{i=1}^{m} a_{ij}S(\mathbf{w}_i) =$$

$$\sum_{i=1}^{m} [a_{ij} \sum_{l=1}^{p} b_{li}\mathbf{x}_l] = \sum_{l=1}^{p}(\sum_{i=1}^{m} b_{li}a_{ij})\mathbf{x}_l, j = 1, \ldots, n.$$

Thus we get that $c_{lj} = \sum_{i=1}^{m} b_{li}a_{ij}, l = 1, \ldots, p, \; j = 1, \ldots, n$, which corresponds exactly to (6.9). $\qquad \Box$

6.4 Change of Basis

A vector space V has many bases. Let $\mathcal{B} = \{\mathbf{b}_1, \ldots, \mathbf{b}_n\}$ and $\mathcal{C} = \{\mathbf{c}_1, \ldots, \mathbf{c}_n\}$ be two bases for V. For $\mathbf{v} \in V$ we have

$$\mathbf{v} = \alpha_1 \mathbf{b}_1 + \cdots + \alpha_n \mathbf{b}_n \quad \leftrightarrow \quad [\mathbf{v}]_{\mathcal{B}} = \begin{bmatrix} \alpha_1 \\ \vdots \\ \alpha_n \end{bmatrix},$$

and

$$\mathbf{v} = \beta_1 \mathbf{c}_1 + \cdots + \beta_n \mathbf{c}_n \quad \leftrightarrow \quad [\mathbf{v}]_{\mathcal{C}} = \begin{bmatrix} \beta_1 \\ \vdots \\ \beta_n \end{bmatrix}.$$

There is a linear relationship between $[\mathbf{v}]_\mathcal{B}$ and $[\mathbf{v}]_\mathcal{C}$, which therefore corresponds to matrix multiplication. The matrix that connect them is in fact $[id]_{\mathcal{C}\leftarrow\mathcal{B}}$:

$$[\mathbf{v}]_\mathcal{C} = [id]_{\mathcal{C}\leftarrow\mathcal{B}}[\mathbf{v}]_\mathcal{B}.$$

This follows from (6.7). The inverse is given via

$$[\mathbf{v}]_\mathcal{B} = [id]_{\mathcal{B}\leftarrow\mathcal{C}}[\mathbf{v}]_\mathcal{C}.$$

Thus $([id]_{\mathcal{B}\leftarrow\mathcal{C}})^{-1} = [id]_{\mathcal{C}\leftarrow\mathcal{B}}$. It follows from (6.8) that

$$[id]_{\mathcal{C}\leftarrow\mathcal{B}} = \begin{bmatrix} [\mathbf{b}_1]_\mathcal{C} & \cdots & [\mathbf{b}_n]_\mathcal{C} \end{bmatrix},$$

since $id(\mathbf{b}_j) = \mathbf{b}_j$.

Let us do an example.

Example 6.4.1. Let $\mathcal{B} = \left\{ \begin{bmatrix} 1 \\ 2 \end{bmatrix}, \begin{bmatrix} 1 \\ 3 \end{bmatrix} \right\}$ and $\mathcal{C} = \left\{ \begin{bmatrix} -2 \\ 1 \end{bmatrix}, \begin{bmatrix} 4 \\ -3 \end{bmatrix} \right\}$ be two bases for $\mathbb{R}^2$. Compute $[id]_{\mathcal{C}\leftarrow\mathcal{B}}$.

We need $[\mathbf{b}_1]_\mathcal{C}$ and $[\mathbf{b}_2]_\mathcal{C}$. In other words, we need to find x_1, x_2, y_1, y_2 so that

$$\mathbf{b}_1 = x_1\mathbf{c}_1 + x_2\mathbf{c}_2, \mathbf{b}_2 = y_1\mathbf{c}_1 + y_2\mathbf{c}_2.$$

Thus

$$\begin{bmatrix} \mathbf{c}_1 & \mathbf{c}_2 \end{bmatrix} \begin{bmatrix} x_1 \\ x_2 \end{bmatrix} = \mathbf{b}_1, \begin{bmatrix} \mathbf{c}_1 & \mathbf{c}_2 \end{bmatrix} \begin{bmatrix} y_1 \\ y_2 \end{bmatrix} = \mathbf{b}_2.$$

We can combine these systems into one augmented matrix

$$\begin{bmatrix} \mathbf{c}_1 & \mathbf{c}_2 & | & \mathbf{b}_1 & \mathbf{b}_2 \end{bmatrix} = \begin{bmatrix} -2 & 4 & | & 1 & 1 \\ 1 & -3 & | & 2 & 3 \end{bmatrix}.$$

Row reducing gives

$$\begin{bmatrix} 1 & 0 & | & -5\frac{1}{2} & -7\frac{1}{2} \\ 0 & 1 & | & -2\frac{1}{2} & -3\frac{1}{2} \end{bmatrix} = \begin{bmatrix} I_2 & | & [id]_{\mathcal{C}\leftarrow\mathcal{B}} \end{bmatrix}.$$

Thus

$$[id]_{\mathcal{C}\leftarrow\mathcal{B}} = \begin{bmatrix} -5\frac{1}{2} & -7\frac{1}{2} \\ -2\frac{1}{2} & -3\frac{1}{2} \end{bmatrix}.$$

□

Example 6.4.2. Let $V = \mathbb{R}^3$ and

$$\mathcal{B} = \left\{ \begin{bmatrix} 3 \\ 5 \\ 11 \end{bmatrix}, \begin{bmatrix} 2 \\ 3 \\ 4 \end{bmatrix}, \begin{bmatrix} 1 \\ 4 \\ 11 \end{bmatrix} \right\}, \quad \mathcal{C} = \left\{ \begin{bmatrix} 0 \\ 2 \\ 4 \end{bmatrix}, \begin{bmatrix} 1 \\ 0 \\ 3 \end{bmatrix}, \begin{bmatrix} 2 \\ 1 \\ 0 \end{bmatrix} \right\}.$$

Find the matrix representation $[id]_{\mathcal{C}\leftarrow\mathcal{B}}$.

We set up the augmented matrix $\begin{bmatrix} \mathbf{c}_1 & \mathbf{c}_2 & \mathbf{c}_3 & | & \mathbf{b}_1 & \mathbf{b}_2 & \mathbf{b}_3 \end{bmatrix}$ and row reduce:

$$\begin{bmatrix} 0 & 1 & 2 & | & 3 & 2 & 1 \\ 2 & 0 & 1 & | & 5 & 3 & 4 \\ 4 & 3 & 0 & | & 11 & 4 & 11 \end{bmatrix} \to \cdots \to \begin{bmatrix} 1 & 0 & 0 & | & 2 & 1 & 2 \\ 0 & 1 & 0 & | & 1 & 0 & 1 \\ 0 & 0 & 1 & | & 1 & 1 & 0 \end{bmatrix}.$$

Thus we find

$$[id_V]_{\mathcal{C}\leftarrow\mathcal{B}} = \begin{bmatrix} 2 & 1 & 2 \\ 1 & 0 & 1 \\ 1 & 1 & 0 \end{bmatrix}.$$

$\square$

In the next corollary, we present an important special case where we change bases in a vector space, and express a linear map with respect to the new basis. Two $n \times n$ matrices A and B are called **similar** if there exists an invertible $n \times n$ matrix P so that

$$A = PBP^{-1}.$$

We have the following corollary.

Corollary 6.4.3. *Let $T : V \to V$ and let $\mathcal{B}$ and $\mathcal{C}$ be two bases in the n-dimensional vector space V. Then*

$$[T]_{\mathcal{B}\leftarrow\mathcal{B}} = [id]_{\mathcal{B}\leftarrow\mathcal{C}}[T]_{\mathcal{C}\leftarrow\mathcal{C}}[id]_{\mathcal{C}\leftarrow\mathcal{B}} = [id]_{\mathcal{C}\leftarrow\mathcal{B}}^{-1}[T]_{\mathcal{C}\leftarrow\mathcal{C}}[id]_{\mathcal{C}\leftarrow\mathcal{B}}. \quad (6.10)$$

In particular, $[T]_{\mathcal{B}\leftarrow\mathcal{B}}$ and $[T]_{\mathcal{C}\leftarrow\mathcal{C}}$ are similar.

In the next chapter we will try to find a basis $\mathcal{C}$ so that $[T]_{\mathcal{C}\leftarrow\mathcal{C}}$ is a diagonal matrix. This leads to the notion of eigenvectors.

6.5 Exercises

Exercise 6.5.1. For the following transformations T from $\mathbb{R}^n$ to $\mathbb{R}^m$, either prove that it is linear or prove that it is not.

(a) $T : \mathbb{R}^2 \to \mathbb{R}^3$ defined via $T(\begin{bmatrix} x_1 \\ x_2 \end{bmatrix}) = \begin{bmatrix} x_1 + 5 \\ 5x_1 - x_2 \\ x_1 \end{bmatrix}.$

(b) $T : \mathbb{R}^3 \to \mathbb{R}^2$ defined via $T(\begin{bmatrix} x_1 \\ x_2 \\ x_3 \end{bmatrix}) = \begin{bmatrix} 2x_1 + 3x_2 \\ x_1 + 5x_2 - 7x_3 \end{bmatrix}$.

(c) $T : \mathbb{R}^3 \to \mathbb{R}^3$ defined via $T(\begin{bmatrix} x_1 \\ x_2 \\ x_3 \end{bmatrix}) = \begin{bmatrix} x_3 \\ x_1 \\ x_2 \end{bmatrix}$.

(d) $T : \mathbb{R}^3 \to \mathbb{R}^2$ defined via $T(\begin{bmatrix} x_1 \\ x_2 \\ x_3 \end{bmatrix}) = \begin{bmatrix} 3x_1 + 2x_2^2 \\ x_1 + 5x_2 - 7x_3 \end{bmatrix}$.

(e) $T : \mathbb{R}^2 \to \mathbb{R}^3$ defined via $T(\begin{bmatrix} x_1 \\ x_2 \end{bmatrix}) = \begin{bmatrix} 2x_1 + 4 \\ 3x_1 - x_2 \\ x_2 \end{bmatrix}$.

(f) $T : \mathbb{R}^3 \to \mathbb{R}^3$ defined via $T(\begin{bmatrix} x_1 \\ x_2 \\ x_3 \end{bmatrix}) = \begin{bmatrix} x_2 \\ x_3 - x_1 \\ x_1 \end{bmatrix}$.

Exercise 6.5.2. For the following maps T, determine whether T is linear or not. Explain your answer.

(a) $T : \mathbb{R}_2[t] \to \mathbb{R}^2$, $T(p_0 + p_1 t + p_2 t^2) = \begin{bmatrix} p_0 + 3p_2 \\ 2p_1 \end{bmatrix}$.

(b) $T : \mathbb{R}_2[t] \to \mathbb{R}_2[t]$, $T(a_0 + a_1 t + a_2 t^2) = a_2 + a_1 t + a_0 t^2$.

(c) $T : \mathbb{R}^{2 \times 2} \to \mathbb{R}^3$, $T(\begin{bmatrix} a & b \\ c & d \end{bmatrix}) = \begin{bmatrix} 3a - 2b \\ a + c - d \\ b - d \end{bmatrix}$.

(d) $T : \mathbb{R}^3 \to \mathbb{R}^4$,

$$T \begin{bmatrix} x_1 \\ x_2 \\ x_3 \end{bmatrix} = \begin{bmatrix} x_1 - 5x_3 \\ 7x_2 + 5 \\ 3x_1 - 6x_2 \\ 8x_3 \end{bmatrix}.$$

(e) $T : \mathbb{R}^3 \to \mathbb{R}^2$,

$$T \begin{bmatrix} x_1 \\ x_2 \\ x_3 \end{bmatrix} = \begin{bmatrix} x_1 - 2x_3 \\ 3x_2 x_3 \end{bmatrix}.$$

(f) $T : \mathbb{R}^{2 \times 2} \to \mathbb{R}^{2 \times 2}$, $T(A) = A - A^T$.

(g) $T : \{f : \mathbb{R} \to \mathbb{R} \ : \ f \text{ is differentiable}\} \to \{f : \mathbb{R} \to \mathbb{R} \ : \ f \text{ is a function}\}$,

$$(T(f))(x) = f'(x)(x^2 + 5).$$

(h) $T : \{f : \mathbb{R} \to \mathbb{R} \ : \ f \text{ is continuous}\} \to \mathbb{R}$,

$$T(f) = \int_{-5}^{10} f(x)dx.$$

(i) $T : \{f : \mathbb{R} \to \mathbb{R} \ : \ f \text{ is differentiable}\} \to \mathbb{R}$,

$$T(f) = f(5)f'(2).$$

Exercise 6.5.3.

(a) Show that T is a linear transformation by finding a matrix that implements the mapping, when T is given by

$$T \begin{bmatrix} x_1 \\ x_2 \\ x_3 \end{bmatrix} = \begin{bmatrix} x_1 - 5x_3 \\ 7x_2 \\ 3x_1 - 6x_2 \\ 8x_3 \end{bmatrix}.$$

(b) Determine if the linear transformation T is one-to-one.

(c) For which h is the vector

$$\begin{bmatrix} 0 \\ 7 \\ 9 \\ h \end{bmatrix}$$

in the range of T?

Exercise 6.5.4. Let $T : V \to W$ and $S : W \to X$ be linear maps. Show that the composition $S \circ T : V \to X$ is also linear.

Exercise 6.5.5. Let $L : \mathbb{R}^{2 \times 2} \to \mathbb{R}^{2 \times 2}$ by $L(A) = 2A - 5A^T$.

(a) Show that L is a linear transformation.

(b) Compute $L(\begin{bmatrix} 2 & 1 \\ 1 & 2 \end{bmatrix})$.

(c) Provide a nonzero element in the range of L.

(d) Is L one-to-one? Explain.

Exercise 6.5.6. Consider the linear map $T : \mathbb{R}_2[t] \to \mathbb{R}^2$ given by $T(p(t)) = \begin{bmatrix} p(1) \\ p(3) \end{bmatrix}$.

(a) Find a basis for the kernel of T.

(b) Find a basis for the range of T.

Exercise 6.5.7. Define $T : \mathbb{R}^{2\times2} \to \mathbb{R}$ via $T \begin{bmatrix} a & b \\ c & d \end{bmatrix} = a+b+c+d$. Determine a basis for the kernel of T.

Exercise 6.5.8. Define $L : \mathbb{R}^{2\times2} \to \mathbb{R}^{2\times2}$ via $L(A) = A - A^T$. Determine a basis for the range of L.

Exercise 6.5.9. Let $T : V \to W$ with $V = \mathbb{R}^4$ and $W = \mathbb{R}^{2\times2}$ be defined by

$$T(\begin{bmatrix} a \\ b \\ c \\ d \end{bmatrix}) = \begin{bmatrix} a+b & b+c \\ c+d & d+a \end{bmatrix}.$$

(a) Find a basis for the kernel of T.

(b) Find a basis for the range of T.

Exercise 6.5.10. Is the linear map $T : \mathbb{R}^3 \longrightarrow \mathbb{R}^2$ described by

$$T = \begin{bmatrix} x_1 \\ x_2 \\ x_3 \end{bmatrix} = \begin{bmatrix} x_1 - 2x_2 + x_3 \\ 2x_1 - 4x_2 + 2x_3 \end{bmatrix}$$

onto? Describe all vectors $\mathbf{u} = \begin{bmatrix} u_1 \\ u_2 \\ u_3 \end{bmatrix} \in \mathbb{R}^3$ with $T(\mathbf{u}) = \begin{bmatrix} 3 \\ 6 \end{bmatrix}$.

Exercise 6.5.11. Show that if $T : V \to W$ is linear and the set $\{T(\mathbf{v}_1), \dots, T(\mathbf{v}_k)\}$ is linearly independent, then the set $\{\mathbf{v}_1, \dots, \mathbf{v}_k\}$ is linearly independent.

Exercise 6.5.12. Show that if $T : V \to W$ is linear and onto, and $\{\mathbf{v}_1 \dots, \mathbf{v}_k\}$ is a basis for V, then the set $\{T(\mathbf{v}_1), \dots, T(\mathbf{v}_k)\}$ spans W. When is $\{T(\mathbf{v}_1), \dots, T(\mathbf{v}_k)\}$ a basis for W?

Exercise 6.5.13. Let V and W be vector spaces, and let $\mathcal{B} = \{\mathbf{b}_1, \dots, \mathbf{b}_n\}$ be a basis for V. Suppose that $T : V \to W$ is a linear one-to-one map. Show that $\dim W \geq n$.

Exercise 6.5.14. Let V be a vector space over $\mathbb{R}$ of dimension n and $\mathcal{B}$ a basis for V. Define $T : V \to \mathbb{R}^n$ by $T(\mathbf{v}) = [\mathbf{v}]_{\mathcal{B}}$.

(a) Show that T is an isomorphism.

(b) Conclude that V and $\mathbb{R}^n$ are isomorphic.

(c) Show that two finite dimensional vector spaces over $\mathbb{R}$ are isomorphic if and only if they have the same dimension.

Exercise 6.5.15. Let $T : V \to W$ be linear, and let $U \subseteq V$ be a subspace of V. Define

$$T[U] := \{\mathbf{w} \in W \; ; \; \text{there exists } \mathbf{u} \in U \text{ so that } \mathbf{w} = T(\mathbf{u})\}.$$

Observe that $T[V] = \text{Ran } T$.

(a) Show that $T[U]$ is a subspace of W.

(b) Assuming dim $U < \infty$, show that dim $T[U] \leq$ dim U.

(c) If $\hat{U}$ is another subspace of V, is it always true that $T[U + \hat{U}] = T[U] + T[\hat{U}]$? If so, provide a proof. If not, provide a counterexample.

(d) If $\hat{U}$ is another subspace of V, is it always true that $T[U \cap \hat{U}] = T[U] \cap T[\hat{U}]$? If so, provide a proof. If not, provide a counterexample.

Exercise 6.5.16. Define the linear map $T : \mathbb{R}^{2 \times 2} \to \mathbb{R}_3[t]$ via

$$T \begin{bmatrix} a & b \\ c & d \end{bmatrix} = (a + b) + (c - b)t + (d - a)t^2 + (a + c)t^3.$$

Let

$$\mathcal{B} = \left\{ \begin{bmatrix} 1 & 0 \\ 0 & 0 \end{bmatrix}, \begin{bmatrix} 0 & 1 \\ 0 & 0 \end{bmatrix}, \begin{bmatrix} 0 & 0 \\ 1 & 0 \end{bmatrix}, \begin{bmatrix} 0 & 0 \\ 0 & 1 \end{bmatrix} \right\}$$

and $\mathcal{C} = \{1, t, t^2, t^3\}$. Find the matrix for T relative to $\mathcal{B}$ and $\mathcal{C}$.

Exercise 6.5.17. Let $\{\mathbf{v}_1, \mathbf{v}_2, \mathbf{v}_3, \mathbf{v}_4\}$ be a basis for a vector space V.

(a) Let $T : V \to V$ be given by $T(\mathbf{v}_i) = \mathbf{v}_{i+1}, i = 1, 2, 3$, and $T(\mathbf{v}_4) = \mathbf{v}_1$. Determine the matrix representation of T with respect to the basis $\{\mathbf{v}_1, \mathbf{v}_2, \mathbf{v}_3, \mathbf{v}_4\}$.

(b) If the matrix representation of a linear map $S : V \to V$ with respect to the basis $\{\mathbf{v}_1, \mathbf{v}_2, \mathbf{v}_3, \mathbf{v}_4\}$ is given by

$$\begin{bmatrix} 1 & 0 & 1 & 1 \\ 0 & 2 & 0 & 2 \\ 1 & 2 & 1 & 3 \\ -1 & 0 & -1 & -1 \end{bmatrix},$$

determine $S(\mathbf{v}_1 - \mathbf{v}_4)$.

(c) Determine bases for Ran S and Ker S.

Exercise 6.5.18. Let $\mathcal{B} = \{b_1, b_2, b_3\}$ and $\mathcal{C} = \{c_1, c_2, c_3\}$ be bases for vector spaces V and W, respectively. Let $T : V \to W$ have the property that

$$T(b_1) = 2c_1 + 3c_3, T(b_2) = 4c_1 - c_2 + 7c_3, T(b_3) = 5c_2 + 10c_3.$$

Find the matrix for T relative to $\mathcal{B}$ and $\mathcal{C}$.

Exercise 6.5.19. Let $T : \mathbb{R}^2 \to \mathbb{R}_2[t]$ be defined by

$$T(\begin{bmatrix} x_1 \\ x_2 \end{bmatrix}) = x_1 t^2 + (x_1 + x_2)t + x_2.$$

(a) Compute $T(\begin{bmatrix} 1 \\ -1 \end{bmatrix})$.

(b) Show that T is linear.

(c) Let $\mathcal{B} = \left\{ \begin{bmatrix} 1 \\ 0 \end{bmatrix}, \begin{bmatrix} 0 \\ 1 \end{bmatrix} \right\}$ and $\mathcal{E} = \{1, t, t^2\}$. Find the matrix representation of T with respect to $\mathcal{B}$ and $\mathcal{E}$.

(d) Find a basis for Ran T.

Exercise 6.5.20. Define $T : \mathbb{R}_2[t] \to \mathbb{R}^2$ by $T(p(t)) = \begin{bmatrix} p(0) + p(5) \\ p(3) \end{bmatrix}$. Find the matrix for T relative to the basis $\{1, t, t^2\}$ for $\mathbb{R}_2[t]$ and the standard basis for $\mathbb{R}^2$.

Exercise 6.5.21. Let $V = \mathbb{R}_3[t]$. Define $T : V \to V$ via

$$T[p(t)] = (3t + 4)p'(t),$$

where p' is the derivative of p.

(a) Find a basis for the kernel of T.

(b) Find a basis for the range of T.

(c) Let $\mathcal{B} = \{1, t, t^2, t^3\}$. Find the matrix representation of T with respect to the basis $\mathcal{B}$.

Exercise 6.5.22. Let $T : \mathbb{R}^{2 \times 2} \to \mathbb{R}_3[t]$ be defined by

$$T \begin{bmatrix} a & b \\ c & d \end{bmatrix} = a + bt + ct^2 + dt^3.$$

Put

$$\mathcal{B} = \left\{ \begin{bmatrix} 1 & 0 \\ 0 & 1 \end{bmatrix}, \begin{bmatrix} 1 & 0 \\ 0 & -1 \end{bmatrix}, \begin{bmatrix} 0 & 1 \\ 1 & 0 \end{bmatrix}, \begin{bmatrix} 0 & 1 \\ -1 & 0 \end{bmatrix} \right\}, \quad \mathcal{C} = \{1, 1 + t, t^2, t^2 + t^3\}.$$

Determine the matrix representation $[T]_{\mathcal{C} \leftarrow \mathcal{B}}$ of T with respect to the bases $\mathcal{B}$ and $\mathcal{C}$.

Exercise 6.5.23. Define $L : \mathbb{R}^{2\times 2} \to \mathbb{R}_2[t]$ via

$$L(\begin{bmatrix} a & b \\ c & d \end{bmatrix}) = (a+b+c) + (b+c+d)t + (a-d)t^2.$$

(a) Determine a basis for the kernel of L.

(b) Determine a basis for the range of L.

(c) Let $\mathcal{B} = \left\{ \begin{bmatrix} 1 & 0 \\ 0 & 1 \end{bmatrix}, \begin{bmatrix} 1 & 0 \\ 0 & -1 \end{bmatrix}, \begin{bmatrix} 0 & 1 \\ 1 & 0 \end{bmatrix}, \begin{bmatrix} 0 & 1 \\ -1 & 0 \end{bmatrix} \right\}$ and $\mathcal{C} = \{1, t, t^2\}$. Determine the matrix representation $[L]_{\mathcal{C} \leftarrow \mathcal{B}}$ of L with respect to the bases $\mathcal{B}$ and $\mathcal{C}$.

Exercise 6.5.24. Let $T : V \to W$ with $V = \mathbb{R}^4$ and $W = \mathbb{R}^{2\times 2}$ be defined by

$$T(\begin{bmatrix} a \\ b \\ c \\ d \end{bmatrix}) = \begin{bmatrix} a+b & b+c \\ c+d & d+a \end{bmatrix}.$$

(a) Find a basis for the kernel of T.

(b) Find a basis for the range of T.

Exercise 6.5.25. For the following $T : V \to W$ with bases $\mathcal{B}$ and $\mathcal{C}$, respectively, determine the matrix representation for T with respect to the bases $\mathcal{B}$ and $\mathcal{C}$. In addition, find bases for the range and kernel of T.

(a) $\mathcal{B} = \mathcal{C} = \{\sin t, \cos t, \sin 2t, \cos 2t\}$, $V = W = \text{Span } \mathcal{B}$, and $T = \frac{d^2}{dt^2} + \frac{d}{dt}$.

(b) $\mathcal{B} = \{1, t, t^2, t^3\}$, $\mathcal{C} = \left\{ \begin{bmatrix} 1 \\ 0 \end{bmatrix}, \begin{bmatrix} 1 \\ -1 \end{bmatrix} \right\}$, $V = \mathbb{R}_3[t]$, and $W = \mathbb{R}^2$, and $T(p) = \begin{bmatrix} p(3) \\ p(5) \end{bmatrix}$.

(c) $\mathcal{B} = \mathcal{C} = \{e^t \cos t, e^t \sin t, e^{3t}, te^{3t}\}$, $V = W = \text{Span } \mathcal{B}$, and $T = \frac{d}{dt}$.

(d) $\mathcal{B} = \{1, t, t^2\}$, $\mathcal{C} = \left\{ \begin{bmatrix} 1 \\ 1 \end{bmatrix}, \begin{bmatrix} 1 \\ 0 \end{bmatrix} \right\}$, $V = \mathbb{R}_2[t]$, and $W = \mathbb{R}^2$, and $T(p) = \begin{bmatrix} \int_0^1 p(t)dt \\ p(1) \end{bmatrix}$.

Exercise 6.5.26. Let $V = \mathbb{R}^{n\times n}$. Define $L : V \to V$ via $L(A) = \frac{1}{2}(A + A^T)$.

(a) Let

$$B = \left\{ \begin{bmatrix} 1 & 0 \\ 0 & 0 \end{bmatrix}, \begin{bmatrix} 0 & 1 \\ 0 & 0 \end{bmatrix}, \begin{bmatrix} 0 & 0 \\ 1 & 0 \end{bmatrix}, \begin{bmatrix} 0 & 0 \\ 0 & 1 \end{bmatrix} \right\}.$$

Determine the matrix representation of L with respect to the basis B.

(b) Determine the dimensions of the subspaces

$$W = \{A \in V \; : \; L(A) = A\}, \text{and}$$

$$\text{Ker } L = \{A \in V \; : \; L(A) = 0\}.$$

Exercise 6.5.27. Let $B = \{1, t, \ldots, t^n\}$, $C = \{1, t, \ldots, t^{n+1}\}$, $V = $ Span B and $W = $ Span C. Define $A : V \to W$ via

$$Af(t) := (2t^2 - 3t + 4)f'(t),$$

where f' is the derivative of f.

(a) Find the matrix representation of A with respect to the bases B and C.

(b) Find bases for Ran A and Ker A.

Exercise 6.5.28. Let $T : \mathbb{R}_2[t] \to \mathbb{R}^3$ be defined by

$$T(p) = \begin{bmatrix} p(0) \\ p(2) \\ p'(1) \end{bmatrix}.$$

(a) Compute $T(t^2 + 5)$.

(b) Let $B = \left\{ \begin{bmatrix} 1 \\ 0 \\ 0 \end{bmatrix}, \begin{bmatrix} 0 \\ 1 \\ 0 \end{bmatrix}, \begin{bmatrix} 0 \\ 0 \\ 1 \end{bmatrix} \right\}$ and $\mathcal{E} = \{1, t, t^2\}$. Find the matrix representation of T with respect to $\mathcal{E}$ and B.

Exercise 6.5.29. Let $T : \mathbb{R}_2[t] \to \mathbb{R}_2[t]$ be the transformation $T(p(t)) = p'(t) - p(t)$. Find the matrix representation $[T]_{\mathcal{E} \leftarrow \mathcal{E}}$ of T with respect to the basis $\mathcal{E} = \{1, t, t^2\}$.

Exercise 6.5.30. Let $B = \left\{ \begin{bmatrix} 2 \\ -1 \end{bmatrix}, \begin{bmatrix} 1 \\ 3 \end{bmatrix} \right\}$ and $C = \left\{ \begin{bmatrix} 1 \\ 1 \end{bmatrix}, \begin{bmatrix} 1 \\ -1 \end{bmatrix} \right\}$.

(a) Find the change of basis matrix $[id]_{C \leftarrow B}$.

(b) Show how you would check your answer under (a).

Exercise 6.5.31. Let $\mathcal{F} = \{\mathbf{f}_1, \mathbf{f}_2, \mathbf{f}_3\}$ and $\mathcal{D} = \{\mathbf{d}_1, \mathbf{d}_2, \mathbf{d}_3\}$ be two bases of a vector space V. Moreover, it is given that $\mathbf{f}_1 = 2\mathbf{d}_1 - \mathbf{d}_2 + 3\mathbf{d}_3, \mathbf{f}_2 = \mathbf{d}_1 - 5\mathbf{d}_3, \mathbf{f}_3 = \mathbf{d}_2 - \mathbf{d}_3$. If $[x]_{\mathcal{F}} = \begin{bmatrix} 1 \\ 2 \\ 3 \end{bmatrix}$, find $[x]_{\mathcal{D}}$.

Exercise 6.5.32. Let $\mathcal{B} = \left\{ \begin{bmatrix} -2 \\ 1 \end{bmatrix}, \begin{bmatrix} 3 \\ 2 \end{bmatrix} \right\}$ and $\mathcal{C} = \left\{ \begin{bmatrix} 1 \\ 2 \end{bmatrix}, \begin{bmatrix} 2 \\ 3 \end{bmatrix} \right\}$ be two bases in $\mathbb{R}^2$. Find the change of coordinates matrix from $\mathcal{B}$ to $\mathcal{C}$.

Exercise 6.5.33. True or False? Justify each answer.

(i) The map $T : \mathbb{R}^2 \to \mathbb{R}^2$ defined by $T \begin{bmatrix} x_1 \\ x_2 \end{bmatrix} = \begin{bmatrix} x_1 + x_2 \\ x_1 x_2 \end{bmatrix}$ is linear.

(ii) The map $T : \mathbb{R}^2 \to \mathbb{R}^2$ defined by $T \begin{bmatrix} x_1 \\ x_2 \end{bmatrix} = \begin{bmatrix} x_1 + x_2 \\ x_2 + 3 \end{bmatrix}$ is linear.

(iii) If T is the map given by $T(\mathbf{x}) = A\mathbf{x}$, then T is one-to-one if and only if the columns of A form a linearly independent set.

Exercise 6.5.34. Suppose we have a sound signal

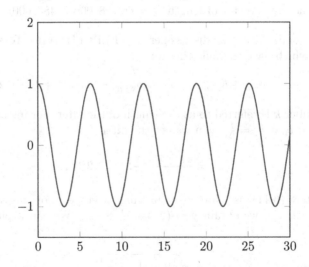

and that it gets corrupted with noise, resulting in the following.

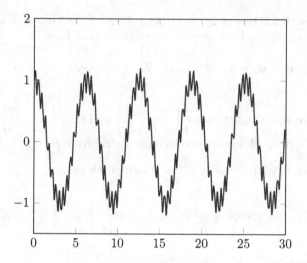

When we receive the corrupted signal, we would like to 'filter out the noise'. The signal that we receive is a digital signal, so it is really a vector of numbers $(x_k)_{k=1}^N$. Here it will be convenient to write the signal as a row vector. Standard for phone calls is a sampling rate of 8000 Hz, which means that the sound is sampled 8000 times per second. Thus a sound signal that is one minute long corresponds to a vector of length $N = 60 \times 8,000 = 480,000$.

The type of filter that we discuss here are **Finite Impulse Response (FIR) filters**, which have the following form

$$y_n = b_0 x_n + b_1 x_{n-1} + \cdots + b_k x_{n-k}, \quad n = k+1, k+2, \ldots.$$

The number k is referred to as the length of the filter. For instance, if we have $k = 1$ and $b_0 = b_1 = \frac{1}{2}$, then we get the filter

$$y_n = \frac{x_n + x_{n-1}}{2}, \quad n = 2, 3, \ldots. \tag{6.11}$$

The filter (6.11) is simply a moving average. For instance, if $\mathbf{x} = (1, 3, 4, 7, 9, \ldots)$, we obtain $\mathbf{y} = (2, 3\frac{1}{2}, 5\frac{1}{2}, 8, \ldots)$. We can depict the filter as follows.

$$\text{input } (x_n)_{n=1,2,\ldots} \longrightarrow \boxed{\text{digital filter}} \longrightarrow \text{output } (y_n)_{n=k+1,k+2,\ldots}$$

Note that the filter causes some delay. Indeed, we need $x_1, \ldots, x_{k+1}$ to be able to compute y_{k+1}. So, for instance, if $k = 10$ there will be a delay of 10 samples, which at the sampling rate of 8000 Hz is $\frac{1}{800}$ of a second. With the right choice of the filter (thus the right choice of the numbers $b_0, \ldots, b_k$), it is possible to recover the original signal above from the corrupted signal. Filter design is all about finding the right filter for your purposes (filtering out noise, separating voices, ...).

Consider the signals

$$\mathbf{a} = (\ldots, 1, 2, 3, 3, 2, 1, 1, 2, 3, 3, 2, 1, 1, 2, 3, \ldots), \mathbf{n} = (\ldots, 1, -1, 1, -1, 1, -1, 1,$$

$$-1, \ldots).$$

For a filter $\mathcal{F}$, with input $\mathbf{x}$, let us denote the output by $\mathcal{F}(\mathbf{x})$.

(a) Compute the output when we let $\mathbf{a}$ be the input of filter (6.11). Do the same with input $\mathbf{n}$. What do you observe?

(b) We would like to make a filter so that it does not change the signal $\mathbf{a}$ (thus $\mathcal{F}(\mathbf{a}) = \mathbf{a}$) and so that $\mathbf{n}$ will have constant zero as the output (thus $\mathcal{F}(\mathbf{n}) = \mathbf{0}$). The signal $\mathbf{a}$ represents a 'good' signal that we want to keep, and the signal $\mathbf{n}$ represents a 'bad' signal that we want to remove. Explain why for such a filter we have that $\mathcal{F}(\mathbf{a} + c\mathbf{n}) = \mathbf{a}$ for all $c \in \mathbb{R}$. What property of $\mathcal{F}$ are you using?

(c) Show that the choice $b_0 = \frac{1}{6}, b_1 = -\frac{1}{3}, b_2 = \frac{1}{3}, b_3 = \frac{5}{6}$ gives a filter satisfying the properties under (b).

(d) Show that no FIR filter of length ≤ 2 exists satisfying the properties under (b).
(Hint: Write out the equations that b_0, b_1, b_2 need to satisfy.)

Exercise 6.5.35. In transmitting information signals errors can occur. In some situations this may lead to significant problems (for instance, with financial or medical information), and thus it is important to make sure to have a way to **detect errors**. A simple rule may be that all the numbers in a signal have to add up to 0. For instance,

$$\begin{bmatrix} 1 \\ -3 \\ 5 \\ -7 \\ 4 \end{bmatrix}$$

is then a valid signal, and

$$\begin{bmatrix} 2 \\ -3 \\ 5 \\ -7 \\ 4 \end{bmatrix}$$

is not. This simple rule provides some protection, but it is limited. Indeed, a common human mistake is that two consecutive entries get transposed. The above rule does not detect that mistake.

A large number of **error detecting** and **error correcting codes** are built on the idea that the only valid signals belong to the kernel of a linear map T. Thus,

$$\mathbf{v} \text{ is a valid signal} \Leftrightarrow T(\mathbf{v}) = \mathbf{0}.$$

Often, these linear maps act on $\mathbb{F}^n$ where $\mathbb{F}$ is a field* other than $\mathbb{R}$ or $\mathbb{C}$, but let us not worry about that here. In our case, let us take $T : \mathbb{R}^6 \to \mathbb{R}^2$.

(a) If T is onto, what is the dimension of Ker T? The dimension of Ker T is the dimension of the subspace containing all valid signals.

(b) Explain how a single entry mistake is always detected if and only if $\mathbf{e}_j \notin$ Ker T, $j = 1, \ldots, 6$.

(c) Explain how a single transposition of two consecutive entries is always detected if and only if $\mathbf{e}_j - \mathbf{e}_{j+1} \notin$ Ker T, $j = 1, \ldots, 5$.

(d) Show that if the standard matrix of T is given by A below, then a single entry mistake and a single transposition of two consecutive entries is always detected.
$$A = \begin{bmatrix} 1 & 2 & -1 & 3 & 1 & -1 \\ 0 & 1 & -2 & 1 & 0 & 1 \end{bmatrix}.$$

(e) If a person transposes two non-consecutive entries, will it always be detected using the linear map T from (d)?

In this exercise we focused on some ideas involving error detection. To make error correcting codes (one that corrects a single entry error or a single transposition, for instance), one has to see to it that the type of errors one wants to correct for result in signals that are 'closer' to Ker T than signals that are the result of any other type of errors. One way to do this is by using the 'Hamming distance' (defined on $\mathbb{Z}_2^n$). For more information, please search for error correcting codes.

---

*For instance, $\mathbb{F} = \mathbb{Z}_p$; see Appendix A.3 for a definition of $\mathbb{Z}_p$.

7

Eigenvectors and Eigenvalues

CONTENTS

7.1   Eigenvectors and Eigenvalues

Let A be a square matrix. For certain vectors $\mathbf{x}$ the effect of the multiplication $A\mathbf{x}$ is very simple: the outcome is $\lambda\mathbf{x}$, where λ is a scalar. These special vectors along with the scalars λ expose important features of the matrix, and therefore have a special name.

> **Definition 7.1.1.** The scalar λ is called an **eigenvalue** of A, if there exists $\mathbf{x} \neq \mathbf{0}$ so that
> $$A\mathbf{x} = \lambda\mathbf{x}.$$
>
> In this case, the nonzero vector $\mathbf{x}$ is called an **eigenvector** of A corresponding to the eigenvalue λ.

If we rewrite the above equation as

$$(A - \lambda I)\mathbf{x} = \mathbf{0}, \quad \mathbf{x} \neq \mathbf{0},$$

it gives us a way to find eigenvalues. Indeed, if $\mathrm{Nul}(A - \lambda I) \neq \{\mathbf{0}\}$ then $A - \lambda I$ is not invertible. Consequently,

$$\det(A - \lambda I) = 0,$$

which gives a polynomial equation for λ. We call this the **characteristic equation** of A.

Definition 7.1.2. The polynomial $p_A(\lambda) = \det(A - \lambda I)$ is called the **characteristic polynomial** of A. Note that the roots of $p_A(\lambda)$ are exactly the eigenvalues of A; in other words, λ is called an eigenvalue of A if and only if $p_A(\lambda) = 0$.

The subspace $\text{Nul}(A - \lambda I_n)$ is called the **eigenspace** of A at λ, and consists of all the eigenvectors of A at λ and the zero vector.

Example 7.1.3. Let $A = \begin{bmatrix} 3 & 3 \\ 4 & 2 \end{bmatrix}$. Then

$$\det(A - \lambda I) = \det \begin{bmatrix} 3 - \lambda & 3 \\ 4 & 2 - \lambda \end{bmatrix} = \lambda^2 - 5\lambda - 6 = (\lambda + 1)(\lambda - 6).$$

Thus A has eigenvalues 6 and -1. Next we determine the eigenspaces.

At $\lambda = 6$, determine the null space of $A - 6I$:

$$\begin{bmatrix} 3 - 6 & 3 & | & 0 \\ 4 & 2 - 6 & | & 0 \end{bmatrix} \rightarrow \begin{bmatrix} -3 & 3 & | & 0 \\ 4 & -4 & | & 0 \end{bmatrix} \rightarrow \begin{bmatrix} 1 & -1 & | & 0 \\ 0 & 0 & | & 0 \end{bmatrix}.$$

So $\text{Nul}(A - 6I) = \text{Span}\left\{ \begin{bmatrix} 1 \\ 1 \end{bmatrix} \right\}$. Next, at $\lambda = -1$,

$$\begin{bmatrix} 3 - (-1) & 3 & | & 0 \\ 4 & 2 - (-1) & | & 0 \end{bmatrix} \rightarrow \begin{bmatrix} 4 & 3 & | & 0 \\ 4 & 3 & | & 0 \end{bmatrix} \rightarrow \begin{bmatrix} 1 & \frac{3}{4} & | & 0 \\ 0 & 0 & | & 0 \end{bmatrix}.$$

So $\text{Nul}(A + I) = \text{Span}\left\{ \begin{bmatrix} -\frac{3}{4} \\ 1 \end{bmatrix} \right\} = \text{Span}\left\{ \begin{bmatrix} -3 \\ 4 \end{bmatrix} \right\}$.

Let us check the answers:

$$\begin{bmatrix} 3 & 3 \\ 4 & 2 \end{bmatrix} \begin{bmatrix} 1 \\ 1 \end{bmatrix} = \begin{bmatrix} 6 \\ 6 \end{bmatrix} = 6 \begin{bmatrix} 1 \\ 1 \end{bmatrix}, \begin{bmatrix} 3 & 3 \\ 4 & 2 \end{bmatrix} \begin{bmatrix} -3 \\ 4 \end{bmatrix} = \begin{bmatrix} 3 \\ -4 \end{bmatrix} = -\begin{bmatrix} -3 \\ 4 \end{bmatrix}.$$

Put $P = \begin{bmatrix} 1 & -3 \\ 1 & 4 \end{bmatrix}$ and $D = \begin{bmatrix} 6 & 0 \\ 0 & -1 \end{bmatrix}$. Then P is invertible, and

$$AP = \begin{bmatrix} 3 & 3 \\ 4 & 2 \end{bmatrix} \begin{bmatrix} 1 & -3 \\ 1 & 4 \end{bmatrix} = \begin{bmatrix} 6 & 3 \\ 6 & -4 \end{bmatrix} = \begin{bmatrix} 1 & -3 \\ 1 & 4 \end{bmatrix} \begin{bmatrix} 6 & 0 \\ 0 & -1 \end{bmatrix} = PD.$$

Thus

$$A = PDP^{-1},$$

where D is a diagonal matrix with the eigenvalues on its main diagonal and P has the corresponding eigenvectors as its columns. $\square$

The determinant of a lower or upper triangular matrix is easy to compute, as in that case the determinant is the product of the diagonal entries. Consequently, we easily find that for triangular matrices the eigenvalues correspond to the diagonal entries.

Theorem 7.1.4.
If $L = (l_{ij})_{i,j=1}^n$ is lower triangular, then its eigenvalues are $l_{11}, \ldots, l_{nn}$.
If $U = (u_{ij})_{i,j=1}^n$ is upper triangular, then its eigenvalues are $u_{11}, \ldots, u_{nn}$.

Proof. By Theorem 4.1.2 we have that $\det(L - \lambda I) = (l_{11} - \lambda) \cdots (l_{nn} - \lambda)$, which has the roots $l_{11}, \ldots, l_{nn}$. The argument is similar for upper triangular matrices. $\square$

The notion of eigenvalues and eigenvectors also apply to linear maps $T : V \to V$. We call $0 \neq \mathbf{x} \in V$ an eigenvector with eigenvalue λ if $T(\mathbf{x}) = \lambda \mathbf{x}$. When we have a matrix representation $A = [T]_{\mathcal{B} \leftarrow \mathcal{B}}$ for T relative to the basis $\mathcal{B}$, one can find the eigenvectors of T by finding the eigenvectors of A. Indeed, we have that

$$T(\mathbf{x}) = \lambda \mathbf{x} \quad \Leftrightarrow \quad A[\mathbf{x}]_{\mathcal{B}} = \lambda [\mathbf{x}]_{\mathcal{B}}.$$

Thus it is important to keep in mind that the eigenvectors of A should be interpreted as the coordinate vectors of the eigenvectors of T with respect to the basis $\mathcal{B}$.

We provide an example.

Example 7.1.5. Find the eigenvectors of $T : \mathbb{R}^{2 \times 2} \to \mathbb{R}^{2 \times 2}$ defined by $T(A) = A + A^T$.

Let $\mathcal{B} = \{E_{11}, E_{12}, E_{21}, E_{22}\}$. Then

$$A = [T]_{\mathcal{B} \leftarrow \mathcal{B}} = \begin{bmatrix} 2 & 0 & 0 & 0 \\ 0 & 1 & 1 & 0 \\ 0 & 1 & 1 & 0 \\ 0 & 0 & 0 & 2 \end{bmatrix}.$$

We find that 2 and 0 are the eigenvalues of A with eigenspaces

$$\text{Nul}(A - 2I) = \text{Span} \left\{ \begin{bmatrix} 1 \\ 0 \\ 0 \\ 0 \end{bmatrix}, \begin{bmatrix} 0 \\ 1 \\ 1 \\ 0 \end{bmatrix}, \begin{bmatrix} 0 \\ 0 \\ 0 \\ 1 \end{bmatrix} \right\}, \text{Nul}(A - 0I) = \text{Span} \left\{ \begin{bmatrix} 0 \\ 1 \\ -1 \\ 0 \end{bmatrix} \right\}.$$

Thus T has eigenvalues 2 and 0, with eigenvectors at $\lambda = 2$

$$\begin{bmatrix} 1 & 0 \\ 0 & 0 \end{bmatrix}, \begin{bmatrix} 0 & 1 \\ 1 & 0 \end{bmatrix}, \begin{bmatrix} 0 & 0 \\ 0 & 1 \end{bmatrix},$$

and at $\lambda = 0$
$$\begin{bmatrix} 0 & 1 \\ -1 & 0 \end{bmatrix}.$$

Notice that indeed

$$T(\begin{bmatrix} 1 & 0 \\ 0 & 0 \end{bmatrix}) = 2\begin{bmatrix} 1 & 0 \\ 0 & 0 \end{bmatrix}, T(\begin{bmatrix} 0 & 1 \\ 1 & 0 \end{bmatrix}) = 2\begin{bmatrix} 0 & 1 \\ 1 & 0 \end{bmatrix},$$

$$T(\begin{bmatrix} 0 & 0 \\ 0 & 1 \end{bmatrix}) = 2\begin{bmatrix} 0 & 0 \\ 0 & 1 \end{bmatrix}, T(\begin{bmatrix} 0 & 1 \\ -1 & 0 \end{bmatrix}) = \begin{bmatrix} 0 & 0 \\ 0 & 0 \end{bmatrix} = 0\begin{bmatrix} 0 & 1 \\ -1 & 0 \end{bmatrix}.$$

□

7.2   Similarity and Diagonalizability

Square matrices A and B of the same size are called **similar** if there exists an invertible matrix P so that $A = PBP^{-1}$. The matrix P is called the **similarity matrix**.

> **Theorem 7.2.1.** *The $n \times n$ matrix A is similar to a diagonal matrix D if and only if A has n linearly independent eigenvectors $\mathbf{v}_1, \ldots, \mathbf{v}_n$ with eigenvalues $\lambda_1, \ldots, \lambda_n$, respectively. If we put $P = \begin{bmatrix} \mathbf{v}_1 & \cdots & \mathbf{v}_n \end{bmatrix}$ and $D =$*
> $$\begin{bmatrix} \lambda_1 & & \\ & \ddots & \\ & & \lambda_n \end{bmatrix}. \ \textit{Then } A = PDP^{-1}.$$

In this case, we say that A is **diagonalizable**.

Note that A in Example 7.1.3 is diagonalizable. In the following example the matrix is not diagonalizable.

Example 7.2.2. Let $A = \begin{bmatrix} 1 & 1 \\ 0 & 1 \end{bmatrix}$. Then $\det(A - \lambda I) = (1-\lambda)^2$. Thus $\lambda = 1$ is the only eigenvalue. For $\mathrm{Nul}(A - I)$, reduce

$$\begin{bmatrix} 1-1 & 1 & | & 0 \\ 0 & 1-1 & | & 0 \end{bmatrix} \rightarrow \begin{bmatrix} 0 & 1 & | & 0 \\ 0 & 0 & | & 0 \end{bmatrix}.$$

So $\mathrm{Nul}(A - I) = \mathrm{Span}\left\{ \begin{bmatrix} 1 \\ 0 \end{bmatrix} \right\}$. There are **not** 2 linearly independent eigenvectors for A. Thus A is not diagonalizable.   □

Theorem 7.2.3. *Eigenvectors at different eigenvalues are linearly independent. Thus, if the $n \times n$ matrix A has n different eigenvalues, then A is diagonalizable.*

For instance $A = \begin{bmatrix} 1 & 2 & 3 \\ 0 & 4 & 5 \\ 0 & 0 & 6 \end{bmatrix}$ is diagonalizable.

Proof. Let $\lambda_1, \ldots, \lambda_n$ be the eigenvalues of A with corresponding eigenvectors $\mathbf{v}_1, \ldots, \mathbf{v}_n$. Suppose that there is a p so that $\{\mathbf{v}_1, \ldots, \mathbf{v}_p\}$ is linearly independent, but $\{\mathbf{v}_1, \ldots, \mathbf{v}_p, \mathbf{v}_{p+1}\}$ is not. Then we must have that $\mathbf{v}_{p+1}$ is a linear combination of $\mathbf{v}_1, \ldots, \mathbf{v}_p$. Thus there exists scalars $c_1, \ldots, c_p$ with

$$\mathbf{v}_{p+1} = c_1 \mathbf{v}_1 + \cdots + c_p \mathbf{v}_p. \tag{7.1}$$

Multiplying with A on the left gives

$$A\mathbf{v}_{p+1} = A(c_1 \mathbf{v}_1 + \cdots + c_p \mathbf{v}_p) = c_1 A\mathbf{v}_1 + \cdots + c_p A\mathbf{v}_p.$$

Using that they are eigenvectors with (different) λ_j gives

$$\lambda_{p+1} \mathbf{v}_{p+1} = c_1 \lambda_1 \mathbf{v}_1 + \cdots + c_p \lambda_p \mathbf{v}_p. \tag{7.2}$$

Multiplying (7.1) with λ_{p+1} and substracting it from (7.2) gives

$$\mathbf{0} = c_1(\lambda_1 - \lambda_{p+1})\mathbf{v}_1 + \cdots + c_p(\lambda_p - \lambda_{p+1})\mathbf{v}_p.$$

Since $\{\mathbf{v}_1, \ldots, \mathbf{v}_p\}$ is linearly independent, we obtain that $c_1(\lambda_1 - \lambda_{p+1}) = 0, \ldots, c_p(\lambda_p - \lambda_{p+1}) = 0$. Since $\lambda_1 - \lambda_{p+1} \neq 0, \ldots, \lambda_p - \lambda_{p+1} \neq 0$, we now get that $c_1 = 0, \ldots, c_p = 0$. This now gives that $\mathbf{v}_{p+1} = \mathbf{0}$, which is a contradiction as $\mathbf{v}_{p+1}$ is an eigenvector (and thus $\neq \mathbf{0}$). $\qquad\square$

When we have written a diagonalizable matrix in the form $A = PDP^{-1}$, it is easy to compute powers A^k of A.

Lemma 7.2.4. *When $A = PDP^{-1}$ is diagonalizable, then $A^k = PD^k P^{-1}$.*

Proof. We have $A^k = PDP^{-1}PDP^{-1}PDP^{-1} \cdots PDP^{-1} = PDD \cdots DP^{-1} = PD^k P^{-1}$. $\qquad\square$

Example 7.2.5. Let $A = \begin{bmatrix} 3 & 4 \\ -2 & -3 \end{bmatrix}$. Then diagonalizing A we find $A = PDP^{-1}$, where

$$P = \begin{bmatrix} -1 & -2 \\ 1 & 1 \end{bmatrix}, D = \begin{bmatrix} -1 & 0 \\ 0 & 1 \end{bmatrix}.$$

Then

$$A^{100} = PD^{100}P^{-1} = P \begin{bmatrix} (-1)^{100} & 0 \\ 0 & 1^{100} \end{bmatrix} P^{-1} = PIP^{-1} = I.$$

$\qquad\square$

7.3   Complex Eigenvalues

Let $A = \begin{bmatrix} 0 & -1 \\ 1 & 0 \end{bmatrix}$ and $\mathbf{x} = \begin{bmatrix} x_1 \\ x_2 \end{bmatrix}$. Then $A\mathbf{x} = \begin{bmatrix} -x_2 \\ x_1 \end{bmatrix}$ is obtained from $\mathbf{x}$ by rotating it 90° counterclockwise around the origin, as depicted below.

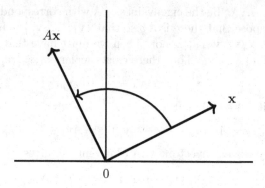

Figure 7.1: Rotation of 90°.

So what eigenvalues and eigenvectors could A possibly have? No vector $A\mathbf{x}$ ends up in the same or opposite direction as $\mathbf{x}$ (nor is $A\mathbf{x} = \mathbf{0}$ for any nonzero $\mathbf{x}$). Indeed, the matrix A does not have any real eigenvalues. However, if we allow complex numbers, then A does have eigenvalues. This section is about that situation. For those who need an introduction or a refresher on complex numbers, please review Appendix A.2, where we cover some basic properties.

Example 7.3.1. Find the eigenvalues and eigenvectors of $A = \begin{bmatrix} 0 & -1 \\ 1 & 0 \end{bmatrix}$.

We find that the characteristic equation is $\det(\lambda I_2 - A) = \lambda^2 + 1 = 0$. So $\lambda^2 = -1$, and thus $\lambda = \pm i$. For $\mathrm{Nul}(A - iI)$, reduce

$$\begin{bmatrix} -i & -1 & | & 0 \\ 1 & -i & | & 0 \end{bmatrix} \rightarrow \begin{bmatrix} 1 & -i & | & 0 \\ -i & -1 & | & 0 \end{bmatrix} \overset{\substack{\text{replace row}_2 \text{ by} \\ \text{row}_2 + i \text{ row}_1}}{\longrightarrow} \begin{bmatrix} 1 & -i & | & 0 \\ 0 & 0 & | & 0 \end{bmatrix}.$$

So x_2 is free, and $x_1 = i x_2$. Consequently, $\mathrm{Nul}(A - iI) = \mathrm{Span}\left\{ \begin{bmatrix} i \\ 1 \end{bmatrix} \right\}$. Similarly, $\mathrm{Nul}(A + iI) = \mathrm{Span}\left\{ \begin{bmatrix} -i \\ 1 \end{bmatrix} \right\}$. Thus the eigenvalues and eigenvectors are $\lambda_1 = i$, $\lambda_2 = -i$ and

$$\mathbf{v}_1 = \begin{bmatrix} i \\ 1 \end{bmatrix}, \mathbf{v}_2 = \begin{bmatrix} -i \\ 1 \end{bmatrix}.$$

Also, A is diagonalizable:

$$\begin{bmatrix} 0 & -1 \\ 1 & 0 \end{bmatrix} = \begin{bmatrix} i & -i \\ 1 & 1 \end{bmatrix} \begin{bmatrix} i & 0 \\ 0 & -i \end{bmatrix} \begin{bmatrix} i & -i \\ 1 & 1 \end{bmatrix}^{-1}.$$

□

Example 7.3.2. Find the eigenvalues and eigenvectors of $A = \begin{bmatrix} 5 & -2 \\ 1 & 3 \end{bmatrix}$.

We find that the characteristic equation is

$$\det(A - \lambda I_2) = (\lambda - 5)(\lambda - 3) + 2 = \lambda^2 - 8\lambda + 17 = (\lambda - 4)^2 + 1 = 0.$$

So $\lambda - 4 = \pm i$, yielding $\lambda = 4 \pm i$. For $\text{Nul}(A - (4+i)I)$, reduce

$$\begin{bmatrix} 1-i & -2 \\ 1 & -1-i \end{bmatrix} \rightarrow \begin{bmatrix} 1 & -1-i \\ 1-i & -2 \end{bmatrix} \overset{\substack{\text{replace row}_2 \text{ by} \\ \text{row}_2 - (1-i)\,\text{row}_1}}{\underset{\rightarrow}{\longrightarrow}} \begin{bmatrix} 1 & -1-i \\ 0 & 0 \end{bmatrix}.$$

(Notice that we left off the augmented zero part.) So x_2 is free, and $x_1 = (1+i)x_2$. Consequently, $\text{Nul}(A-(4+i)I) = \text{Span}\left\{ \begin{bmatrix} 1+i \\ 1 \end{bmatrix} \right\}$. Similarly, $\text{Nul}(A-(4-i)I) = \text{Span}\left\{ \begin{bmatrix} 1-i \\ 1 \end{bmatrix} \right\}$. Thus the eigenvalues and eigenvectors are $\lambda_1 = 4+i$, $\lambda_2 = 4-i$ and

$$\mathbf{v}_1 = \begin{bmatrix} 1+i \\ 1 \end{bmatrix}, \mathbf{v}_2 = \begin{bmatrix} 1-i \\ 1 \end{bmatrix}.$$

Also, A is diagonalizable:

$$\begin{bmatrix} 5 & -2 \\ 1 & 3 \end{bmatrix} = \begin{bmatrix} 1+i & 1-i \\ 1 & 1 \end{bmatrix} \begin{bmatrix} 4+i & 0 \\ 0 & 4-i \end{bmatrix} \begin{bmatrix} 1+i & 1-i \\ 1 & 1 \end{bmatrix}^{-1}.$$

□

For a matrix $A = (a_{ij})_{i,j=1}^n$ and a vector $\mathbf{v} = (v_i)_{i=1}^n$ we define

$$\overline{A} = (\overline{a}_{ij})_{i,j=1}^n, \quad \overline{\mathbf{v}} = \begin{bmatrix} \overline{v}_1 \\ \vdots \\ \overline{v}_n \end{bmatrix},$$

where $\overline{z}$ denotes the complex conjugate of z. It follows easily that $\overline{A\mathbf{v}} = \overline{A}\,\overline{\mathbf{v}}$.

Notice that in both examples above we have

$$\lambda_2 = \overline{\lambda_1}, \quad \mathbf{v}_2 = \overline{\mathbf{v}_1}.$$

This is not by chance. This is due to the following general rule.

Theorem 7.3.3. *If A is a real $n \times n$ matrix with eigenvalue $\lambda = a+bi$, then $\overline{\lambda} = a - bi$ is also an eigenvalue of A. In addition, if $\mathbf{v}_1$ is an eigenvector of A at λ, then $\mathbf{v}_2 = \overline{\mathbf{v}_1}$ is an eigenvector of A at $\overline{\lambda}$.*

Proof. Let us write

$$p(t) = \det(tI - A) = t^n + p_{n-1}t^{n-1} + \cdots + p_1 t + p_0.$$

Since A has real entries, the coefficients p_j of $p(t)$ are all real. Thus $\overline{p_j} = p_j$, $j = 0, \ldots, n-1$. Now suppose that $p(\lambda) = 0$. Then

$$0 = \overline{p(\lambda)} = \overline{\lambda^n + p_{n-1}\lambda^{n-1} + \cdots + p_1\lambda + p_0} = \overline{\lambda}^n + \overline{p_{n-1}}\overline{\lambda}^{n-1} + \cdots + \overline{p_1}\overline{\lambda} + \overline{p_0}$$

$$= \overline{\lambda}^n + p_{n-1}\overline{\lambda}^{n-1} + \cdots + p_1\overline{\lambda} + p_0 = p(\overline{\lambda}).$$

Thus $p(\overline{\lambda}) = 0$, yielding that $\overline{\lambda}$ is also an eigenvalue.

Next, let $\mathbf{v}_1$ be an eigenvector of A at λ. Thus $A\mathbf{v}_1 = \lambda\mathbf{v}_1$. Then $\overline{A}\,\overline{\mathbf{v}_1} = \overline{\lambda}\overline{\mathbf{v}_1}$. Since A is real, we have $\overline{A} = A$, and thus $A\overline{\mathbf{v}_1} = \overline{\lambda}\overline{\mathbf{v}_1}$, giving that $\overline{\mathbf{v}_1}$ is an eigenvector of A at $\overline{\lambda}$. $\qquad\square$

7.4 Systems of Differential Equations: the Diagonalizable Case

A system of linear differential equations with initial conditions has the form

$$\begin{cases} x_1'(t) = a_{11}x_1(t) + \cdots + a_{1n}x_n(t), & x_1(0) = c_1, \\ \quad\vdots & \vdots \\ x_n'(t) = a_{n1}x_1(t) + \cdots + a_{nn}x_n(t), & x_n(0) = c_n, \end{cases}$$

which in shorthand we can write as

$$\begin{cases} \mathbf{x}'(t) = A\mathbf{x}(t), \\ \mathbf{x}(0) = \mathbf{c}, \end{cases} \tag{7.3}$$

where $A = (a_{ij})_{i,j=1}^n$, and

$$\mathbf{x}(t) = \begin{bmatrix} x_1(t) \\ \vdots \\ x_n(t) \end{bmatrix}, \mathbf{c} = \begin{bmatrix} c_1 \\ \vdots \\ c_n \end{bmatrix}.$$

Here $x_j'(t)$ is the derivative of $x_j(t)$. The objective is to find differentiable functions $x_1(t), \ldots, x_n(t)$, satisfying the above equations. When $n = 1$, we simply have the scalar valued equation

$$x'(t) = \lambda x(t), \quad x(0) = c.$$

The solution in this case is $x(t) = ce^{\lambda t}$. Indeed, recalling that $\frac{d}{dt} e^{\lambda t} = \lambda e^{\lambda t}$, we get that $x'(t) = c\lambda e^{\lambda t} = \lambda x(t)$ and $x(0) = ce^0 = c$.

When the coefficient matrix A is diagonalizable, the solution to the system is easily expressed using the eigenvalues and eigenvalues of A. Here is the result.

Theorem 7.4.1. *Consider the system* (7.3) *with A diagonalizable. Let $\lambda_1, \ldots, \lambda_n$ be the eigenvalues of A and let $\mathcal{B} = \{v_1, \ldots, v_n\}$ be a basis of corresponding eigenvectors. Then*

$$x(t) = d_1 e^{\lambda_1 t} v_1 + \cdots + d_n e^{\lambda_n t} v_n,$$

where $\mathbf{d} = [c]_{\mathcal{B}}$*, is the solution to* (7.3)*.*

Proof. With $x(t)$ as above we find that

$$x'(t) = d_1 \lambda_1 e^{\lambda_1 t} v_1 + \cdots + d_n \lambda_n e^{\lambda_n t} v_n.$$

Also,

$$Ax(t) = \sum_{j=1}^{n} d_j e^{\lambda_j t} A v_j = \sum_{j=1}^{n} d_j e^{\lambda_j t} \lambda_j v_j,$$

where we used that $A v_j = \lambda_j v_j$. Thus $x'(t) = Ax(t)$. Finally, observe that $x(0) = \sum_{j=1}^{n} d_j v_j = c$, so indeed $x(t)$ is the solution. $\qquad\square$

Example 7.4.2. Consider the system

$$\begin{cases} x_1'(t) = 3x_1(t) + 3x_2(t), & x_1(0) = -11, \\ x_2'(t) = 4x_1(t) + 2x_2(t), & x_2(0) = 10. \end{cases}$$

In Example 7.1.3 we computed the eigenvalues $\lambda = 6$, $\lambda_2 = -1$, and the corresponding eigenvectors

$$v_1 = \begin{bmatrix} 1 \\ 1 \end{bmatrix}, v_2 = \begin{bmatrix} -3 \\ 4 \end{bmatrix}.$$

It remains to write the vector $\mathbf{c}$ of initial values as a linear combination of v_1 and v_2:

$$\mathbf{c} = \begin{bmatrix} -11 \\ 10 \end{bmatrix} = -2 \begin{bmatrix} 1 \\ 1 \end{bmatrix} + 3 \begin{bmatrix} -3 \\ 4 \end{bmatrix}, \quad \text{thus } \mathbf{d} = [c]_{\{v_1, v_2\}} = \begin{bmatrix} -2 \\ 3 \end{bmatrix}.$$

Apply now Theorem 7.4.1 to get the solution

$$\begin{bmatrix} x_1(t) \\ x_2(t) \end{bmatrix} = -2e^{6t}\begin{bmatrix} 1 \\ 1 \end{bmatrix} + 3e^{-t}\begin{bmatrix} -3 \\ 4 \end{bmatrix} = \begin{bmatrix} -2e^{6t} - 9e^{-t} \\ -2e^{6t} + 12e^{-t} \end{bmatrix}.$$

□

We know that $\frac{d}{dt}\cos t = -\sin(t)$ and $\frac{d}{dt}\sin t = \cos(t)$. Thus, if we let $x_1(t) = \alpha\cos t + \beta\sin t$ and $x_2(t) = -\beta\cos t + \alpha\sin t$, they satisfy the system of differential equations

$$\begin{bmatrix} x_1'(t) \\ x_2'(t) \end{bmatrix} = \begin{bmatrix} 0 & -1 \\ 1 & 0 \end{bmatrix}\begin{bmatrix} x_1(t) \\ x_2(t) \end{bmatrix}.$$

So what happens when we apply Theorem 7.4.1 to a system with this coefficient matrix?

Example 7.4.3. Consider

$$\begin{cases} x_1'(t) = & -x_2(t), & x_1(0) = 2, \\ x_2'(t) = x_1(t) & , & x_2(0) = 3. \end{cases}$$

By Example 7.3.1 we find $\lambda_1 = i$, $\lambda_2 = -i$, and the corresponding eigenvectors

$$\mathbf{v}_1 = \begin{bmatrix} i \\ 1 \end{bmatrix}, \mathbf{v}_2 = \begin{bmatrix} -i \\ 1 \end{bmatrix}.$$

Next, write the vector $\mathbf{c}$ of initial values as a linear combination of $\mathbf{v}_1$ and $\mathbf{v}_2$:

$$\mathbf{c} = \begin{bmatrix} 2 \\ 3 \end{bmatrix} = (\frac{3}{2} - i)\begin{bmatrix} i \\ 1 \end{bmatrix} + (\frac{3}{2} + i)\begin{bmatrix} -i \\ 1 \end{bmatrix}, \quad \text{thus } \mathbf{d} = [\mathbf{c}]_{\{\mathbf{v}_1,\mathbf{v}_2\}} = \begin{bmatrix} \frac{3}{2} - i \\ \frac{3}{2} + i \end{bmatrix}.$$

Apply now Theorem 7.4.1 to get the solution

$$\begin{bmatrix} x_1(t) \\ x_2(t) \end{bmatrix} = (\frac{3}{2} - i)e^{it}\begin{bmatrix} i \\ 1 \end{bmatrix} + (\frac{3}{2} + i)e^{-it}\begin{bmatrix} -i \\ 1 \end{bmatrix}.$$

If we now use that $e^{\pm it} = \cos t \pm i\sin t$, then we obtain the solution

$$\begin{bmatrix} x_1(t) \\ x_2(t) \end{bmatrix} = \begin{bmatrix} 2\cos t - 3\sin t \\ 3\cos t + 2\sin t \end{bmatrix}.$$

□

More general, when you start with a real system of differential equations with real initial conditions, and find eigenvalues $\lambda = a \pm bi$, $a, b \in \mathbb{R}$, the functions $e^{at}\cos(bt)$ and $e^{at}\sin(bt)$ appear in the solution. Indeed, we have

$$e^{(a\pm bi)t} = e^{at}e^{\pm ibt} = e^{at}(\cos(bt) \pm i\sin(bt)).$$

7.5 Exercises

Exercise 7.5.1. Find the eigenvalues and eigenvectors of the following matrices.

(a) $\begin{bmatrix} 4 & -2 \\ -2 & 4 \end{bmatrix}$.

(b) $\begin{bmatrix} 3 & 0 & -3 \\ 1 & 2 & 5 \\ 3 & 0 & 3 \end{bmatrix}$.

(c) $\begin{bmatrix} 2 & 1 & 0 \\ 0 & -3 & 1 \\ 0 & 9 & -3 \end{bmatrix}$.

Exercise 7.5.2. Find the eigenspace of the matrix $\begin{bmatrix} 2 & 0 & -3 \\ 1 & 1 & 5 \\ 3 & 0 & -4 \end{bmatrix}$ at the eigenvalue $\lambda = -1$.

Exercise 7.5.3. Is $\lambda = 5$ an eigenvalue of $A = \begin{bmatrix} 4 & 0 & -2 \\ 2 & 5 & 4 \\ 0 & 0 & 5 \end{bmatrix}$? If so, find the eigenspace of A at $\lambda = 5$.

Exercise 7.5.4. Is $\lambda = -3$ an eigenvalue of $A = \begin{bmatrix} 0 & 0 & -1 \\ 1 & -4 & 0 \\ 4 & -13 & 0 \end{bmatrix}$? If so, find the eigenspace of A at $\lambda = -3$.

Exercise 7.5.5. Find a basis for the eigenspace of $A = \begin{bmatrix} 3 & 5 & 0 \\ 4 & 6 & 5 \\ 2 & 2 & 4 \end{bmatrix}$ corresponding to the eigenvalue $\lambda = 1$.

Exercise 7.5.6. Let $B = \begin{bmatrix} -1 & 1 & -2 \\ 2 & -5 & 1 \\ -3 & 4 & -3 \end{bmatrix}$. Find the eigenspace of B at the eigenvalue $\lambda = -2$.

Exercise 7.5.7. Find the eigenvalues and eigenvectors of $\begin{bmatrix} 2 & i \\ -i & 2 \end{bmatrix}$.

Exercise 7.5.8. Which of the following matrices are diagonalizable? If diagonalizable, give P and D such that $A = PDP^{-1}$.

(a) $\begin{bmatrix} 1 & 2 & -1 \\ -2 & -4 & 2 \\ 3 & 6 & -3 \end{bmatrix}$.

(b) $\begin{bmatrix} 2 & 1 \\ 0 & 2 \end{bmatrix}$.

Exercise 7.5.9. Let A be $n \times n$.

(a) Use Exercise 3.7.6 to show that if $A = PDP^{-1}$ is diagonalizable, then $\operatorname{tr} A = \operatorname{tr} D = $ sum of the eigenvalues of A.

(b) Write the characteristic polynomial $p_A(t) = \det(A - tI)$ as $(-1)^n(t^n + p_{n-1}t^{n-1} + \cdots + p_0)$. Show that $\operatorname{tr} A = -p_{n-1}$.

(c) Write also $p_A(t) = (-1)^n \prod_{j=1}^{n}(t - \lambda_j)$. Use (b) to conclude that $\operatorname{tr} A = \sum_{j=1}^{n} \lambda_j = $ sum of the eigenvalues of A, even when A is not diagonalizable.

Exercise 7.5.10. Suppose that a 3×3 matrix C has the following eigenvalues and eigenvectors:

$$2, \begin{bmatrix} 1 \\ 5 \\ 0 \end{bmatrix}, \begin{bmatrix} 4 \\ 1 \\ 0 \end{bmatrix} \quad \text{and} \quad 3, \begin{bmatrix} 0 \\ 0 \\ 2 \end{bmatrix}.$$

What is $\det C$? And $\operatorname{tr} C$?

Exercise 7.5.11. Let A be a square matrix so that $A^2 = A$. Show that A only has eigenvalues 0 or 1.

Hint: Suppose $A\mathbf{x} = \lambda\mathbf{x}, \mathbf{x} \neq 0$. Compute $A^2\mathbf{x}$ in terms of λ and $\mathbf{x}$, and use $A^2 = A$ to get an equality for λ. A square matrix A so that $A^2 = A$ is called a **projection**.

Exercise 7.5.12. Let A be a square matrix.

(a) Show that $\det(A - \lambda I) = \det(A^T - \lambda I)$.

(b) Conclude that A and A^T have the same eigenvalues.

Exercise 7.5.13. Let A be an invertible matrix. Show that λ is an eigenvalue of A with eigenvector $\mathbf{x}$ if and only if λ^{-1} is an eigenvalue of A^{-1} with eigenvector $\mathbf{x}$.
(Hint: Start with $A\mathbf{x} = \lambda\mathbf{x}$ and multiply on both sides with A^{-1} on the left.)

Exercise 7.5.14. Let A be a square matrix satisfying $A^2 = -A$. Determine the possible eigenvalues of A.

Exercise 7.5.15. Let $A = PDP^{-1}$, where $P = \begin{bmatrix} 1 & 2 \\ 3 & 4 \end{bmatrix}$ and $D = \begin{bmatrix} 2 & 0 \\ 0 & -2 \end{bmatrix}$. Compute A^4.

Exercise 7.5.16. Compute A^{24} for

$$A = \begin{bmatrix} 0 & 1 & 0 \\ -1 & 0 & 0 \\ 1 & 1 & 1 \end{bmatrix}.$$

Exercise 7.5.17. Given that $A = PDP^{-1}$, with $P = \begin{bmatrix} -2 & 0 & -1 \\ 0 & 1 & 2 \\ 1 & 0 & 0 \end{bmatrix}$ and

$D = \begin{bmatrix} 1 & 0 & 0 \\ 0 & -1 & 0 \\ 0 & 0 & 0 \end{bmatrix}$, compute A^{100}.

Exercise 7.5.18.

(a) Find the eigenvectors and eigenvalues of $A = \begin{bmatrix} \frac{3}{4} & \frac{1}{4} \\ \frac{1}{4} & \frac{3}{4} \end{bmatrix}$.

(Hint: One of the eigenvalues is 1.)

(b) Compute A^k.

(c) Show that A^k approaches $\begin{bmatrix} \frac{1}{2} & \frac{1}{2} \\ \frac{1}{2} & \frac{1}{2} \end{bmatrix}$ as $k \to \infty$.

Exercise 7.5.19.

(a) Find the eigenvectors and eigenvalues of $A = \begin{bmatrix} \frac{1}{2} & \frac{1}{2} \\ \frac{1}{4} & \frac{3}{4} \end{bmatrix}$. (Hint: One of the

eigenvalues is 1.)

(b) Compute A^k.

(c) Show that A^k approaches $\begin{bmatrix} \frac{1}{3} & \frac{2}{3} \\ \frac{1}{3} & \frac{2}{3} \end{bmatrix}$ as $k \to \infty$.

Exercise 7.5.20. Let $V = \mathbb{R}^{2 \times 2}$. Define $L : V \to V$ via $L(A) = A - A^T$.

(a) Let

$$\mathcal{B} = \left\{ \begin{bmatrix} 1 & 0 \\ 0 & 0 \end{bmatrix}, \begin{bmatrix} 0 & 1 \\ 0 & 0 \end{bmatrix}, \begin{bmatrix} 0 & 0 \\ 1 & 0 \end{bmatrix}, \begin{bmatrix} 0 & 0 \\ 0 & 1 \end{bmatrix} \right\}.$$

Determine the matrix representation of L with respect to the basis $\mathcal{B}$.

(b) Determine the eigenvalues and eigenvectors of L.

Exercise 7.5.21. Let the linear map $L : \mathbb{R}^{2 \times 2} \to \mathbb{R}^{2 \times 2}$ be given by $L(A) = 2A + 3A^T$. Show that $\begin{bmatrix} 1 & 2 \\ 2 & 1 \end{bmatrix}$ is an eigenvector for L. What is the corresponding eigenvalue?

Exercise 7.5.22. Let C be a square matrix.

(a) Show that for $k \geq 1$,

$$(I-C)(I+C+C^2+\cdots+C^{k-1}) = I-C^k = (I+C+C^2+\cdots+C^{k-1})(I-C).$$

For the remainder of this exercise, we assume that C is diagonalizable and all its eigenvalues λ satisfy $|\lambda| < 1$.

(b) Show that $\lim_{k\to\infty} C^k = 0$.

(c) Using (a) and (b) show that for a matrix C as in (b) we have that $(I - C)^{-1} = \sum_{k=0}^{\infty} C^k$.

(d) If, in addition, all the entries of C are nonnegative, show that $(I - C)^{-1}$ has all nonnegative entries.

The observation under (d) is useful when C represents a cost matrix in a Leontief input-output model (see Exercise 3.7.27).

Exercise 7.5.23. Determine the (complex!) eigenvalues and eigenvectors of the following matrices.

(a) $\begin{bmatrix} -2 & -2 \\ 2 & -2 \end{bmatrix}$.

(b) $\begin{bmatrix} 6 & -5 \\ 5 & 0 \end{bmatrix}$.

Exercise 7.5.24. Solve the following systems of linear differential equations.

(a) $\begin{cases} x_1'(t) = x_2(t), & x_1(0) = 1, \\ x_2'(t) = 6x_1(t) - x_2(t), & x_2(0) = 2. \end{cases}$

(b) $\begin{cases} x_1'(t) = 2x_1(t) + 3x_2(t), & x_1(0) = 1, \\ x_2'(t) = 4x_1(t) + 3x_2(t), & x_2(0) = -1. \end{cases}$

(c) $\begin{cases} x_1'(t) = 3x_1(t) - 2x_2(t), & x_1(0) = 1, \\ x_2'(t) = -2x_1(t) + 3x_2(t), & x_2(0) = 2. \end{cases}$

(d) $\begin{cases} x_1'(t) = -x_2(t), & x_1(0) = 1, \\ x_2'(t) = x_1(t), & x_2(0) = 2. \end{cases}$

Exercise 7.5.25. In a **Markov chain** one transitions from state to state where the probability of each transition depends only on the previous state. Here is an example with four states. The transitions are depicted as arrows

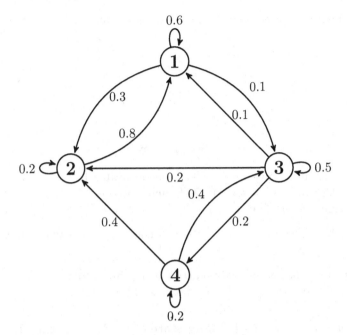

Figure 7.2: A transition graph.

with their probabilities next to them. For instance, to transition from state 1 to state 1 has a 0.6 probability. Also, to transition from state 4 to state 3 has probability 0.4.

When we put the probabilities in a matrix we obtain

$$A = \begin{bmatrix} 0.6 & 0.3 & 0.1 & 0 \\ 0.8 & 0.2 & 0 & 0 \\ 0.1 & 0.2 & 0.5 & 0.2 \\ 0 & 0.4 & 0.4 & 0.2 \end{bmatrix},$$

where a_{ij} is the probability of transitioning from state i to j. This matrix is **row stochastic** as it is a square matrix in which each entry is nonnegative, and the entries in each row add up to 1.

(a) Show that a row stochastic matrix A has 1 as an eigenvalue.
   (Hint: Compute $A\mathbf{e}$, where $\mathbf{e}$ is the vector in which each entry equals 1.)

(b) Show that A^T also has 1 as an eigenvalue.

(c) For the remainder, suppose that the other eigenvalues λ of A satisfy $|\lambda| < 1^*$, and that A is diagonalizable. Diagonalize A as $A = SDS^{-1}$, where

---

*When all the entries are positive this holds, due to the Perron-Frobenius Theorem.

$D = \text{diag}(\lambda_i)_{i=1}^n$ and $\lambda_1 = 1$, and $\mathbf{e}$ the first column of S. Show that

$$\lim_{k \to \infty} D^k = \begin{bmatrix} 1 & 0 & \cdots & 0 \\ 0 & 0 & \cdots & 0 \\ \vdots & \vdots & & \vdots \\ 0 & 0 & \cdots & 0 \end{bmatrix}.$$

(d) Show that

$$\lim_{k \to \infty} A^k = \mathbf{e}\mathbf{y}^T,$$

where $\mathbf{y}$ is an eigenvector of A^T at the eigenvalue 1.
(Hint: Use that $A^T = (S^{-1})^T D S^T$, and thus the first column of $(S^{-1})^T$ is an eigenvector of A^T at 1.)

(e) Show that $A^k \mathbf{e} = \mathbf{e}$ for all $k \in \mathbb{N}$, and conclude that $\mathbf{e}\mathbf{y}^T\mathbf{e} = \mathbf{e}$ and thus $\mathbf{y}^T\mathbf{e} = 1$.

(f) Show that $\mathbf{y}$ has nonnegative entries that sum up to 1.
(Hint: The matrices A^k all have nonnegative entries.)

The vector $\mathbf{y}$ is the so-called **steady state** of the Markov chain. If we start with some population distributed over the different states, the vector $\mathbf{y}$ represents how the population will eventually be distributed among the states. In the example above

$$\mathbf{y} = \begin{bmatrix} 0.5614 \\ 0.2632 \\ 0.1404 \\ 0.0351 \end{bmatrix}.$$

If the above picture represents an internet with 4 webpages and the transition matrix describes the probabilities how people move from page to page, then eventually 56% will end up on webpage 1, 26% on webpage 2, 14% on webpage 3, and 4% on webpage 4. This makes webpage 1 'more important' than the others. This is a crucial idea in Google's PageRank algorithm.

Exercise 7.5.26. Diagonalization of matrices can be useful in determining formulas for sequences given by linear recurrence relations. A famous recurrence relation is $F_n = F_{n-1} + F_{n-2}$, $n \geq 2$, which together with the initial conditions $F_0 = 0$, $F_1 = 1$, defines the **Fibonacci sequence**. The sequence starts off as

$$0, 1, 1, 2, 3, 5, 8, 13, 21, 34, 55, \ldots .$$

How can we use matrices to determine a direct formula for F_n?

(a) Show that we have

$$\begin{bmatrix} F_n \\ F_{n-1} \end{bmatrix} = \begin{bmatrix} 1 & 1 \\ 1 & 0 \end{bmatrix} \begin{bmatrix} F_{n-1} \\ F_{n-2} \end{bmatrix}, \quad n \geq 2.$$

(b) Show that

$$\begin{bmatrix} F_n \\ F_{n-1} \end{bmatrix} = \begin{bmatrix} 1 & 1 \\ 1 & 0 \end{bmatrix}^{n-1} \begin{bmatrix} F_1 \\ F_0 \end{bmatrix} = \begin{bmatrix} 1 & 1 \\ 1 & 0 \end{bmatrix}^{n-1} \begin{bmatrix} 1 \\ 0 \end{bmatrix}.$$

(c) Show that by diagonalization we obtain

$$\begin{bmatrix} 1 & 1 \\ 1 & 0 \end{bmatrix} = \begin{bmatrix} 1 & \frac{1-\sqrt{5}}{2} \\ \frac{\sqrt{5}-1}{2} & 1 \end{bmatrix} \begin{bmatrix} \frac{1+\sqrt{5}}{2} & 0 \\ 0 & \frac{1-\sqrt{5}}{2} \end{bmatrix} \begin{bmatrix} 1 & \frac{1-\sqrt{5}}{2} \\ \frac{\sqrt{5}-1}{2} & 1 \end{bmatrix}^{-1}.$$

(d) Use (b) and (c) to obtain the formula $F_n = \frac{1}{\sqrt{5}} \left((\frac{1+\sqrt{5}}{2})^n - (\frac{1-\sqrt{5}}{2})^n \right)$.

Exercise 7.5.27. Consider the recurrence relation $a_n = 7a_{n-1} - 10a_{n-2}$, $n \geq 2$, with initial conditions $a_0 = 2$, $a_1 = 3$.

(a) Show that

$$\begin{bmatrix} a_n \\ a_{n-1} \end{bmatrix} = \begin{bmatrix} 7 & -10 \\ 1 & 0 \end{bmatrix} \begin{bmatrix} a_{n-1} \\ a_{n-2} \end{bmatrix}, \quad n \geq 2.$$

(b) Show that

$$\begin{bmatrix} a_n \\ a_{n-1} \end{bmatrix} = \begin{bmatrix} 7 & -10 \\ 1 & 0 \end{bmatrix}^{n-1} \begin{bmatrix} 3 \\ 2 \end{bmatrix}.$$

(c) Diagonalize the matrix $\begin{bmatrix} 7 & -10 \\ 1 & 0 \end{bmatrix}$.

(d) Use (b) and (c) to find a formula for a_n.

Exercise 7.5.28. Given a monic polynomial $p(t) = t^k + p_{k-1}t^{k-1} + \cdots + p_2 t^2 + p_1 t + p_0$, we introduce the matrix

$$C_p = \begin{bmatrix} -p_{k-1} & -p_{k-2} & -p_{k-3} & \cdots & -p_1 & -p_0 \\ 1 & 0 & 0 & \cdots & 0 & 0 \\ 0 & 1 & 0 & \cdots & 0 & 0 \\ \vdots & \vdots & \ddots & \ddots & & \vdots \\ 0 & 0 & \cdots & 1 & 0 & 0 \\ 0 & 0 & \cdots & 0 & 1 & 0 \end{bmatrix}.$$

The matrix C_p is called the **companion matrix** of the polynomial $p(t)$. For instance, if $p(t) = t^3 + 2t^2 + 3t + 4$, then

$$C_p = \begin{bmatrix} -2 & -3 & -4 \\ 1 & 0 & 0 \\ 0 & 1 & 0 \end{bmatrix}.$$

(a) Show that the characteristic polynomial of C_p is $(-1)^n p(t)$; equivalently, show that $\det(tI - C_p) = p(t)$.

(b) Given a recurrence relation $x_n = -2x_{n-1} - 3x_{n-2} - 4x_{n-3}$, $n \geq 3$, show that

$$\begin{bmatrix} x_n \\ x_{n-1} \\ x_{n-2} \end{bmatrix} = \begin{bmatrix} -2 & -3 & -4 \\ 1 & 0 & 0 \\ 0 & 1 & 0 \end{bmatrix} \begin{bmatrix} x_{n-1} \\ x_{n-2} \\ x_{n-3} \end{bmatrix}, n \geq 3.$$

(c) For a general **linear recurrence relation** of the form $x_n + \sum_{i=1}^{k} p_{k-i} x_{n-i} = 0$, show that

$$\begin{bmatrix} x_n \\ x_{n-1} \\ \vdots \\ x_{n-k+1} \end{bmatrix} = C_p \begin{bmatrix} x_{n-1} \\ n_{n-2} \\ \vdots \\ x_{n-k} \end{bmatrix}.$$

Thus analyzing the matrix C_p provides information about the sequence $(x_n)_{n=0}^{\infty}$ defined via the corresponding recurrence relation.

Exercise 7.5.29. True or False? Justify each answer.

(i) $\begin{bmatrix} 1 \\ 1 \\ 1 \end{bmatrix}$ is an eigenvector of $\begin{bmatrix} 1 & 2 & 3 \\ 6 & 0 & 0 \\ 1 & 5 & 1 \end{bmatrix}$.

(ii) The sum of two eigenvectors of A both at eigenvalue λ, is again an eigenvector of A at eigenvalue λ.

(iii) The difference of two different eigenvectors of A both at eigenvalue λ, is again an eigenvector of A at eigenvalue λ.

(iv) If A is an $n \times n$ matrix with eigenvalue 0, then $\operatorname{rank} A < n$.

(v) A diagonalizable matrix is necessarily invertible.

(vi) If A is diagonalizable, then so is A^4.

(vii) If $A \in \mathbb{R}^{n \times n}$ has fewer than n different eigenvalues, then A is not diagonalizable.

(viii) The standard basis vector $e_1 \in \mathbb{R}^n$ is an eigenvector of every upper-triangular $n \times n$-matrix.

(ix) The standard basis vector $e_n \in \mathbb{R}^n$ is an eigenvector of every upper-triangular $n \times n$-matrix.

(x) If λ is an eigenvalue of two $n \times n$-matrices A and B, then λ is an eigenvalue of $A + B$ as well.

(xi) If $\mathbf{v}$ is an eigenvector of two $n \times n$-matrices A and B, then $\mathbf{v}$ is an eigenvector of $A + B$ as well.

(xii) If two $n \times n$-matrices A and B have the same characteristic polynomial, then $\det A = \det B$.

8

Orthogonality

CONTENTS

8.1 Dot Product and the Euclidean Norm

Definition 8.1.1. We define the **dot product** of $\mathbf{u}$ and $\mathbf{v}$ in $\mathbb{R}^n$ by

$$\langle \mathbf{u}, \mathbf{v} \rangle = \mathbf{v}^T \mathbf{u} = u_1 v_1 + \cdots + u_n v_n.$$

For instance

$$\langle \begin{bmatrix} 1 \\ 3 \\ -2 \end{bmatrix}, \begin{bmatrix} 5 \\ 2 \\ -1 \end{bmatrix} \rangle = 5 + 6 + 2 = 13.$$

Proposition 8.1.2. *The dot product on $\mathbb{R}^n$ satisfies*

(i) Linearity in the first component: $\langle a\,\mathbf{u} + b\,\mathbf{v}, \mathbf{w} \rangle = a\langle \mathbf{u}, \mathbf{w} \rangle + b\langle \mathbf{v}, \mathbf{w} \rangle$.

(ii) Symmetry: $\langle \mathbf{u}, \mathbf{v} \rangle = \langle \mathbf{v}, \mathbf{u} \rangle$.

(iii) Definiteness: $\langle \mathbf{u}, \mathbf{u} \rangle \geq 0$ *and* $[\langle \mathbf{u}, \mathbf{u} \rangle = 0 \Leftrightarrow \mathbf{u} = \mathbf{0}]$.

Proof. (i) follows since $\mathbf{w}^T(a\,\mathbf{u} + b\,\mathbf{v}) = a\,\mathbf{w}^T\mathbf{u} + b\,\mathbf{w}^T\mathbf{v}$. (ii) is clear. Finally, for (iii) observe that $\langle \mathbf{u}, \mathbf{u} \rangle = u_1^2 + \cdots + u_n^2 \geq 0$, since squares of real numbers are nonnegative. Furthermore, if $u_1^2 + \cdots + u_n^2 = 0$, then each term has to be 0, and thus $u_1 = \cdots = u_n = 0$, yielding $\mathbf{u} = \mathbf{0}$. Also, $\langle \mathbf{0}, \mathbf{0} \rangle = 0$. $\square$

Note that if we combine (i) and (ii) we also get linearity in the second component:

$$\langle \mathbf{u}, a\,\mathbf{v} + b\,\mathbf{w} \rangle = a\langle \mathbf{u}, \mathbf{v} \rangle + b\langle \mathbf{u}, \mathbf{w} \rangle.$$

Due to Proposition 8.1.2(iii), we can introduce the following.

Definition 8.1.3. We define the **norm** (or **length**) of $\mathbf{u} \in \mathbb{R}^n$ by

$$\|\mathbf{u}\| := \sqrt{\langle \mathbf{u}, \mathbf{u} \rangle} = \sqrt{u_1^2 + \cdots + u_n^2}.$$

For instance

$$\left\| \begin{bmatrix} 1 \\ 3 \\ -2 \end{bmatrix} \right\| = \sqrt{1 + 9 + 4} = \sqrt{14}.$$

Geometrically, $\|\mathbf{u}\|$ represents the Euclidean distance from $\mathbf{u}$ to the origin.

Proposition 8.1.4. *The norm on $\mathbb{R}^n$ satisfies*

(i) $\|\mathbf{u}\| \geq 0$ *and* $[\ \|\mathbf{u}\| = 0 \Leftrightarrow \mathbf{u} = \mathbf{0}\ ]$,

(ii) $\|a\,\mathbf{u}\| = |a|\,\|\mathbf{u}\|, a \in \mathbb{R}$,

(iii) $\|\mathbf{u} + \mathbf{v}\| \leq \|\mathbf{u}\| + \|\mathbf{v}\|$. *(triangle inequality)*

The proofs of (i) and (ii) are direct. To prove the triangle inequality, we first need to derive the Cauchy–Schwarz inequality. The geometric interpretation of the triangle inequality is depicted in Figure 8.1.

In this chapter, we also allow vectors with complex entries. A **vector** in $\mathbb{C}^n$ has the form

$$\mathbf{u} = \begin{bmatrix} u_1 \\ u_2 \\ \vdots \\ u_n \end{bmatrix}, \text{ with } u_1, \ldots, u_n \in \mathbb{C}.$$

We define addition and scalar multiplication in the same way as we did in $\mathbb{R}^n$, with the only difference that the scalars are now complex numbers:

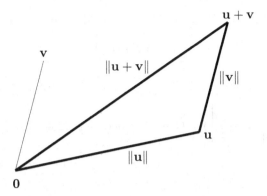

Figure 8.1: The triangle inequality.

(i) $\mathbf{u} + \mathbf{v} = \begin{bmatrix} u_1 \\ u_2 \\ \vdots \\ u_n \end{bmatrix} + \begin{bmatrix} v_1 \\ v_2 \\ \vdots \\ v_n \end{bmatrix} := \begin{bmatrix} u_1 + v_1 \\ u_2 + v_2 \\ \vdots \\ u_n + v_n \end{bmatrix} \in \mathbb{C}^n.$

(ii) $\alpha \mathbf{u} = \alpha \begin{bmatrix} u_1 \\ u_2 \\ \vdots \\ u_n \end{bmatrix} := \begin{bmatrix} \alpha u_1 \\ \alpha u_2 \\ \vdots \\ \alpha u_n \end{bmatrix} \in \mathbb{C}^n.$ Here $\alpha \in \mathbb{C}.$

With these operations $\mathbb{C}^n$ is a vector space over $\mathbb{C}$. When we use concepts like linear independence, span, basis, coordinate system, etc., in $\mathbb{C}^n$ we have to remember that the scalars are now complex.

For vectors $\mathbf{u} \in \mathbb{C}^n$ we define the **conjugate transpose** as

$$\mathbf{u}^* := (\overline{\mathbf{u}})^T = \begin{bmatrix} \overline{u_1} & \cdots & \overline{u_n} \end{bmatrix}.$$

It is easy to that $(\alpha \mathbf{u} + \beta \mathbf{v})^* = \overline{\alpha}\mathbf{u}^* + \overline{\beta}\mathbf{v}^*$. Let us also introduce the operation on matrices: $A^* = (\overline{A})^T$, which is the **conjugate transpose** of A. In other words, if $A = (a_{jk})_{j=1,k=1}^{m\ \ n} \in \mathbb{C}^{m \times n}$, then $A^* = (\overline{a_{kj}})_{j=1,k=1}^{n\ \ m} \in \mathbb{C}^{n \times m}$. For instance,

$$\begin{bmatrix} 1+2i & 3-4i & 5+6i \\ 7-8i & 9+10i & 11-12i \end{bmatrix}^* = \begin{bmatrix} 1-2i & 7+8i \\ 3+4i & 9-10i \\ 5-6i & 11+12i \end{bmatrix}.$$

We have the following rules

$$(A^*)^* = A, (A+B)^* = A^* + B^*, (cA)^* = \overline{c}A^*, (AB)^* = B^*A^*.$$

Definition 8.1.5. We define the **dot product** of $\mathbf{u}$ and $\mathbf{v}$ in $\mathbb{C}^n$ by

$$\langle \mathbf{u}, \mathbf{v} \rangle = \mathbf{v}^* \mathbf{u} = u_1 \overline{v_1} + \cdots + u_n \overline{v_n}.$$

For instance

$$\left\langle \begin{bmatrix} 1-i \\ 3+2i \\ 2-3i \end{bmatrix}, \begin{bmatrix} -4i \\ 2-6i \\ -1-7i \end{bmatrix} \right\rangle = (1-i)(4i) + (3+2i)(2+6i) + (2-3i)(-1+7i)$$

$$= 4i + 4 + 6 + 4i + 18i - 12 - 2 + 3i + 14i + 21 = 17 + 43i.$$

Proposition 8.1.6. *The dot product on* $\mathbb{C}^n$ *satisfies*

(i) *Linearity in the first component:* $\langle a\, \mathbf{u} + b\, \mathbf{v}, \mathbf{w} \rangle = a\langle \mathbf{u}, \mathbf{w} \rangle + b\langle \mathbf{v}, \mathbf{w} \rangle.$

(ii) *Conjugate symmetry:* $\langle \mathbf{u}, \mathbf{v} \rangle = \overline{\langle \mathbf{v}, \mathbf{u} \rangle}.$

(iii) *Definiteness:* $\langle \mathbf{u}, \mathbf{u} \rangle \geq 0$ *and* $[\langle \mathbf{u}, \mathbf{u} \rangle = 0 \iff \mathbf{u} = \mathbf{0}].$

Proof. (i) follows since $\mathbf{w}^*(a\, \mathbf{u} + b\, \mathbf{v}) = a\, \mathbf{w}^* \mathbf{u} + b\, \mathbf{w}^* \mathbf{v}$. (ii) is clear. Finally, for (iii) observe that $\langle \mathbf{u}, \mathbf{u} \rangle = |u_1|^2 + \cdots + |u_n|^2 \geq 0$, since squares of absolute values are nonnegative. Furthermore, if $|u_1|^2 + \cdots + |u_n|^2 = 0$, then each term has to be 0, and thus $u_1 = \cdots = u_n = 0$, yielding $\mathbf{u} = \mathbf{0}$. Also, $\langle \mathbf{0}, \mathbf{0} \rangle = 0$. $\square$

Note that if we combine (i) and (ii) we also get 'skew-linearity' in the second component:

$$\langle \mathbf{u}, a\, \mathbf{v} + b\, \mathbf{w} \rangle = \overline{a}\langle \mathbf{u}, \mathbf{v} \rangle + \overline{b}\langle \mathbf{u}, \mathbf{w} \rangle.$$

So the main difference of the dot product on $\mathbb{C}^n$ with the one on $\mathbb{R}^n$ is that we perform complex conjugation on the entries in the second component. We use the same notation for the dot product on $\mathbb{C}^n$ and $\mathbb{R}^n$, as when all the entries are real the two definitions actually coincide (since complex conjugation does not do anything to a real number). So there should be no confusion. We just have to remember that whenever there are complex numbers involved, we have to take the complex conjugates of the entries appearing in the second component.

Definition 8.1.7. We define the **norm** (or **length**) of $\mathbf{u} \in \mathbb{C}^n$ by

$$\|\mathbf{u}\| := \sqrt{\langle \mathbf{u}, \mathbf{u} \rangle} = \sqrt{|u_1|^2 + \cdots + |u_n|^2}.$$

For instance

$$\left\| \begin{bmatrix} 1+i \\ 3+2i \\ 4i \end{bmatrix} \right\| = \sqrt{(1+i)(1-i) + (3+2i)(3-2i) + (4i)(-4i)} = \sqrt{31}.$$

Proposition 8.1.8. *The norm on $\mathbb{C}^n$ satisfies*

(i) $\|\mathbf{u}\| \geq 0$ and $[\ \|\mathbf{u}\| = 0\ \Leftrightarrow \mathbf{u} = \mathbf{0}\ ]$,

(ii) $\|a\ \mathbf{u}\| = |a|\ \|\mathbf{u}\|, a \in \mathbb{C}$,

(iii) $\|\mathbf{u} + \mathbf{v}\| \leq \|\mathbf{u}\| + \|\mathbf{v}\|$. *(triangle inequality)*

As you see above, Propositions 8.1.4 and 8.1.8 are almost the same. The only way they differ is that the vectors and scalars are real or complex. Similarly, Propositions 8.1.2 and 8.1.6 are almost the same (although, one needs to be careful with the (conjugate) symmetry rule). Going forward, it often makes sense to combine the statements, whether the vectors are real or complex. The way we do this is by saying that $\mathbf{u} \in \mathbb{F}^n$ and $a \in \mathbb{F}$, where $\mathbb{F} = \mathbb{R}$ or $\mathbb{C}$. Hopefully this new notation does not distract from the main points.

The first main result is the **Cauchy–Schwarz inequality**.

Theorem 8.1.9. *(Cauchy–Schwarz inequality) Let $\mathbb{F} = \mathbb{R}$ or $\mathbb{C}$. For $\mathbf{x}, \mathbf{y} \in \mathbb{F}^n$ we have*

$$|\langle \mathbf{x}, \mathbf{y} \rangle| \leq \|\mathbf{x}\|\|\mathbf{y}\|. \tag{8.1}$$

Moreover, equality in (8.1) holds if and only if $\{\mathbf{x}, \mathbf{y}\}$ is linearly dependent.

Proof. When $\mathbf{x} = \mathbf{0}$, inequality (8.1) clearly holds since both sides equal 0. Next, suppose that $\mathbf{x} \neq \mathbf{0}$. Then $\langle \mathbf{x}, \mathbf{x} \rangle > 0$ and we let $\alpha = \frac{\langle \mathbf{y}, \mathbf{x} \rangle}{\langle \mathbf{x}, \mathbf{x} \rangle}$ and $\mathbf{z} = \mathbf{y} - \alpha \mathbf{x}$. We have $\langle \mathbf{z}, \mathbf{z} \rangle \geq 0$. This gives that

$$0 \leq \langle \mathbf{y} - \alpha \mathbf{x}, \mathbf{y} - \alpha \mathbf{x} \rangle = \langle \mathbf{y}, \mathbf{y} \rangle - \alpha \langle \mathbf{x}, \mathbf{y} \rangle - \overline{\alpha} \langle \mathbf{y}, \mathbf{x} \rangle + |\alpha|^2 \langle \mathbf{x}, \mathbf{x} \rangle. \tag{8.2}$$

Now

$$\alpha \langle \mathbf{x}, \mathbf{y} \rangle = \frac{\langle \mathbf{y}, \mathbf{x} \rangle \langle \mathbf{x}, \mathbf{y} \rangle}{\langle \mathbf{x}, \mathbf{x} \rangle} = \frac{|\langle \mathbf{x}, \mathbf{y} \rangle|^2}{\langle \mathbf{x}, \mathbf{x} \rangle} = \overline{\alpha} \langle \mathbf{y}, \mathbf{x} \rangle = |\alpha|^2 \langle \mathbf{x}, \mathbf{x} \rangle.$$

Thus (8.2) simplifies to

$$0 \leq \langle \mathbf{z}, \mathbf{z} \rangle = \langle \mathbf{y}, \mathbf{y} \rangle - \frac{|\langle \mathbf{x}, \mathbf{y} \rangle|^2}{\langle \mathbf{x}, \mathbf{x} \rangle}, \quad \text{which gives } |\langle \mathbf{x}, \mathbf{y} \rangle|^2 \leq \langle \mathbf{x}, \mathbf{x} \rangle \langle \mathbf{y}, \mathbf{y} \rangle.$$

Taking square roots on both sides, yields (8.1).

If $\{\mathbf{x}, \mathbf{y}\}$ is linearly dependent, it is easy to check that equality in (8.1) holds (as $\mathbf{x} = \mathbf{0}$ or $\mathbf{y}$ is a multiple of $\mathbf{x}$). Conversely, suppose that equality holds in (8.1). If $\mathbf{x} = \mathbf{0}$, then clearly $\{\mathbf{x}, \mathbf{y}\}$ is linearly dependent. Next, let us suppose that $\mathbf{x} \neq \mathbf{0}$. As before, put $\mathbf{z} = \mathbf{y} - \frac{\langle \mathbf{y}, \mathbf{x} \rangle}{\langle \mathbf{x}, \mathbf{x} \rangle} \mathbf{x}$. Equality in (8.1) yields that $\langle \mathbf{z}, \mathbf{z} \rangle = 0$. Thus $\mathbf{z} = \mathbf{0}$, showing that $\{\mathbf{x}, \mathbf{y}\}$ is linearly dependent. $\square$

Proof of Propositions 8.1.4 and 8.1.8. Condition (i) follows directly from Proposition 8.1.6(iii) (or Proposition 8.1.2(iii)).

For (ii) compute that $\|a\mathbf{x}\|^2 = \langle a\mathbf{x}, a\mathbf{x}\rangle = a\bar{a}\langle\mathbf{x},\mathbf{x}\rangle = |a|^2\|\mathbf{x}\|^2$. Now take square roots on both sides.

For (iii) we observe that

$$\|\mathbf{x}+\mathbf{y}\|^2 = \langle\mathbf{x}+\mathbf{y},\mathbf{x}+\mathbf{y}\rangle = \langle\mathbf{x},\mathbf{x}\rangle + \langle\mathbf{x},\mathbf{y}\rangle + \langle\mathbf{y},\mathbf{x}\rangle + \langle\mathbf{y},\mathbf{y}\rangle$$

$$\leq \langle\mathbf{x},\mathbf{x}\rangle + 2|\langle\mathbf{x},\mathbf{y}\rangle| + \langle\mathbf{y},\mathbf{y}\rangle \leq \langle\mathbf{x},\mathbf{x}\rangle + 2\|\mathbf{x}\|\|\mathbf{y}\| + \langle\mathbf{y},\mathbf{y}\rangle = (\|\mathbf{x}\|+\|\mathbf{y}\|)^2, \quad (8.3)$$

where we used the Cauchy–Schwarz inequality (8.1) in the last inequality. Taking square roots on both sides proves (iii). ☐

The norm also satisfies the following variation of the triangle inequality.

Lemma 8.1.10. *Let* $\mathbb{F} = \mathbb{R}$ *or* $\mathbb{C}$. *For* $\mathbf{x},\mathbf{y} \in \mathbb{F}^n$ *we have*

$$\big|\, \|\mathbf{x}\| - \|\mathbf{y}\| \,\big| \leq \|\mathbf{x}-\mathbf{y}\|. \qquad (8.4)$$

Proof. Note that the triangle inequality implies

$$\|\mathbf{x}\| = \|\mathbf{x}-\mathbf{y}+\mathbf{y}\| \leq \|\mathbf{x}-\mathbf{y}\| + \|\mathbf{y}\|,$$

and thus
$$\|\mathbf{x}\| - \|\mathbf{y}\| \leq \|\mathbf{x}-\mathbf{y}\|. \qquad (8.5)$$

Reversing the roles of $\mathbf{x}$ and $\mathbf{y}$, we also obtain that

$$\|\mathbf{y}\| - \|\mathbf{x}\| \leq \|\mathbf{y}-\mathbf{x}\| = \|\mathbf{x}-\mathbf{y}\|. \qquad (8.6)$$

Combining (8.5) and (8.6) yields (8.4). ☐

8.2 Orthogonality and Distance to Subspaces

Definition 8.2.1. We say that $\mathbf{v}$ and $\mathbf{w}$ are **orthogonal** if $\langle\mathbf{v},\mathbf{w}\rangle = 0$, and we will denote this as $\mathbf{v} \perp \mathbf{w}$. Notice that $\mathbf{0}$ is orthogonal to any vector, and it is the only vector that is orthogonal to itself.

The **Pythagoras rule** is the following.

Proposition 8.2.2. *If* $\mathbf{v} \perp \mathbf{w}$ *then* $\|\mathbf{v} + \mathbf{w}\|^2 = \|\mathbf{v}\|^2 + \|\mathbf{w}\|^2$.

Proof. $\|\mathbf{v} + \mathbf{w}\|^2 = \langle \mathbf{v} + \mathbf{w}, \mathbf{v} + \mathbf{w} \rangle = \langle \mathbf{v}, \mathbf{v} \rangle + \langle \mathbf{w}, \mathbf{v} \rangle + \langle \mathbf{v}, \mathbf{w} \rangle + \langle \mathbf{w}, \mathbf{w} \rangle =$
$\langle \mathbf{v}, \mathbf{v} \rangle + 0 + 0 + \langle \mathbf{w}, \mathbf{w} \rangle = \|\mathbf{v}\|^2 + \|\mathbf{w}\|^2$. □

For $\emptyset \neq W \subseteq \mathbb{F}^n$ we define

$$W^{\perp} = \{\mathbf{v} \in \mathbb{F}^n : \langle \mathbf{v}, \mathbf{w} \rangle = 0 \text{ for all } \mathbf{w} \in W\} = \{\mathbf{v} : \mathbf{v} \perp \mathbf{w} \text{ for all } \mathbf{w} \in W\}.$$

Notice that in this definition we do not require that W is a subspace; W can be any set of vectors of $\mathbb{F}^n$.

Lemma 8.2.3. *For* $\emptyset \neq W \subseteq V$ *we have that* $W^{\perp}$ *is a subspace of* $\mathbb{F}^n$.

Proof. Clearly $\mathbf{0} \in W^{\perp}$ as $\mathbf{0}$ is orthogonal to any vector, in particular to those in W. Next, let $\mathbf{x}, \mathbf{y} \in W^{\perp}$ and $c, d \in \mathbb{F}$. Then for every $\mathbf{w} \in W$ we have that $\langle c\mathbf{x} + d\mathbf{y}, \mathbf{w} \rangle = c\langle \mathbf{x}, \mathbf{w} \rangle + d\langle \mathbf{y}, \mathbf{w} \rangle = c\,0 + d\,0 = 0$. Thus $c\mathbf{x} + d\mathbf{y} \in W^{\perp}$, showing that $W^{\perp}$ is a subspace. □

Example 8.2.4. Let $W = \text{Span} \left\{ \begin{bmatrix} 1 \\ 1 \\ 1 \\ 2 \end{bmatrix}, \begin{bmatrix} 0 \\ -1 \\ 1 \\ 0 \end{bmatrix} \right\} \subset \mathbb{R}^4$. Find a basis for $W^{\perp}$.

Let $\mathbf{x} = (x_i)_{i=1}^4 \in W^{\perp}$. This holds if and only if $x_1 + x_2 + x_3 + 2x_4 = 0$, $-x_2 + x_3 = 0$. In other words, $\mathbf{x}$ is in the null space of $\begin{bmatrix} 1 & 1 & 1 & 2 \\ 0 & -1 & 1 & 0 \end{bmatrix}$, which in row-reduced echelon form is the matrix $\begin{bmatrix} 1 & 0 & 2 & 2 \\ 0 & 1 & -1 & 0 \end{bmatrix}$. This leads to the basis

$$\left\{ \begin{bmatrix} -2 \\ 1 \\ 1 \\ 0 \end{bmatrix}, \begin{bmatrix} -2 \\ 0 \\ 0 \\ 1 \end{bmatrix} \right\}$$

for $W^{\perp}$. □

Example 8.2.5. Let $W = \text{Span} \left\{ \begin{bmatrix} 2 \\ i \\ 1+i \end{bmatrix}, \begin{bmatrix} 1 \\ 0 \\ i \end{bmatrix} \right\} \subset \mathbb{C}^3$. Find a basis for $W^{\perp}$.

Let $\mathbf{z} = (z_j)_{j=1}^3 \in W^{\perp}$. This holds if and only if $2z_1 - iz_2 + (1-i)z_3 = 0$, $z_1 - iz_3 = 0$. In other words, $\mathbf{z}$ is in the null space of $\begin{bmatrix} 2 & -i & 1-i \\ 1 & 0 & -i \end{bmatrix}$. Row reducing, switching rows first, we get

$$\begin{bmatrix} 1 & 0 & -i \\ 2 & -i & 1-i \end{bmatrix} \rightarrow \begin{bmatrix} 1 & 0 & -i \\ 0 & -i & 1+i \end{bmatrix} \rightarrow \begin{bmatrix} 1 & 0 & -i \\ 0 & 1 & -1+i \end{bmatrix}.$$

We find that z_3 is the free variable, and

$$\begin{bmatrix} z_1 \\ z_2 \\ z_3 \end{bmatrix} = z_3 \begin{bmatrix} i \\ 1-i \\ 1 \end{bmatrix}.$$

This leads to the basis

$$\left\{ \begin{bmatrix} i \\ 1-i \\ 1 \end{bmatrix} \right\}$$

for $W^\perp$.   □

Definition 8.2.6. The **distance** between two vectors is given by the norm of the difference vector; that is

$$\text{distance}(\mathbf{u}, \mathbf{v}) = \|\mathbf{u} - \mathbf{v}\|.$$

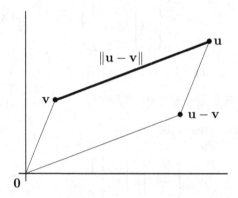

Figure 8.2: Distance between two vectors.

Example 8.2.7. Find the distance between $\begin{bmatrix} 1 \\ 2 \\ 3 \end{bmatrix}$ and $\begin{bmatrix} 0 \\ 4 \\ -3 \end{bmatrix}$.

The distance equals

$$\left\| \begin{bmatrix} 1 \\ 2 \\ 3 \end{bmatrix} - \begin{bmatrix} 0 \\ 4 \\ -3 \end{bmatrix} \right\| = \left\| \begin{bmatrix} 1 \\ -2 \\ 6 \end{bmatrix} \right\| = \sqrt{1^2 + (-2)^2 + 6^2} = \sqrt{41}.$$

□

Example 8.2.8. Find the distance between $\begin{bmatrix} 1+i \\ 2+3i \end{bmatrix}$ and $\begin{bmatrix} 4-2i \\ -1+i \end{bmatrix}$.

The distance equals

$$\left\| \begin{bmatrix} 1+i \\ 2+3i \end{bmatrix} - \begin{bmatrix} 4-2i \\ -1+i \end{bmatrix} \right\| = \left\| \begin{bmatrix} -3+3i \\ 3+2i \end{bmatrix} \right\| = \sqrt{|-3+3i|^2 + |3+2i|^2} =$$

$$\sqrt{(-3)^2 + 3^2 + 3^2 + 2^2} = \sqrt{31}.$$

$\square$

When W is a subspace and $\mathbf{u}$ is a vector, then the distance between $\mathbf{u}$ and W defined as the minimum over all numbers $\|\mathbf{u} - \mathbf{w}\|$, where $\mathbf{w} \in W$:

$$\text{distance}(\mathbf{u}, W) = \min_{\mathbf{w} \in W} \|\mathbf{u} - \mathbf{w}\|.$$

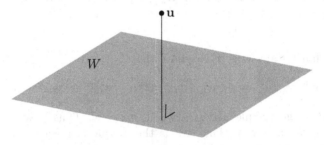

Figure 8.3: Distance between a vector $\mathbf{u}$ and a subspace W.

Orthogonality appears in a natural way when computing the distance from a vector $\mathbf{u}$ to a subspace W, as it is the distance from $\mathbf{u}$ to a point $\mathbf{w} \in W$ so that $\mathbf{u} - \mathbf{w}$ is orthogonal to W. The statement is as follows.

Theorem 8.2.9. *Let* $\mathbb{F} = \mathbb{R}$ *or* $\mathbb{C}$. *Let* $\mathbf{u} \in \mathbb{F}^n$ *and* $W \subseteq \mathbb{F}^n$ *a subspace. Then there exists a unique element* $\mathbf{w}_0 \in W$ *so that*

$$\text{distance}(\mathbf{u}, W) = \|\mathbf{u} - \mathbf{w}_0\|.$$

This element $\mathbf{w}_0$ *is the unique element of* W *with the property that* $\mathbf{u} - \mathbf{w}_0 \in W^\perp$.

In order to prove Theorem 8.2.9 we need the following auxiliary result.

Lemma 8.2.10. *Let* $\mathbb{F} = \mathbb{R}$ *or* $\mathbb{C}$. *Let* $\{\mathbf{w}_1, \ldots, \mathbf{w}_k\}$ *be linearly independent. Then the matrix* $G = (\langle \mathbf{w}_j, \mathbf{w}_i \rangle)_{i,j=1}^k$ *is invertible.*

The matrix G is called the **Gram matrix** of the vectors $\{\mathbf{w}_1, \ldots, \mathbf{w}_k\}$.

Proof. Since the matrix G is square, it suffices to prove that $\text{Nul } G = \{\mathbf{0}\}$. Let $\mathbf{c} = (c_i)_{i=1}^k \in \text{Nul } G$. Then $G\mathbf{c} = \mathbf{0}$. Thus also $\mathbf{c}^*G\mathbf{c} = 0$. A straightforward computation show that

$$0 = \mathbf{c}^*G\mathbf{c} = \langle \sum_{i=1}^k c_i \mathbf{w}_i, \sum_{i=1}^k c_i \mathbf{w}_i \rangle = \| \sum_{i=1}^k c_i \mathbf{w}_i \|^2.$$

Thus we get that $\sum_{i=1}^k c_i \mathbf{w}_i = \mathbf{0}$. As $\{\mathbf{w}_1, \ldots, \mathbf{w}_k\}$ is linearly independent, this implies that $c_1 = \cdots = c_k = 0$. Consequently $\mathbf{c} = \mathbf{0}$. $\square$

Proof of Theorem 8.2.9. First let us find $\mathbf{w}_0 \in W$ so that $\mathbf{u} - \mathbf{w}_0 \in W^\perp$. Letting $\{\mathbf{w}_1, \ldots, \mathbf{w}_k\}$ be a basis of W, we have that $\mathbf{w}_0 \in W$ implies that $\mathbf{w}_0 = \sum_{j=1}^k c_j \mathbf{w}_j$ for some $c_1, \ldots, c_k \in \mathbb{F}$. The equations $\langle \mathbf{u} - \mathbf{w}_0, \mathbf{w}_i \rangle = 0$, $i = 1, \ldots, k$, lead to the matrix equation $G\mathbf{c} = (\langle \mathbf{u}, \mathbf{w}_i \rangle)_{i=1}^n$, where G is the Gram matrix of the vectors $\{\mathbf{w}_1, \ldots, \mathbf{w}_k\}$. By Lemma 8.2.10 the matrix G is invertible, so this equation has a unique solution. Thus there is a unique $\mathbf{w}_0 \in W$ so that $\mathbf{u} - \mathbf{w}_0 \in W^\perp$.

Next notice that for any $\mathbf{w} \in W$, we have that

$$\|\mathbf{u} - \mathbf{w}\|^2 = \|\mathbf{u} - \mathbf{w}_0 + \mathbf{w}_0 - \mathbf{w}\|^2 = \|\mathbf{u} - \mathbf{w}_0\|^2 + \|\mathbf{w}_0 - \mathbf{w}\|^2,$$

due to the Pythagoras rule and $\mathbf{u} - \mathbf{w}_0 \perp \mathbf{w}_0 - \mathbf{w}$. Thus if $\mathbf{w} \neq \mathbf{w}_0$, we have $\|\mathbf{u} - \mathbf{w}\| > \|\mathbf{u} - \mathbf{w}_0\|$. Thus $\mathbf{w}_0$ is the unique element in W so that $\text{distance}(\mathbf{u}, W) = \|\mathbf{u} - \mathbf{w}_0\|$. $\square$

The proof gives a way to compute the distance from a vector $\mathbf{u}$ and a subspace W with basis $\{\mathbf{w}_1, \ldots, \mathbf{w}_k\}$, as follows.

1. Compute $G = (\langle \mathbf{w}_j, \mathbf{w}_i \rangle)_{i,j=1}^k$.

2. Compute $\mathbf{c} = (c_j)_{j=1}^k = G^{-1}(\langle \mathbf{u}, \mathbf{w}_i \rangle)_{i=1}^n$.

3. Put $\mathbf{w}_0 = \sum_{j=1}^k c_j \mathbf{w}_j$.

4. Compute $\text{distance}(\mathbf{u}, W) = \|\mathbf{u} - \mathbf{w}_0\|$.

The vector $\mathbf{w}_0$ is also referred to as the **orthonogonal projection** of $\mathbf{u}$ onto W.

Example 8.2.11. Find the distance of $\mathbf{u} = \begin{bmatrix} 6 \\ -2 \end{bmatrix}$ to the line $2x_1 + x_2 = 0$.

Note that the line is the same as $W = \text{Span}\{\mathbf{w}_1\}$, where $\mathbf{w}_1 = \begin{bmatrix} 1 \\ -2 \end{bmatrix}$. To find $\mathbf{w}_0$, we need to find the scalar c so that $\mathbf{u} - \mathbf{w}_0 = \mathbf{u} - c\mathbf{w}_1 \perp \mathbf{w}_1$. This gives

$c = \frac{\langle \mathbf{u}, \mathbf{w}_1 \rangle}{\langle \mathbf{w}_1, \mathbf{w}_1 \rangle} = \frac{10}{5} = 2$. Thus $\mathbf{w}_0 = 2\mathbf{w}_1 = \begin{bmatrix} 2 \\ -4 \end{bmatrix}$. And therefore the distance of $\mathbf{u}$ to the line is

$$\text{distance}(\mathbf{u}, W) = \|\mathbf{u} - \mathbf{w}_0\| = \| \begin{bmatrix} 4 \\ 2 \end{bmatrix} \| = \sqrt{4^2 + 2^2} = 2\sqrt{5}.$$

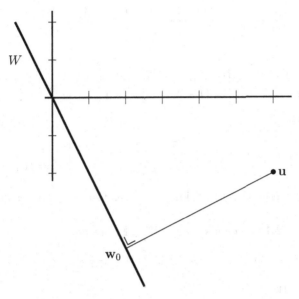

Figure 8.4: Distance from $\mathbf{u}$ to W.

$\square$

Example 8.2.12. Find the distance of $\mathbf{u}$ to the subspace $W = \text{Span} \{\mathbf{w}_1, \mathbf{w}_2\}$, where

$$\mathbf{u} = \begin{bmatrix} 3 \\ -1 \\ 2 \\ 2 \end{bmatrix}, \mathbf{w}_1 = \begin{bmatrix} 1 \\ 1 \\ 1 \\ 1 \end{bmatrix}, \mathbf{w}_2 = \begin{bmatrix} -1 \\ 3 \\ 0 \\ 0 \end{bmatrix}.$$

We have

$$\mathbf{c} = \begin{bmatrix} \langle \mathbf{w}_1, \mathbf{w}_1 \rangle & \langle \mathbf{w}_2, \mathbf{w}_1 \rangle \\ \langle \mathbf{w}_1, \mathbf{w}_2 \rangle & \langle \mathbf{w}_2, \mathbf{w}_2 \rangle \end{bmatrix}^{-1} \begin{bmatrix} \langle \mathbf{u}, \mathbf{w}_1 \rangle \\ \langle \mathbf{u}, \mathbf{w}_2 \rangle \end{bmatrix} = \begin{bmatrix} 4 & 2 \\ 2 & 10 \end{bmatrix}^{-1} \begin{bmatrix} 6 \\ -6 \end{bmatrix} = \begin{bmatrix} 2 \\ -1 \end{bmatrix}.$$

Thus

$$\mathbf{w}_0 = 2\mathbf{w}_1 - \mathbf{w}_2 = \begin{bmatrix} 3 \\ -1 \\ 2 \\ 2 \end{bmatrix},$$

and consequently $\text{distance}(\mathbf{u}, W) = \|\mathbf{u} - \mathbf{w}_0\| = \|\mathbf{0}\| = 0$. In this case, $\mathbf{u}$ is an element of W. $\square$

Example 8.2.13. Find the distance of $\mathbf{u}$ to the subspace $W = \mathrm{Span}\,\{\mathbf{w}_1\}$, where

$$\mathbf{u} = \begin{bmatrix} 1+i \\ -1 \end{bmatrix}, \mathbf{w}_1 = \begin{bmatrix} 2-i \\ i \end{bmatrix}.$$

We have

$$\mathbf{c} = \frac{\langle \mathbf{u}, \mathbf{w}_1 \rangle}{\langle \mathbf{w}_1, \mathbf{w}_1 \rangle} = \frac{(2+i)(1+i) + (-i)(-1)}{|2-i|^2 + |i|^2} = \frac{1+4i}{6}.$$

Thus

$$\mathbf{w}_0 = \frac{1+4i}{6}\mathbf{w}_1 = \frac{1}{6}\begin{bmatrix} (1+4i)(2-i) \\ (1+4i)i \end{bmatrix} = \frac{1}{6}\begin{bmatrix} 6+7i \\ -4+i \end{bmatrix} = \begin{bmatrix} 1+\frac{7}{6}i \\ -\frac{2}{3}+\frac{1}{6}i \end{bmatrix}$$

and consequently

$$\mathrm{distance}(\mathbf{u}, W) = \|\mathbf{u} - \mathbf{w}_0\| = \left\| \begin{bmatrix} -\frac{1}{6}i \\ -\frac{1}{3}-\frac{1}{6}i \end{bmatrix} \right\| = \frac{1}{6}\left\| \begin{bmatrix} -i \\ -2-i \end{bmatrix} \right\| = \frac{\sqrt{6}}{6}.$$

It is easy to determine the dimension of $W^\perp$, as the following result shows.

Proposition 8.2.14. *For a subspace $W \subseteq \mathbb{F}^n$ we have*

$$W + W^\perp = \mathbb{F}^n, \quad W \cap W^\perp = \{\mathbf{0}\}.$$

Also $(W^\perp)^\perp = W$ and

$$\dim W + \dim W^\perp = n.$$

Proof. Let $\mathbf{u} \in \mathbb{F}^n$. Let $\mathbf{w}_0$ be the orthogonal projection of $\mathbf{u}$ onto W. Then $\mathbf{u}-\mathbf{w}_0 \in W^\perp$. Thus $\mathbf{u} = \mathbf{w}_0+(\mathbf{u}-\mathbf{w}_0) \in W+W^\perp$. This shows $\mathbb{F}^n \subseteq W+W^\perp$. The other inclusion is trivial (since both W and $W^\perp$ are subspaces of $\mathbb{F}^n$), and thus equality holds.

Next let $\mathbf{x} \in W \cap W^\perp$. But then $\mathbf{x} \perp \mathbf{x}$, and thus $\mathbf{x} = \mathbf{0}$. Thus $W \dotplus W^\perp = \mathbb{F}^n$ is a direct sum, and $\dim W + \dim W^\perp = n$ follows from Theorem 5.3.14.

Finally, it is clear that any vector in W is orthogonal to vectors in $W^\perp$. Thus $W \subseteq (W^\perp)^\perp$. Since $\dim(W^\perp)^\perp = n - \dim W^\perp = n - (n - \dim W) = \dim W$, we obtain $(W^\perp)^\perp = W$. □

When $W = \mathrm{Col}A$ we have a way of describing $W^\perp$.

Proposition 8.2.15. *Let $A \in \mathbb{F}^{m \times n}$. Then*

(i) $(\mathrm{Col}\,A)^\perp = \mathrm{Nul}\,A^*$.

(ii) $(\text{Nul } A)^\perp = \text{Col } A^*$.

(iii) $\dim \text{Nul } A + \text{rank } A = n$.

(iv) $\dim \text{Nul } A^* + \text{rank } A = m$.

Proof. Let $\mathbf{x} \in (\text{Col } A)^\perp$. Then for any $\mathbf{y}$ we have that $\mathbf{x} \perp A\mathbf{y}$. Thus $0 = (A\mathbf{y})^*\mathbf{x} = \mathbf{y}^*(A^*\mathbf{x})$. Thus $A^*\mathbf{x} \perp \mathbf{y}$ for every $\mathbf{y}$. The only vector that is orthogonal to all $\mathbf{y}$ is the zero vector. Thus $A^*\mathbf{x} = \mathbf{0}$, or equivalently $\mathbf{x} \in \text{Nul } A^*$. This shows $(\text{Col } A)^\perp \subseteq \text{Nul } A^*$.

For the other inclusion, let $\mathbf{x} \in \text{Nul } A^*$. Then $A^*\mathbf{x} = \mathbf{0}$. Thus for every $\mathbf{y}$ we have $0 = \mathbf{y}^*(A^*\mathbf{x}) = (A\mathbf{y})^*\mathbf{x}$. Thus $\mathbf{x} \perp A\mathbf{y}$ for every $\mathbf{y}$. Thus $\mathbf{x} \in (\text{Col } A)^\perp$. This finishes (i).

For (ii) apply (i) to the matrix A^* instead of A, giving $(\text{Col } A^*)^\perp = \text{Nul } A$. Next use the rule $(W^\perp)^\perp$ to obtain from this that $\text{Col } A^* = ((\text{Col } A^*)^\perp)^\perp = (\text{Nul } A)^\perp$.

Using Proposition 8.2.14, (iii) and (iv) follow. $\qquad\square$

In Example 8.2.4 we had that $W = \text{Col } A$, where

$$A = \begin{bmatrix} 1 & 0 \\ 1 & -1 \\ 1 & 1 \\ 2 & 0 \end{bmatrix}.$$

We indeed discovered there that a basis for $\text{Nul } A^*$ gave a basis for $W^\perp$. We found the basis

$$\left\{ \begin{bmatrix} -2 \\ 1 \\ 1 \\ 0 \end{bmatrix}, \begin{bmatrix} -2 \\ 0 \\ 0 \\ 1 \end{bmatrix} \right\}$$

for $W^\perp = \text{Nul } A^*$.

8.3   Orthonormal Bases and Gram–Schmidt

Definition 8.3.1. We say that a set $\{\mathbf{w}_1, \ldots, \mathbf{w}_m\}$ is an **orthogonal set** if $\mathbf{w}_k \perp \mathbf{w}_j$ when $k \neq j$. If, in addition, $\|\mathbf{w}_\ell\| = 1$, $\ell = 1, \ldots, m$, then it is an **orthonormal set**. Given a subspace W, we call $\mathcal{B} = \{\mathbf{w}_1, \ldots, \mathbf{w}_m\}$

an **orthogonal/orthonormal basis** of W, if $\mathcal{B}$ is a basis and an orthogonal/orthonormal set.

Notice that the Gram matrix of an orthogonal set is diagonal. Moreover, the Gram matrix of an orthonormal set is the identity matrix. This indicates that computations with orthogonal and orthonormal sets are much easier. This raises the question whether any subspace has an orthogonal and/or an orthonormal basis. The answer to this question is 'yes', and its proof relies on the **Gram–Schmidt process**, which we now introduce.

Theorem 8.3.2 (*Gram–Schmidt process*). *Let* $\{\mathbf{v}_1, \ldots, \mathbf{v}_p\} \subset \mathbb{F}^n$ *be linearly independent. Construct* $\{\mathbf{z}_1, \ldots, \mathbf{z}_p\}$ *as follows:*

$$\mathbf{z}_1 = \mathbf{v}_1$$

$$\mathbf{z}_k = \mathbf{v}_k - \frac{\langle \mathbf{v}_k, \mathbf{z}_{k-1} \rangle}{\langle \mathbf{z}_{k-1}, \mathbf{z}_{k-1} \rangle} \mathbf{z}_{k-1} - \cdots - \frac{\langle \mathbf{v}_k, \mathbf{z}_1 \rangle}{\langle \mathbf{z}_1, \mathbf{z}_1 \rangle} \mathbf{z}_1, \ k = 2, \ldots, p.$$

$$(8.7)$$

Then for $k = 1, \ldots, p$, *we have that* $\{\mathbf{z}_1, \ldots, \mathbf{z}_p\}$ *is an orthogonal linearly independent set satisfying*

$$\mathrm{Span}\, \{\mathbf{v}_1, \ldots, \mathbf{v}_k\} = \mathrm{Span}\, \{\mathbf{z}_1, \ldots, \mathbf{z}_k\} = \mathrm{Span}\, \left\{ \frac{\mathbf{z}_1}{\|\mathbf{z}_1\|}, \ldots, \frac{\mathbf{z}_k}{\|\mathbf{z}_k\|} \right\}. \quad (8.8)$$

The set $\left\{ \frac{\mathbf{z}_1}{\|\mathbf{z}_1\|}, \ldots, \frac{\mathbf{z}_p}{\|\mathbf{z}_p\|} \right\}$ *is an orthonormal set.*

Proof. We prove this by induction. Clearly when $p = 1$, then $\mathbf{z}_1 = \mathbf{v}_1$, and the statements are trivial.

Now suppose that the theorem has been proven for sets with up to $p - 1$ vectors. Next, we are given $\{\mathbf{v}_1, \ldots, \mathbf{v}_p\}$ and we construct $\{\mathbf{z}_1, \ldots, \mathbf{z}_p\}$. Notice that $\{\mathbf{z}_1, \ldots, \mathbf{z}_{p-1}\}$ are obtained by applying the Gram–Schmidt process to $\{\mathbf{v}_1, \ldots, \mathbf{v}_{p-1}\}$, and thus by the induction assumption $\{\mathbf{z}_1, \ldots, \mathbf{z}_{p-1}\}$ is a linearly independent set and (8.8) holds for $k = 1, \ldots, p-1$. Let $\mathbf{z}_p$ be defined via (8.7). Observe that for $k \leq p - 1$,

$$\langle \mathbf{z}_p, \mathbf{z}_k \rangle = \langle \mathbf{v}_p, \mathbf{z}_k \rangle - \langle \sum_{j=1}^{p-1} \frac{\langle \mathbf{v}_p, \mathbf{z}_j \rangle}{\langle \mathbf{z}_j, \mathbf{z}_j \rangle} \mathbf{z}_j, \mathbf{z}_k \rangle = \langle \mathbf{v}_p, \mathbf{z}_k \rangle - \frac{\langle \mathbf{v}_p, \mathbf{z}_k \rangle}{\langle \mathbf{z}_k, \mathbf{z}_k \rangle} \langle \mathbf{z}_k, \mathbf{z}_k \rangle = 0,$$

where we used that $\langle \mathbf{z}_j, \mathbf{z}_k \rangle = 0$ for $j \neq k$, $1 \leq j, k \leq p - 1$. This proves the orthogonality. Also, we see that

$$\mathbf{z}_p \in \mathrm{Span}\, \{\mathbf{v}_p\} + \mathrm{Span}\, \{\mathbf{z}_1, \ldots, \mathbf{z}_{p-1}\} =$$

$$\text{Span}\{\mathbf{v}_p\} + \text{Span}\{\mathbf{v}_1, \dots, \mathbf{v}_{p-1}\} = \text{Span}\{\mathbf{v}_1, \dots, \mathbf{v}_p\}.$$

Next, since $\mathbf{v}_p = \mathbf{z}_p + \sum_{j=1}^{p-1} \frac{\langle \mathbf{v}_p, \mathbf{z}_j \rangle}{\langle \mathbf{z}_j, \mathbf{z}_j \rangle} \mathbf{z}_j$, we have that

$$\mathbf{v}_p \in \text{Span}\{\mathbf{z}_1, \dots, \mathbf{z}_p\}.$$

Combining these observations with the induction assumption yields

$$\text{Span}\{\mathbf{v}_1, \dots, \mathbf{v}_p\} = \text{Span}\{\mathbf{z}_1, \dots, \mathbf{z}_p\}.$$

Since $\{\mathbf{v}_1, \dots, \mathbf{v}_p\}$ is linearly independent, they span a p dimensional space. Then $\{\mathbf{z}_1, \dots, \mathbf{z}_p\}$ also span a p dimensional space (the same one), and thus this set of vectors is also linearly independent. Finally, dividing each $\mathbf{z}_i$ by its length does not change the span, and makes the vectors orthonormal. $\quad\square$

Corollary 8.3.3. *Let W be a subspace of $\mathbb{F}^n$. Then W has an orthonormal basis.*

Proof. Start with a basis $\{\mathbf{v}_1, \dots, \mathbf{v}_p\}$ for W. Perform the Gram–Schmidt process on this set of vectors. We then obtain that $\left\{ \frac{\mathbf{z}_1}{\|\mathbf{z}_1\|}, \dots, \frac{\mathbf{z}_p}{\|\mathbf{z}_p\|} \right\}$ is an orthonormal basis for W. $\quad\square$

Example 8.3.4. Let $W = \text{Span}\left\{ \begin{bmatrix} 2 \\ 0 \\ -2 \end{bmatrix}, \begin{bmatrix} 1 \\ 2 \\ 3 \end{bmatrix} \right\}$. Find an orthonormal basis for W. Applying the Gram–Schmidt process we obtain,

$$\mathbf{z}_1 = \begin{bmatrix} 2 \\ 0 \\ -2 \end{bmatrix},$$

$$\mathbf{z}_2 = \begin{bmatrix} 1 \\ 2 \\ 3 \end{bmatrix} - \frac{-4}{8} \begin{bmatrix} 2 \\ 0 \\ -2 \end{bmatrix} = \begin{bmatrix} 2 \\ 2 \\ 2 \end{bmatrix}. \tag{8.9}$$

Dividing these vectors by their lengths $\|\mathbf{z}_1\| = \sqrt{2^2 + (-2)^2} = 2\sqrt{2}$ and $\|\mathbf{z}_2\| = \sqrt{2^2 + 2^2 + 2^2} = 2\sqrt{3}$, we find that

$$\left\{ \frac{1}{\sqrt{2}} \begin{bmatrix} 1 \\ 0 \\ -1 \end{bmatrix}, \frac{1}{\sqrt{3}} \begin{bmatrix} 1 \\ 1 \\ 1 \end{bmatrix} \right\}$$

is an orthonormal basis for W. $\quad\square$

In Theorem 8.3.2 we required $\{\mathbf{v}_1, \dots, \mathbf{v}_p\}$ to be linearly independent. One can also perform the Gram–Schmidt process to a set that is not necessarily linearly independent. In that case one may reach a point where $\mathbf{z}_k = 0$. This happens exactly when $\mathbf{v}_k \in \mathrm{Span}\,\{\mathbf{v}_1, \ldots, \mathbf{v}_{k-1}\}$. At this point one would continue the process pretending $\mathbf{v}_k$ was not part of the original set; or equivalently, leaving $\mathbf{z}_k = 0$ out of the subsequent calculations. In the end one would still end up with an orthogonal set spanning the same space. To state this formally is somewhat cumbersome, but let us see how it works out in an example.

Example 8.3.5. Let $\mathbf{v}_1 = \begin{bmatrix} 1 \\ 1 \\ 1 \\ 1 \end{bmatrix}, \mathbf{v}_2 = \begin{bmatrix} 0 \\ -2 \\ 0 \\ -2 \end{bmatrix}, \mathbf{v}_3 = \begin{bmatrix} 1 \\ 0 \\ 1 \\ 0 \end{bmatrix}, \mathbf{v}_4 = \begin{bmatrix} 2 \\ -2 \\ 0 \\ 0 \end{bmatrix}$. Apply-

ing the Gram–Schmidt process we obtain,

$$\mathbf{z}_1 = \begin{bmatrix} 1 \\ 1 \\ 1 \\ 1 \end{bmatrix},$$

$$\mathbf{z}_2 = \begin{bmatrix} 0 \\ -2 \\ 0 \\ -2 \end{bmatrix} - \frac{-4}{4} \begin{bmatrix} 1 \\ 1 \\ 1 \\ 1 \end{bmatrix} = \begin{bmatrix} 1 \\ -1 \\ 1 \\ -1 \end{bmatrix},$$

$$\mathbf{z}_3 = \begin{bmatrix} 1 \\ 0 \\ 1 \\ 0 \end{bmatrix} - \frac{2}{4} \begin{bmatrix} 1 \\ 1 \\ 1 \\ 1 \end{bmatrix} - \frac{2}{4} \begin{bmatrix} 1 \\ -1 \\ 1 \\ -1 \end{bmatrix} = \begin{bmatrix} 0 \\ 0 \\ 0 \\ 0 \end{bmatrix}. \tag{8.10}$$

Thus $\mathbf{v}_3 \in \mathrm{Span}\,\{\mathbf{z}_1, \mathbf{z}_2\} = \mathrm{Span}\,\{\mathbf{v}_1, \mathbf{v}_2\}$, and thus the original set of vectors was not linearly independent. Let us just continue the process pretending $\mathbf{v}_3$ was never there. We then get

$$\mathbf{z}_4 = \begin{bmatrix} 2 \\ -2 \\ 0 \\ 0 \end{bmatrix} - 0 \begin{bmatrix} 1 \\ 1 \\ 1 \\ 1 \end{bmatrix} - \frac{4}{4} \begin{bmatrix} 1 \\ -1 \\ 1 \\ -1 \end{bmatrix} = \begin{bmatrix} 1 \\ -1 \\ -1 \\ 1 \end{bmatrix}.$$

Thus we find

$$\mathrm{Span}\,\{\mathbf{v}_1, \mathbf{v}_2, \mathbf{v}_3, \mathbf{v}_4\} = \mathrm{Span}\,\{\mathbf{v}_1, \mathbf{v}_2, \mathbf{v}_4\} = \mathrm{Span}\,\{\mathbf{z}_1, \mathbf{z}_2, \mathbf{z}_4\},$$

and $\{\mathbf{z}_1, \mathbf{z}_2, \mathbf{z}_4\}$ is an orthogonal linearly independent set. When we divide $\mathbf{z}_1, \mathbf{z}_2, \mathbf{z}_4$ by their lengths we obtain the orthonormal basis

$$\begin{bmatrix} \frac{1}{2} \\ \frac{1}{2} \\ \frac{1}{2} \\ \frac{1}{2} \end{bmatrix}, \begin{bmatrix} \frac{1}{2} \\ -\frac{1}{2} \\ \frac{1}{2} \\ -\frac{1}{2} \end{bmatrix}, \begin{bmatrix} \frac{1}{2} \\ -\frac{1}{2} \\ -\frac{1}{2} \\ \frac{1}{2} \end{bmatrix}$$

for $\mathrm{Span}\,\{\mathbf{v}_1, \mathbf{v}_2, \mathbf{v}_3, \mathbf{v}_4\} = \mathrm{Span}\,\{\mathbf{v}_1, \mathbf{v}_2, \mathbf{v}_4\}$.   □

When we have an orthonormal basis for W it is easy to compute the orthogonal projection of a vector $\mathbf{u}$ onto W.

Proposition 8.3.6. *Let W have orthogonal basis $\{\mathbf{w}_1, \ldots, \mathbf{w}_k\}$. Introduce the matrix*

$$P_W = \sum_{i=1}^{k} \frac{1}{\mathbf{w}_i^* \mathbf{w}_i} \mathbf{w}_i \mathbf{w}_i^*.$$

Then the orthogonal projection $\mathbf{w}_0$ of $\mathbf{u}$ onto W is given by

$$\mathbf{w}_0 = P_W \mathbf{u} = \frac{\langle \mathbf{u}, \mathbf{w}_1 \rangle}{\langle \mathbf{w}_1, \mathbf{w}_1 \rangle} \mathbf{w}_1 + \cdots + \frac{\langle \mathbf{u}, \mathbf{w}_k \rangle}{\langle \mathbf{w}_k, \mathbf{w}_k \rangle} \mathbf{w}_k.$$

Proof. We have that the Gram matrix is a diagonal matrix with diagonal entries $\langle \mathbf{w}_j, \mathbf{w}_j \rangle$, so applying the algorithm from the previous section we find $\mathbf{c} = (\frac{\langle \mathbf{u}, \mathbf{w}_i \rangle}{\langle \mathbf{w}_i, \mathbf{w}_i \rangle})_{i=1}^{k}$. Thus $\mathbf{w}_0 = \sum_{i=j}^{k} c_j \mathbf{w}_j = \frac{\langle \mathbf{u}, \mathbf{w}_1 \rangle}{\langle \mathbf{w}_1, \mathbf{w}_1 \rangle} \mathbf{w}_1 + \cdots + \frac{\langle \mathbf{u}, \mathbf{w}_k \rangle}{\langle \mathbf{w}_k, \mathbf{w}_k \rangle} \mathbf{w}_k.$ □

Using this notation, the iteration in the Gram–Schmidt process can be summarized as

$$\mathbf{z}_k = \mathbf{v}_k - P_{\text{Span}\{\mathbf{z}_1, \ldots, \mathbf{z}_{k-1}\}} \mathbf{v}_k.$$

Example 8.3.7. Find the distance of $\mathbf{u}$ to the subspace $W = \text{Span}\{\mathbf{v}_1, \mathbf{v}_2\}$, where

$$\mathbf{u} = \begin{bmatrix} -1 \\ 2 \\ 2 \end{bmatrix}, \mathbf{v}_1 = \begin{bmatrix} 2 \\ 0 \\ -2 \end{bmatrix}, \mathbf{v}_2 = \begin{bmatrix} 1 \\ 2 \\ 3 \end{bmatrix}.$$

In Example 8.3.4 we found

$$\left\{ \begin{bmatrix} 2 \\ 0 \\ -2 \end{bmatrix}, \begin{bmatrix} 2 \\ 2 \\ 2 \end{bmatrix} \right\} =: \{\mathbf{w}_1, \mathbf{w}_2\}$$

to be an orthogonal basis for W. Thus

$$\mathbf{w}_0 = P_W \mathbf{u} = \frac{\langle \mathbf{u}, \mathbf{w}_1 \rangle}{\langle \mathbf{w}_1, \mathbf{w}_1 \rangle} \mathbf{w}_1 + \frac{\langle \mathbf{u}, \mathbf{w}_2 \rangle}{\langle \mathbf{w}_2, \mathbf{w}_2 \rangle} \mathbf{w}_2 = \frac{-6}{8} \begin{bmatrix} 2 \\ 0 \\ -2 \end{bmatrix} + \frac{6}{12} \begin{bmatrix} 2 \\ 2 \\ 2 \end{bmatrix} = \begin{bmatrix} -\frac{1}{2} \\ 1 \\ \frac{5}{2} \end{bmatrix},$$

and we find

$$\text{distance}(\mathbf{u}, W) = \|\mathbf{u} - \mathbf{w}_0\| = \left\| \begin{bmatrix} -\frac{1}{2} \\ 1 \\ -\frac{1}{2} \end{bmatrix} \right\| = \frac{\sqrt{6}}{2}.$$

□

Another reason why it is easy to work with an orthonormal basis, is that it is easy to find the coordinates of a vector with respect to an orthonormal basis.

Lemma 8.3.8. *Let* $\mathcal{B} = \{\mathbf{v}_1, \ldots, \mathbf{v}_n\}$ *be an orthonormal basis of a subspace* W. *Let* $\mathbf{x} \in W$. *Then*

$$[\mathbf{x}]_{\mathcal{B}} = \begin{bmatrix} \langle \mathbf{x}, \mathbf{v}_1 \rangle \\ \vdots \\ \langle \mathbf{x}, \mathbf{v}_n \rangle \end{bmatrix}.$$

Proof. Let $\mathbf{x} = \sum_{i=1}^{n} c_i \mathbf{v}_i$. Then $\langle \mathbf{x}, \mathbf{v}_j \rangle = \sum_{i=1}^{n} c_i \langle \mathbf{v}_i, \mathbf{v}_j \rangle = c_j$, proving the lemma. $\qquad\square$

8.4 Isometries, Unitary Matrices and QR Factorization

When $\{\mathbf{w}_1, \ldots, \mathbf{w}_k\}$ is an orthonormal set in $\mathbb{F}^n$, then its Gram matrix equals the identity:

$$G = \begin{bmatrix} \mathbf{w}_1^* \mathbf{w}_1 & \cdots & \mathbf{w}_1^* \mathbf{w}_k \\ \vdots & & \vdots \\ \mathbf{w}_1^* \mathbf{w}_k & \cdots & \mathbf{w}_k^* \mathbf{w}_1 \end{bmatrix} = \begin{bmatrix} \mathbf{w}_1^* \\ \vdots \\ \mathbf{w}_k^* \end{bmatrix} \begin{bmatrix} \mathbf{w}_1 & \cdots & \mathbf{w}_k \end{bmatrix} = I_k.$$

When $k = n$ this orthonormal set is in fact a basis of $\mathbb{F}^n$, and $A = \begin{bmatrix} \mathbf{w}_1 & \cdots & \mathbf{w}_k \end{bmatrix}$ is invertible with inverse equal to A^*. This leads to the following definition.

Definition 8.4.1. *We call a matrix* $A \in \mathbb{F}^{n \times k}$ *an* **isometry** *if* $A^* A = I_k$. *We call a matrix* $A \in \mathbb{F}^{n \times n}$ **unitary** *if both* A *and* A^* *are isometries. That is,* A *is unitary if and only if* $A^* A = I_n = AA^*$.

The above observation gives the following.

Lemma 8.4.2. *A matrix* $A \in \mathbb{F}^{n \times k}$ *is an isometry if and only its columns form an orthonormal set. A matrix* $A \in \mathbb{F}^{n \times n}$ *is unitary if and only its columns form an orthonormal basis of* $\mathbb{F}^n$.

It is easy to check the following.

Lemma 8.4.3. *If matrices $A \in \mathbb{F}^{n \times k}$ and $B \in \mathbb{F}^{k \times m}$ are isometries, then so is AB. Also, if $A \in \mathbb{F}^{n \times n}$ and $B \in \mathbb{F}^{n \times n}$ are unitary, then so are A^* and AB.*

Proof. For the first statement we observe that if A and B are isometries, then $(AB)^*AB = B^*A^*AB = B^*I_kB = B^*B = I_m$. This shows that AB is an isometry.

If A and B are unitaries, then A^* and B^* are unitary as well. But then $(AB)^*AB = I_n = AB(AB)^*$ follows easily, and thus AB is unitary. $\quad\square$

Proposition 8.4.4. *If A is an isometry, then $\langle A\mathbf{u}, A\mathbf{v} \rangle = \langle \mathbf{u}, \mathbf{v} \rangle$ for all vectors $\mathbf{u}, \mathbf{v}$. In particular, $\|A\mathbf{u}\| = \|\mathbf{u}\|$ for all vectors $\mathbf{u}$.*

Thus multiplying a vector with an isometry does not change the length. It is this property that gave isometries their name.

Proof. $\langle A\mathbf{u}, A\mathbf{v} \rangle = (A\mathbf{v})^* A\mathbf{u} = \mathbf{v}^* A^* A\mathbf{u} = \mathbf{v}^* I_k \mathbf{u} = \langle \mathbf{u}, \mathbf{v} \rangle$. $\quad\square$

It is easy to check that

$$\frac{1}{\sqrt{2}} \begin{bmatrix} 1 & 1 \\ i & -i \end{bmatrix}, \frac{1}{\sqrt{3}} \begin{bmatrix} 1 & 1 & 1 \\ 1 & e^{\frac{2i\pi}{3}} & e^{\frac{4i\pi}{3}} \\ 1 & e^{\frac{4i\pi}{3}} & e^{\frac{8i\pi}{3}} \end{bmatrix}, \frac{1}{2} \begin{bmatrix} 1 & 1 & 1 & 1 \\ 1 & -1 & 1 & -1 \\ 1 & 1 & -1 & -1 \\ 1 & -1 & -1 & 1 \end{bmatrix},$$

are examples of unitary matrices. When we take a unitary matrix and remove some of its columns, then we obtain an isometry. For instance,

$$\frac{1}{\sqrt{2}} \begin{bmatrix} 1 \\ -i \end{bmatrix}, \frac{1}{\sqrt{3}} \begin{bmatrix} 1 & 1 \\ 1 & e^{\frac{4i\pi}{3}} \\ 1 & e^{\frac{8i\pi}{3}} \end{bmatrix}, \frac{1}{2} \begin{bmatrix} 1 & 1 \\ 1 & -1 \\ -1 & -1 \\ -1 & 1 \end{bmatrix},$$

are examples of isometries.

From the Gram–Schmidt process we can deduce the following.

Theorem 8.4.5. *(QR factorization) Let $A \in \mathbb{F}^{m \times n}$ with $m \geq n$. Then there exists an isometry $Q \in \mathbb{F}^{m \times n}$ and an upper triangular matrix $R \in \mathbb{F}^{n \times n}$ with nonnegative entries on the diagonal, so that*

$$A = QR.$$

If A has rank equal to n, then the diagonal entries of R are positive, and R is invertible. If $m = n$, then Q is unitary.

Proof. First we consider the case when $\text{rank} A = n$. Let $\mathbf{v}_1, \ldots, \mathbf{v}_n$ denote the columns of A, and let $\mathbf{z}_1, \ldots, \mathbf{z}_n$ denote the resulting vectors when we apply the Gram–Schmidt process to $\mathbf{v}_1, \ldots, \mathbf{v}_n$ as in Theorem 8.3.2. Let now Q be the matrix with columns $\frac{\mathbf{z}_1}{\|\mathbf{z}_1\|}, \ldots, \frac{\mathbf{z}_n}{\|\mathbf{z}_n\|}$. Then $Q^*Q = I_n$ as the columns of Q are orthonormal. Moreover, we have that

$$\mathbf{v}_k = \|\mathbf{z}_k\| \frac{\mathbf{z}_k}{\|\mathbf{z}_k\|} + \sum_{j=1}^{k-1} r_{kj} \mathbf{z}_j,$$

for some $r_{kj} \in \mathbb{F}$, $k > j$. Putting $r_{kk} = \|\mathbf{z}_k\|$, and $r_{kj} = 0$, $k < j$, and letting $R = (r_{kj})_{k,j=1}^n$, we get the desired upper triangular matrix R yielding $A = QR$.

When rank $< n$, apply the Gram–Schmidt process with those columns of A that do not lie in the span of the preceding columns. Place the vectors $\frac{\mathbf{z}}{\|\mathbf{z}\|}$ that are found in this way in the corresponding columns of Q. Next, one can fill up the remaining columns of Q with any vectors making the matrix an isometry. The upper triangular entries in R are obtained from writing the columns of A as linear combinations of the $\frac{\mathbf{z}}{\|\mathbf{z}\|}$'s found in the process above. $\square$

Example 8.4.6. Find a QR factorization of $\begin{bmatrix} 1 & 1 & 0 \\ 1 & 0 & 1 \\ 0 & 1 & 1 \end{bmatrix}$.

Applying the Gram–Schmidt process to the columns of A we obtain,

$$\mathbf{z}_1 = \begin{bmatrix} 1 \\ 1 \\ 0 \end{bmatrix},$$

$$\mathbf{z}_2 = \begin{bmatrix} 1 \\ 0 \\ 1 \end{bmatrix} - \frac{1}{2} \begin{bmatrix} 1 \\ 1 \\ 0 \end{bmatrix} = \begin{bmatrix} \frac{1}{2} \\ -\frac{1}{2} \\ 1 \end{bmatrix},$$

$$\mathbf{z}_3 = \begin{bmatrix} 0 \\ 1 \\ 1 \end{bmatrix} - \frac{1}{2} \begin{bmatrix} 1 \\ 1 \\ 0 \end{bmatrix} - \frac{\frac{1}{2}}{\frac{6}{4}} \begin{bmatrix} \frac{1}{2} \\ -\frac{1}{2} \\ 1 \end{bmatrix} = \begin{bmatrix} -\frac{2}{3} \\ \frac{2}{3} \\ \frac{2}{3} \end{bmatrix}. \tag{8.11}$$

Dividing by their lengths we obtain the columns of Q and thus

$$Q = \begin{bmatrix} \frac{\sqrt{2}}{2} & \frac{1}{\sqrt{6}} & -\frac{1}{\sqrt{3}} \\ \frac{\sqrt{2}}{2} & -\frac{1}{\sqrt{6}} & \frac{1}{\sqrt{3}} \\ 0 & \frac{\sqrt{6}}{3} & \frac{1}{\sqrt{3}} \end{bmatrix}.$$

To compute R, we multiply the equation $A = QR$ with Q^* on the left, and obtain $Q^*A = Q^*QR = I_kR$, and thus

$$R = Q^*A = \begin{bmatrix} \sqrt{2} & \frac{\sqrt{2}}{2} & \frac{\sqrt{2}}{2} \\ 0 & \frac{\sqrt{6}}{2} & \frac{1}{\sqrt{6}} \\ 0 & 0 & \frac{2\sqrt{3}}{3} \end{bmatrix}.$$

$\square$

Let us illustrate the QR factorization on an example where the columns of A are linearly dependent.

Example 8.4.7. Let $A = \begin{bmatrix} 1 & 0 & 1 & 2 \\ 1 & -2 & 0 & -2 \\ 1 & 0 & 1 & 0 \\ 1 & -2 & 0 & 0 \end{bmatrix}$. Applying the Gram–Schmidt

process we obtain,

$$\mathbf{z}_1 = \begin{bmatrix} 1 \\ 1 \\ 1 \\ 1 \end{bmatrix},$$

$$\mathbf{z}_2 = \begin{bmatrix} 0 \\ -2 \\ 0 \\ -2 \end{bmatrix} - \frac{-4}{4}\begin{bmatrix} 1 \\ 1 \\ 1 \\ 1 \end{bmatrix} = \begin{bmatrix} 1 \\ -1 \\ 1 \\ -1 \end{bmatrix},$$

$$\mathbf{z}_3 = \begin{bmatrix} 1 \\ 0 \\ 1 \\ 0 \end{bmatrix} - \frac{2}{4}\begin{bmatrix} 1 \\ 1 \\ 1 \\ 1 \end{bmatrix} - \frac{2}{4}\begin{bmatrix} 1 \\ -1 \\ 1 \\ -1 \end{bmatrix} = \begin{bmatrix} 0 \\ 0 \\ 0 \\ 0 \end{bmatrix}. \tag{8.12}$$

We thus notice that the third column of A is a linear combination of the first two columns of A, so we continue to compute $\mathbf{z}_4$ without using $\mathbf{z}_3$:

$$\mathbf{z}_4 = \begin{bmatrix} 2 \\ -2 \\ 0 \\ 0 \end{bmatrix} - 0\begin{bmatrix} 1 \\ 1 \\ 1 \\ 1 \end{bmatrix} - \frac{4}{4}\begin{bmatrix} 1 \\ -1 \\ 1 \\ -1 \end{bmatrix} = \begin{bmatrix} 1 \\ -1 \\ -1 \\ 1 \end{bmatrix}.$$

Dividing $\mathbf{z}_1, \mathbf{z}_2, \mathbf{z}_4$ by their respective lengths, and putting them in the matrix Q, we get

$$Q = \begin{bmatrix} \frac{1}{2} & \frac{1}{2} & ? & \frac{1}{2} \\ \frac{1}{2} & -\frac{1}{2} & ? & -\frac{1}{2} \\ \frac{1}{2} & \frac{1}{2} & ? & -\frac{1}{2} \\ \frac{1}{2} & -\frac{1}{2} & ? & \frac{1}{2} \end{bmatrix},$$

where it remains to fill in the third column of Q. To make the columns of

Q orthonormal, we choose the third column to be a unit vector in $\{\mathbf{z}_1, \mathbf{z}_2, \mathbf{z}_4\}^{\perp}$. Here we choose $\begin{bmatrix} \frac{1}{2} & \frac{1}{2} & -\frac{1}{2} & -\frac{1}{2} \end{bmatrix}^*$, so we get

$$Q = \begin{bmatrix} \frac{1}{2} & \frac{1}{2} & \frac{1}{2} & \frac{1}{2} \\ \frac{1}{2} & -\frac{1}{2} & \frac{1}{2} & -\frac{1}{2} \\ \frac{1}{2} & \frac{1}{2} & -\frac{1}{2} & -\frac{1}{2} \\ \frac{1}{2} & -\frac{1}{2} & -\frac{1}{2} & \frac{1}{2} \end{bmatrix}.$$

To compute R, we multiply the equation $A = QR$ with Q^* on the left, and obtain $Q^*A = Q^*QR = I_kR$, and thus

$$R = Q^*A = \begin{bmatrix} -2 & 2 & -1 & 0 \\ 0 & 2 & 1 & 2 \\ 0 & 0 & 0 & 0 \\ 0 & 0 & 0 & 2 \end{bmatrix}.$$

□

Let us give a pseudo code for QR factorization.

Algorithm 4 QR factorization

1: **procedure** QR(A)   ▷ Finds QR factorization of $m \times n$ matrix A
2:     $k \leftarrow 1, Q \leftarrow 0_{m\times n}, R \leftarrow 0_{n\times n}$
3:     **while** $k \leq n$ **do**
4:         $\mathbf{q}_k \leftarrow \mathbf{a_k} - \sum_{s=1}^{j-1} \langle \mathbf{a_k}, \mathbf{q}_s \rangle \mathbf{q}_s$
5:         **if** $\mathbf{q}_k \neq 0$ **then** $\mathbf{q}_k \leftarrow \mathbf{q}_k/\|\mathbf{q}_k\|$
6:         $k \leftarrow k + 1$
7:     **end**
8:     $R \leftarrow Q^*A$
9:     Replace zero columns of Q to make its columns an orthonormal set
10:     **return** Q, R ▷ $A = QR$ is a QR factorization

To end this section, we will do an example with a complex matrix.

Example 8.4.8. Find a QR factorization of $\begin{bmatrix} 1 & 2 & i \\ i & -1+i & 1 \\ i & 1+i & -1 \\ 1 & 0 & -i \end{bmatrix}$.

Applying the Gram–Schmidt process to the columns of A we obtain,

$$\mathbf{z}_1 = \begin{bmatrix} 1 \\ i \\ i \\ 1 \end{bmatrix},$$

$$\mathbf{z}_2 = \begin{bmatrix} 2 \\ -1+i \\ 1+i \\ 0 \end{bmatrix} - \frac{2 + (-1+i)(-i) + (1+i)(-i) + 0 \cdot 1}{4} \begin{bmatrix} 1 \\ i \\ i \\ 1 \end{bmatrix} = \begin{bmatrix} 1 \\ -1 \\ 1 \\ -1 \end{bmatrix},$$

$$\mathbf{z}_3 = \begin{bmatrix} i \\ 1 \\ -1 \\ -i \end{bmatrix} - \frac{0}{4} \begin{bmatrix} 1 \\ i \\ i \\ 1 \end{bmatrix} - \frac{-2+2i}{4} \begin{bmatrix} 1 \\ -1 \\ 1 \\ -1 \end{bmatrix} = \begin{bmatrix} \frac{1}{2} + \frac{1}{2}i \\ \frac{1}{2} + \frac{1}{2}i \\ -\frac{1}{2} - \frac{1}{2}i \\ -\frac{1}{2} - \frac{1}{2}i \end{bmatrix}. \tag{8.13}$$

Dividing by their lengths we obtain the columns of Q and thus

$$Q = \begin{bmatrix} \frac{1}{2} & \frac{1}{2} & \frac{1+i}{\sqrt{8}} \\ \frac{1}{2}i & -\frac{1}{2} & \frac{1+i}{\sqrt{8}} \\ \frac{1}{2}i & \frac{1}{2} & \frac{-1-i}{\sqrt{8}} \\ \frac{1}{2} & -\frac{1}{2} & \frac{-1-i}{\sqrt{8}} \end{bmatrix}.$$

Finally, $R = Q^*A$ and thus

$$R = \begin{bmatrix} \frac{1}{2} & -\frac{1}{2}i & -\frac{1}{2}i & \frac{1}{2} \\ \frac{1}{2} & -\frac{1}{2} & \frac{1}{2} & -\frac{1}{2} \\ \frac{1-i}{\sqrt{8}} & \frac{1-i}{\sqrt{8}} & \frac{-1+i}{\sqrt{8}} & \frac{-1+i}{\sqrt{8}} \end{bmatrix} \begin{bmatrix} 1 & 2 & i \\ i & -1+i & 1 \\ i & 1+i & -1 \\ 1 & 0 & -i \end{bmatrix} = \begin{bmatrix} 2 & 2 & 0 \\ 0 & 2 & -1+i \\ 0 & 0 & \sqrt{2} \end{bmatrix}.$$

□

8.5 Least Squares Solution and Curve Fitting

When the equation $A\mathbf{x} = \mathbf{b}$ does not have a solution, one may be interested in finding an $\mathbf{x}$ so that $\|A\mathbf{x} - \mathbf{b}\|$ is minimal. Such an $\mathbf{x}$ is called a **least squares solution** to $A\mathbf{x} = \mathbf{b}$.

To find a least squares solution, we need to find a vector $A\mathbf{x} \in \text{Col } A$ that is closest to $\mathbf{b}$. In other words, we would like $\mathbf{x}$ so that $A\mathbf{x}$ corresponds to the orthogonal projection of $\mathbf{b}$ onto Col A. In other words, $A\mathbf{x} - \mathbf{b}$ needs to be orthogonal to Col$A$. The vector space that is orthogonal to Col$A$ is exactly Nul A^*. Thus we want $A\mathbf{x} - \mathbf{b} \in \text{Nul} A^*$. Thus we need

$$A^*(A\mathbf{x} - \mathbf{b}) = \mathbf{0} \iff \mathbf{x} = (A^*A)^{-1}A^*\mathbf{b},$$

where we assumed that A has linearly independent columns (implying that A^*A is invertible by Lemma 8.2.10).

Theorem 8.5.1. *Let* $A \in \mathbb{F}^{m \times n}$ *with* rank $A = n$. *Then the least squares solution to the equation* $A\mathbf{x} = \mathbf{b}$ *is given by*

$$\mathbf{x} = (A^*A)^{-1}A^*\mathbf{b}.$$

When $A = QR$ *with* Q *an isometry, then the least squares solution is given by* $\mathbf{x} = R^{-1}Q^*\mathbf{b}$.

Example 8.5.2. Find the least squares solution to

$$\begin{bmatrix} 1 & 2 \\ -1 & 4 \\ 1 & 2 \end{bmatrix} \mathbf{x} = \begin{bmatrix} 3 \\ -1 \\ 5 \end{bmatrix}.$$

We find

$$\mathbf{x} = \begin{bmatrix} 3 & 0 \\ 0 & 24 \end{bmatrix}^{-1} \begin{bmatrix} 1 & -1 & 1 \\ 2 & 4 & 2 \end{bmatrix} \begin{bmatrix} 3 \\ -1 \\ 5 \end{bmatrix} = \begin{bmatrix} 3 \\ \frac{1}{2} \end{bmatrix}.$$

$\square$

One important application where we use least squares solutions is when we do curve fitting. For instance, fitting a (regression) line between several points.

Example 8.5.3. Fit a line $y = mx + b$ through the points $(1,3)$, $(2,1)$, $(3,4)$ and $(4,3)$.

We set up the equations

$$3 = m + b, \quad 1 = 2m + b, \quad 4 = 3m + b, \quad 3 = 4m + b.$$

Clearly it is impossible to solve this system as no line fits exactly through these points. Thus we will use the least squares solution. Writing the equations in matrix form, we get

$$\begin{bmatrix} 1 & 1 \\ 2 & 1 \\ 3 & 1 \\ 4 & 1 \end{bmatrix} \begin{bmatrix} m \\ b \end{bmatrix} = \begin{bmatrix} 3 \\ 1 \\ 4 \\ 3 \end{bmatrix}.$$

The least squares solution to this equation is $\begin{bmatrix} m \\ b \end{bmatrix} = \begin{bmatrix} \frac{3}{10} \\ 2 \end{bmatrix}$. So the least squares

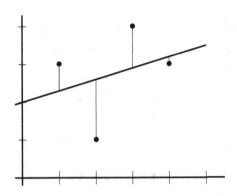

Figure 8.5: Least squares fit.

fit is $y = \frac{3}{10}x + 2$. This line minimizes the sum of the squares of the lengths of the vertical lines between the points and the line. □

This curve fitting can be applied to any equation of the form

$$y = c_1 f_1(x) + \cdots + c_k f_k(x),$$

where $f_1(x), \ldots, f_k(x)$ are some functions. In the above example, we used $f_1(x) = x$, $f_2(x) \equiv 1$, so that we get the equation of a line. If we want to fit a parabola through a set of points, we will use $f_1(x) = x^2$, $f_2(x) = x$, $f_3(x) \equiv 1$. Here is such an example.

Example 8.5.4. Fit a parabola through the points $(1,3)$, $(2,0)$, $(3,4)$ and $(4,2)$.

The general form of a parabola is $y = ax^2 + bx + c$. Set up the equations:

$$3 = a + b + c, \quad 0 = 4a + 2b + c, \quad 4 = 9a + 3b + c, \quad 2 = 16a + 4b + c.$$

Writing this in matrix form, we get

$$\begin{bmatrix} 1 & 1 & 1 \\ 4 & 2 & 1 \\ 9 & 3 & 1 \\ 16 & 4 & 1 \end{bmatrix} \begin{bmatrix} a \\ b \\ c \end{bmatrix} = \begin{bmatrix} 3 \\ 0 \\ 4 \\ 2 \end{bmatrix}.$$

The least squares solution is $\begin{bmatrix} a \\ b \\ c \end{bmatrix} = \begin{bmatrix} 0.25 \\ -1.15 \\ 3.25 \end{bmatrix}$. So the least squares parabola

fit is $y = 0.25x^2 - 1.15x + 3.25$. This parabola minimizes the sum of the squares of the lengths of the vertical lines between the points and the parabola.

□

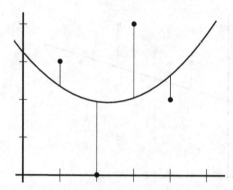

Figure 8.6: Parabola least squares fit.

8.6 Real Symmetric and Hermitian Matrices

Let $\mathbb{F} = \mathbb{R}$ or $\mathbb{C}$. We call a matrix $A \in \mathbb{F}^{n \times n}$ **Hermitian** if $A^* = A$. For instance,

$$\begin{bmatrix} 2 & 1-3i \\ 1+3i & 5 \end{bmatrix} \text{ and } \begin{bmatrix} 1 & 3 & 7 \\ 3 & -10 & 9 \\ 7 & 9 & -2 \end{bmatrix}$$

are Hermitian. If A is real and Hermitian then A satisfies $A = A^T$ and is also referred to as a **symmetric** matrix.

Proposition 8.6.1. *Let A be Hermitian. Then A has only real eigenvalues. Moreover, eigenvectors belonging to different eigenvalues are orthogonal.*

Proof. Suppose that $A\mathbf{v}_1 = \lambda_1\mathbf{v}_1$, $\mathbf{v}_1 \neq \mathbf{0}$. Then $\mathbf{v}_1^* A\mathbf{v}_1 = \mathbf{v}_1^* \lambda_1\mathbf{v}_1 = \lambda_1\mathbf{v}_1^*\mathbf{v}_1$. Also $(\mathbf{v}_1^* A\mathbf{v}_1)^* = \mathbf{v}_1^* A^*\mathbf{v}_1^{**} = \mathbf{v}_1^* A\mathbf{v}_1$. Thus this scalar is real. But then $\lambda_1 = \frac{\mathbf{v}_1^* A\mathbf{v}_1}{\mathbf{v}_1^*\mathbf{v}_1}$ is real as well.

Next, suppose that in addition $A\mathbf{v}_2 = \lambda_2\mathbf{v}_2$, $\mathbf{v}_2 \neq \mathbf{0}$, with $\lambda_1 \neq \lambda_2$. Then

$$\mathbf{v}_2^*(\lambda_1\mathbf{v}_1) = \mathbf{v}_2^*(A\mathbf{v}_1) = \mathbf{v}_2^* A^*\mathbf{v}_1 = (A\mathbf{v}_2)^*\mathbf{v}_1 = (\lambda_2\mathbf{v}_2)^*\mathbf{v}_1$$

and thus $(\lambda_1 - \lambda_2)(\mathbf{v}_2^*\mathbf{v}_1) = 0$. Since $\lambda_1 - \lambda_2 \neq 0$, we get $\mathbf{v}_2^*\mathbf{v}_1 = 0$. Thus $\mathbf{v}_1 \perp \mathbf{v}_2$. □

In fact we have the following stronger result.

Theorem 8.6.2. *A square matrix A is Hermitian if and only if $A = UDU^*$ where D is a real diagonal matrix and U is unitary.*
In other words, $A = A^*$ if and only if A has real eigenvalues and corresponding eigenvectors which form an orthonormal basis of $\mathbb{F}^n$.

Proof. We prove this by induction on the size n. When $n = 1$, then $A = [a]$ is Hermitian if and only if $a = \bar{a}$. Thus a is real, and the result follows (with $D = [a], U = [1]$).

Next, suppose that the theorem holds for matrices of size $(n-1) \times (n-1)$, and let A be $n \times n$. Let $\mathbf{u}_1$ be an eigenvector of A, which we can take to be of length 1 (otherwise, replace $\mathbf{u}_1$ by $\mathbf{u}_1/\|\mathbf{u}_1\|$), with eigenvalue λ. Let U be a unitary matrix with first column $\mathbf{u}_1$ (one can make U by taking a basis $\{\mathbf{v}_1, \ldots, \mathbf{v}_n\}$ of $\mathbb{F}^n$ with $\mathbf{v}_1 = \mathbf{u}_1$, and performing Gram-Schmidt on this basis to get an orthonormal basis that will be the columns of U). Now we get that $A\mathbf{u}_1 = \lambda\mathbf{u}_1$ and thus $\mathbf{u}_i^* A\mathbf{u}_1 = \lambda\mathbf{u}_i^*\mathbf{u}_1$, which equals λ when $i = 1$ and 0 otherwise. Then

$$U^*AU = \begin{bmatrix} \mathbf{u}_1^* \\ \vdots \\ \mathbf{u}_n^* \end{bmatrix} \begin{bmatrix} A\mathbf{u}_1 & A\mathbf{u}_2 & \cdots & A\mathbf{u}_n \end{bmatrix} = \begin{bmatrix} \lambda & * & \cdots & * \\ 0 & * & \cdots & * \\ \vdots & \vdots & & \vdots \\ 0 & * & \cdots & * \end{bmatrix}.$$

Since $(U^*AU)^* = U^*A^*U^{**} = U^*AU$ is Hermitian, we must have 0's in the entries $(1, 2), \ldots, (1, n)$ as well. In fact,

$$U^*AU = \begin{bmatrix} \lambda & 0 \\ 0 & B \end{bmatrix},$$

with $\lambda = \bar{\lambda}$ and $B = B^*$. In particular $\lambda \in \mathbb{R}$. By the induction assumption $B = V\hat{D}V^*$ for some unitary V and real diagonal $\hat{D}$. Then

$$A = U \begin{bmatrix} \lambda & 0 \\ 0 & V\hat{D}V^* \end{bmatrix} U^* = U \begin{bmatrix} 1 & 0 \\ 0 & V \end{bmatrix} \begin{bmatrix} \lambda & 0 \\ 0 & \hat{D} \end{bmatrix} \begin{bmatrix} 1 & 0 \\ 0 & V^* \end{bmatrix} U^*.$$

Since $W := U \begin{bmatrix} 1 & 0 \\ 0 & V \end{bmatrix}$ is a product of unitaries, we get that W is unitary. Also $D = \begin{bmatrix} \lambda & 0 \\ 0 & \hat{D} \end{bmatrix}$ is a real diagonal matrix. Thus A has the required factorization $A = WDW^*$. $\qquad\square$

Example 8.6.3. Write $A = \begin{bmatrix} 2 & -5 \\ -5 & 2 \end{bmatrix} = UDU^*$, with D diagonal and U unitary.

We find $p_A(t) = \det(A - tI) = (t-2)^2 - 25 = 0$ if and only if $t - 2 = \pm 5$.

Thus A has eigenvalues $-3, 7$. Next

$$A - (-3)I = \begin{bmatrix} 5 & -5 \\ -5 & 5 \end{bmatrix} \rightarrow \begin{bmatrix} 1 & -1 \\ 0 & 0 \end{bmatrix},$$

and thus $\begin{bmatrix} 1 & 1 \end{bmatrix}^T$ is an eigenvector at -3. At eigenvalue 7 we get

$$A - 7I = \begin{bmatrix} -5 & -5 \\ -5 & -5 \end{bmatrix} \rightarrow \begin{bmatrix} 1 & 1 \\ 0 & 0 \end{bmatrix},$$

and thus $\begin{bmatrix} -1 & 1 \end{bmatrix}^T$ is an eigenvector at 7. Dividing the eigenvectors by their length, we obtain

$$U = \begin{bmatrix} \frac{1}{\sqrt{2}} & \frac{-1}{\sqrt{2}} \\ \frac{1}{\sqrt{2}} & \frac{1}{\sqrt{2}} \end{bmatrix}, D = \begin{bmatrix} -3 & 0 \\ 0 & 7 \end{bmatrix}.$$

□

Example 8.6.4. Write $A = \begin{bmatrix} 3 & -2 & -1 \\ -2 & 3 & -1 \\ -1 & -1 & 2 \end{bmatrix} = UDU^*$, with D diagonal and U unitary.

We compute

$$-p_A(t) = \det(tI - A) = t^3 - 8t^2 + 15t = t(t-3)(t-5).$$

Thus A has eigenvalues $0, 3$ and 5. Next

$$A - 0I \rightarrow \begin{bmatrix} -1 & -1 & 2 \\ 0 & -5 & 5 \\ 0 & 5 & -5 \end{bmatrix} \rightarrow \begin{bmatrix} -1 & 0 & 1 \\ 0 & 1 & -1 \\ 0 & 0 & 0 \end{bmatrix},$$

and thus $\begin{bmatrix} 1 & 1 & 1 \end{bmatrix}^T$ is an eigenvector at 0. For $A - 3I$ we get

$$\begin{bmatrix} 0 & -2 & -1 \\ -2 & 0 & -1 \\ -1 & -1 & -1 \end{bmatrix} \rightarrow \begin{bmatrix} -1 & -1 & -1 \\ 0 & -2 & -1 \\ 0 & 2 & 1 \end{bmatrix} \rightarrow \begin{bmatrix} 1 & 0 & \frac{1}{2} \\ 0 & 1 & \frac{1}{2} \\ 0 & 0 & 0 \end{bmatrix},$$

and thus $\begin{bmatrix} -\frac{1}{2} & -\frac{1}{2} & 1 \end{bmatrix}^T$ is an eigenvector at 3. For $A - 5I$ we get

$$\begin{bmatrix} -2 & -2 & -1 \\ -2 & -2 & -1 \\ -1 & -1 & -3 \end{bmatrix} \rightarrow \begin{bmatrix} -1 & -1 & -3 \\ 0 & 0 & 5 \\ 0 & 0 & 0 \end{bmatrix} \rightarrow \begin{bmatrix} 1 & 1 & 0 \\ 0 & 0 & 1 \\ 0 & 0 & 0 \end{bmatrix},$$

and thus $\begin{bmatrix} -1 & 1 & 0 \end{bmatrix}^T$ is an eigenvector at 5. Dividing the eigenvectors by their length, we find

$$U = \begin{bmatrix} \frac{1}{\sqrt{3}} & -\frac{1}{\sqrt{6}} & -\frac{1}{\sqrt{2}} \\ \frac{1}{\sqrt{3}} & -\frac{1}{\sqrt{6}} & \frac{1}{\sqrt{2}} \\ \frac{1}{\sqrt{3}} & \frac{2}{\sqrt{6}} & 0 \end{bmatrix}, D = \begin{bmatrix} 0 & 0 & 0 \\ 0 & 3 & 0 \\ 0 & 0 & 5 \end{bmatrix}.$$

□

Definition 8.6.5. A $n \times n$ Hermitian matrix A is called **positive semidef-
inite** if $\langle A\mathbf{x}, \mathbf{x} \rangle \geq 0$ for all vectors $\mathbf{x} \in \mathbb{C}^n$. If in addition, $\langle A\mathbf{x}, \mathbf{x} \rangle > 0$ for
all vectors $\mathbf{0} \neq \mathbf{x} \in \mathbb{C}^n$, then A is called **positive definite**.

Theorem 8.6.6. *A Hermitian $n \times n$ matrix A is positive semidefinite if
and only if A has all nonnegative eigenvalues.*
*A Hermitian $n \times n$ matrix A is positive definite if and only if A has all
positive eigenvalues.*

Proof. Since A is Hermitian, it follows from Theorem 8.6.2 that $A = UDU^*$
with D a real diagonal matrix and U is unitary. The diagonal entries $\lambda_1, \dots, \lambda_n$
of D are the eigenvalues of A. Suppose that A only has positive eigenvalues. Let
$\mathbf{x} \in \mathbb{C}^n$. Let $u_1, \dots, u_n$ be the entries of the vector $U^*\mathbf{x}$; that is, $U^*\mathbf{x} = (u_i)_{i=1}^n$.
Then

$$\langle A\mathbf{x}, \mathbf{x} \rangle = \mathbf{x}^* A\mathbf{x} = \mathbf{x}^* UDU^*\mathbf{x} = (U^*\mathbf{x})^* D(U^*\mathbf{x}) = \sum_{i=1}^n \lambda_i |u_i|^2 \geq 0,$$

and thus A is positive semidefinite.

Conversely, assume that A is positive semidefinite. If λ is an eigenvalue of
A with eigenvector $\mathbf{u}$, then we have that $\langle A\mathbf{u}, \mathbf{u} \rangle \geq 0$, since A is positive
semidefinite. Now $0 \leq \langle A\mathbf{u}, \mathbf{u} \rangle = \langle \lambda\mathbf{u}, \mathbf{u} \rangle = \lambda \|\mathbf{u}\|^2$, implies that $\lambda \geq 0$ (since
$\|\mathbf{u}\|^2 > 0$).

The proof of the second statement is similar. One just needs to observe that if
$\lambda_1, \dots, \lambda_n$ are all positive and $u_1, \dots, u_n$ are not all zero, then $\sum_{i=1}^n \lambda_i |u_i|^2 >$
0. $\qquad \square$

Notice that the matrix in Example 8.6.4 is positive semidefinite, but not positive definite.

8.7 Singular Value Decomposition

The **singular value decomposition** gives a way to write a general (typically,
non-square) matrix A as the product $A = V\Sigma W^*$, where V and W are unitary
and Σ is a matrix of the same size as A with nonnegative entries in positions
(i, i) and zeros elsewhere. As before, we let $\mathbb{F}$ be $\mathbb{R}$ or $\mathbb{C}$.

Theorem 8.7.1. *Let* $A \in \mathbb{F}^{m \times n}$ *have rank* k. *Then there exist unitary matrices* $V \in \mathbb{F}^{m \times m}$, $W \in \mathbb{F}^{n \times n}$, *and a matrix* $\Sigma \subset \mathbb{F}^{m \times n}$ *of the form*

$$\Sigma = \begin{bmatrix} \sigma_1 & 0 & \cdots & 0 & \cdots & 0 \\ 0 & \sigma_2 & \cdots & 0 & \cdots & 0 \\ \vdots & \vdots & \ddots & \vdots & & \vdots \\ 0 & 0 & \cdots & \sigma_k & \cdots & 0 \\ \vdots & \vdots & & \vdots & \cdot 0 & \vdots \\ 0 & 0 & \cdots & 0 & \cdots & 0 \end{bmatrix}, \quad \sigma_1 \geq \sigma_2 \geq \ldots \geq \sigma_k > 0, \quad (8.14)$$

so that $A = V \Sigma W^*$.

One of the main applications of the singular value decomposition is that it gives an easy way to approximate a matrix with a low rank one. The advantage of a low rank matrix is that it requires less memory to store it (see Exercise 8.8.56 for details). Let us illustrate this on the digital black and white image in Figure 8.7, which is stored as a matrix containing grayscale pixel values. When we apply the singular value decomposition to this matrix and do an approximation with a low rank matrix, we get Figure 8.8 as the result.

Figure 8.7: The original image (of size 768×1024).

In Figures 8.9 and 8.10 we depict the singular value decomposition of an $m \times n$ matrix for the cases when the matrix is tall (when $m \geq n$) and when the matrix is wide (when $m \leq n$), respectively.

Proof. From Exercise 8.8.12(a) it follows that $k = \text{rank} A^* A$. We observe that $A^* A$ is positive semidefinite. Indeed,

$$\langle A^* A \mathbf{x}, \mathbf{x} \rangle = \mathbf{x}^* A^* A \mathbf{x} = (A\mathbf{x})^* (A\mathbf{x}) = \|A\mathbf{x}\|^2 \geq 0.$$

Figure 8.8: The image after approximation using svd at about 23% of storage.

Figure 8.9: Depiction of the singular value decomposition of a tall matrix.

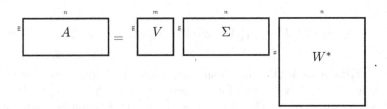

Figure 8.10: Depiction of the singular value decomposition of a wide matrix.

Thus there exists a $n \times n$ unitary W and a diagonal matrix $\Lambda = (\lambda_i)_{i=1}^n$, with $\lambda_1 \geq \cdots \geq \lambda_k > 0 = \lambda_{k+1} = \cdots = \lambda_n$, so that $A^*A = W\Lambda W^*$. Put $\sigma_j = \sqrt{\lambda_j}$, $j = 1, \ldots, k$, and write $W = \begin{bmatrix} \mathbf{w}_1 & \cdots & \mathbf{w}_n \end{bmatrix}$. Next, put $\mathbf{v}_j = \frac{1}{\sigma_j} A\mathbf{w}_j$, $j = 1, \ldots, k$, and let $\{\mathbf{v}_{k+1}, \ldots, \mathbf{v}_m\}$ be an orthonormal basis for KerA^*. Put $V = \begin{bmatrix} \mathbf{v}_1 & \cdots & \mathbf{v}_m \end{bmatrix}$. First, let us show that V is unitary. When $i, j \in \{1, \ldots, k\}$, then

$$\mathbf{v}_j^* \mathbf{v}_i = \frac{1}{\sigma_i \sigma_j} \mathbf{w}_j^* A^* A \mathbf{w}_i = \frac{1}{\sigma_i \sigma_j} \mathbf{w}_j^* W \Lambda W^* \mathbf{w}_i =$$

$$\frac{1}{\sigma_i \sigma_j} \mathbf{e}_j^* \Lambda \mathbf{e}_i = \begin{cases} 0 & \text{when } i \neq j, \\ \frac{\lambda_j}{\sigma_j^2} = 1 & \text{when } i = j. \end{cases}$$

Next, when $j \in \{1, \ldots, k\}$ and $i \in \{k+1, \ldots, m\}$, we get that $\mathbf{v}_j^* \mathbf{v}_i = \frac{1}{\sigma_j} \mathbf{w}_j A^* \mathbf{v}_i = 0$ as $\mathbf{v}_i \in $ Ker A^*. Similarly, $\mathbf{v}_j^* \mathbf{v}_i = 0$ when $i \in \{1, \ldots, k\}$ and $j \in \{k+1, \ldots, m\}$. Finally, we use that $\{\mathbf{v}_{k+1}, \ldots, \mathbf{v}_m\}$ is an orthonormal set, to conclude that $V^*V = I_m$.

It remains to show that $A = V\Sigma W^*$, or equivalently, $AW = V\Sigma$. The equality in the first k columns follows from the definition of $\mathbf{v}_j$, $j = 1, \ldots, k$. In columns $k+1, \ldots, n$ we have $\mathbf{0}$ on both sides, and thus $AW = V\Sigma$ follows. $\square$

The values σ_j are called the **singular values** of A, and they are uniquely determined by A. We also denote them by $\sigma_j(A)$. It is important to observe that

$$\sigma_j(A) = \sqrt{\lambda_j(A^*A)} = \sqrt{\lambda_j(AA^*)}, \ j = 1, \ldots, k(= \text{rank}A).$$

The first equality follows from the proof. For the second equality one can use that $AA^* = V(\Sigma\Sigma^*)V^*$ and that $\Sigma\Sigma^*$ is a $m \times m$ diagonal matrix with diagonal entries $\sigma_1(A)^2, \ldots, \sigma_k(A)^2, 0, \ldots, 0$.

Proposition 8.7.2. *Let $A \in \mathbb{F}^{m \times n}$, and let $\| \cdot \|$ be the Euclidean norm. Then*

$$\sigma_1(A) = \max_{\|\mathbf{x}\|=1} \|A\mathbf{x}\|. \tag{8.15}$$

In addition,

$$\sigma_1(A+B) \leq \sigma_1(A) + \sigma_1(B), \ \sigma_1(cA) = |c| \ \sigma_1(A). \tag{8.16}$$

Proof. Write $A = V\Sigma W^*$ in its singular value decomposition. For U unitary we have that $\|U\mathbf{v}\| = \|\mathbf{v}\|$ for all vectors $\mathbf{v}$. Thus $\|A\mathbf{x}\| = \|V\Sigma W^*\mathbf{x}\| = \|\Sigma W^*\mathbf{x}\|$. Let $\mathbf{u} = W^*\mathbf{x}$. Then $\|\mathbf{x}\| = \|W\mathbf{u}\| = \|\mathbf{u}\|$. Combining these observations, we have that

$$\max_{\|\mathbf{x}\|=1} \|A\mathbf{x}\| = \max_{\|\mathbf{u}\|=1} \|\Sigma\mathbf{u}\| = \max_{\|\mathbf{u}\|=1} \sqrt{\sigma_1^2|u_1|^2 + \cdots + \sigma_k^2|u_k|^2},$$

which is clearly bounded above by $\sqrt{\sigma_1^2 |u_1|^2 + \cdots + \sigma_1^2 |u_k|^2} \leq \sigma_1 \|\mathbf{u}\| = \sigma_1$. When $\mathbf{u} = \mathbf{e}_1$, then we get that $\|\Sigma \mathbf{u}\| = \sigma_1$. Thus $\max_{\|\mathbf{u}\|=1} \|\Sigma \mathbf{u}\| = \sigma_1$ follows.

Next, we have

$$\sigma_1(A+B) = \max_{\|\mathbf{x}\|=1} \|(A+B)\mathbf{x}\| \leq \max_{\|\mathbf{x}\|=1} \|A\mathbf{x}\| + \max_{\|\mathbf{x}\|=1} \|B\mathbf{x}\| = \sigma_1(A) + \sigma_1(B),$$

and

$$\sigma_1(cA) = \max_{\|\mathbf{x}\|=1} \|(cA)\mathbf{x}\| = \max_{\|\mathbf{x}\|=1} (|c| \|A\mathbf{x}\|) = |c| \, \sigma_1(A).$$

$\square$

As mentioned before, an important application of the singular value decomposition is low rank approximation of matrices.

Proposition 8.7.3. *Let $A \in \mathbb{F}^{m \times n}$ have singular value decomposition $A = V\Sigma W^*$ with Σ as in (8.14). Let $l \leq k$. Put $\hat{A} = V\hat{\Sigma}W^*$ with*

$$\hat{\Sigma} = \begin{bmatrix} \sigma_1 & 0 & \cdots & 0 & \cdots & 0 \\ 0 & \sigma_2 & \cdots & 0 & \cdots & 0 \\ \vdots & \vdots & \ddots & \vdots & & \vdots \\ 0 & 0 & \cdots & \sigma_l & \cdots & 0 \\ \vdots & \vdots & & \vdots & 0 & \vdots \\ 0 & 0 & \cdots & 0 & \cdots & 0 \end{bmatrix}. \tag{8.17}$$

Then rank $\hat{A} = l$, $\sigma_1(A - \hat{A}) = \sigma_{l+1}$, *and for any matrix B with* rank$B \leq l$ *we have* $\sigma_1(A - B) \geq \sigma_1(A - \hat{A})$.

Proof. Clearly rank $\hat{A} = l$, $\sigma_1(A - \hat{A}) = \sigma_{l+1}$. Next, let B with rank$B \leq l$. Put $C = V^* B W$. Then rank$C = $ rank$B \leq l$, and $\sigma_1(A - B) = \sigma_1(\Sigma - C)$. Notice that dim Ker $C \geq n - l$, and thus Ker $C \cap$ Span $\{\mathbf{e}_1, \ldots, \mathbf{e}_{l+1}\}$ has dimension ≥ 1. Thus we can find a $\mathbf{v} \in$ Ker $C \cap$ Span $\{\mathbf{e}_1, \ldots, \mathbf{e}_{l+1}\}$ with $\|\mathbf{v}\| = 1$. Then

$$\sigma_1(\Sigma - C) \geq \|(\Sigma - C)\mathbf{v}\| = \|\Sigma \mathbf{v}\| \geq \sigma_{l+1},$$

where in the last step we used that $\mathbf{v} \in$ Span $\{\mathbf{e}_1, \ldots, \mathbf{e}_{l+1}\}$. This proves the statement. $\square$

As mentioned, a low rank matrix requires less storage than a full rank matrix, so that could be one reason to approximate with a low rank matrix. Another reason is if you are trying to discern the major trends in your matrix. In other words, you might want to change your matrix to extract the important parts (achieved above by making some of the smaller singular values equal to 0). An example of where this general idea is used is in principal component analysis (PCA); see Exercise 8.8.57 for more details.

We end this section with an example where we compute the singular value decomposition of a matrix. For this it is useful to notice that if $A = V\Sigma W^*$, then $AA^* = V\Sigma\Sigma^* V^*$ and $A^*A = W\Sigma^*\Sigma W^*$. Thus the columns of V are eigenvectors of AA^*, and the diagonal elements σ_j^2 of the diagonal matrix $\Sigma\Sigma^*$ are the eigenvalues of AA^*. Thus the singular values can be found by computing the square roots of the nonzero eigenvalues of AA^*. Similarly, the columns of W are eigenvectors of A^*A, and the diagonal elements σ_j^2 of the diagonal matrix $\Sigma^*\Sigma$ are the nonzero eigenvalues of A^*A, as we have seen in the proof of Theorem 8.7.1.

Example 8.7.4. Let $A = \begin{bmatrix} 6 & 4 & 4 \\ 4 & 6 & -4 \end{bmatrix}$. Find the singular value decomposition of A.

Compute

$$AA^* = \begin{bmatrix} 68 & 32 \\ 32 & 68 \end{bmatrix},$$

which has eigenvalues 36 and 100. So the singular values of A are 6 and 10, and we get

$$\Sigma = \begin{bmatrix} 10 & 0 & 0 \\ 0 & 6 & 0 \end{bmatrix}.$$

To find V, we find unit eigenvectors of AA^*, giving

$$V = \begin{bmatrix} 1/\sqrt{2} & 1/\sqrt{2} \\ 1/\sqrt{2} & -1/\sqrt{2} \end{bmatrix}.$$

For W, observe that $V^*A = \Sigma W^*$. Writing $W = \begin{bmatrix} \mathbf{w}_1 & \mathbf{w}_2 & \mathbf{w}_3 \end{bmatrix}$, we get

$$\begin{bmatrix} 10/\sqrt{2} & 10/\sqrt{2} & 0 \\ 2/\sqrt{2} & -2/\sqrt{2} & 8/\sqrt{2} \end{bmatrix} = \begin{bmatrix} 10\mathbf{w}_1^* \\ 6\mathbf{w}_2^* \end{bmatrix}.$$

This yields $\mathbf{w}_1$ and $\mathbf{w}_2$. To find $\mathbf{w}_3$, we need to make sure that W is unitary, and thus $\mathbf{w}_3$ needs to be a unit vector orthogonal to $\mathbf{w}_1$ and $\mathbf{w}_2$. We find

$$W = \begin{bmatrix} 1/\sqrt{2} & 1/3\sqrt{2} & 2/3 \\ 1/\sqrt{2} & -1/3\sqrt{2} & -2/3 \\ 0 & 4/3\sqrt{2} & -1/3 \end{bmatrix}.$$

$\square$

If the matrix $A \in \mathbb{F}^{m \times n}$ has rank k, then the last $n - k$ columns in the matrix Σ are zero, as well as the last $m - k$ rows of $\Sigma \in \mathbb{F}^{m \times n}$. If we remove these zero rows and columns, and also remove the last $m - k$ columns of V

and the last $n - k$ rows of W^*, we obtain the so-called **compact singular value decomposition**. Equivalently, we say that $A = \hat{V}\hat{\Sigma}\hat{W}^*$ is the **compact singular value decomposition** if $\hat{V} \in \mathbb{F}^{m \times k}, \hat{W} \in \mathbb{F}^{n \times k}$ are isometries and $\hat{\Sigma} = \mathrm{diag}(\sigma_i(A))_{i=1}^{k}$ is a positive definite diagonal matrix. For instance, for the matrix A in Example 8.7.5 we have that

$$A = \begin{bmatrix} 6 & 4 & 4 \\ 4 & 6 & -4 \end{bmatrix} = \begin{bmatrix} 1/\sqrt{2} & 1/\sqrt{2} \\ 1/\sqrt{2} & -1/\sqrt{2} \end{bmatrix} \begin{bmatrix} 10 & 0 \\ 0 & 6 \end{bmatrix} \begin{bmatrix} 1/\sqrt{2} & 1/3\sqrt{2} \\ 1/\sqrt{2} & -1/3\sqrt{2} \\ 0 & 4/3\sqrt{2} \end{bmatrix}^*$$

is the compact singular value decomposition of A.

Example 8.7.5. Let $A = \begin{bmatrix} -1 & 5 & -1 & 5i \\ 5 & -1 & 5 & -i \\ -1 & 5 & -1 & 5i \\ 5 & -1 & 5 & -i \end{bmatrix}$. Find the compact singular value decomposition of A.

Compute

$$AA^* = \begin{bmatrix} 52 & -20 & 52 & -20 \\ -20 & 52 & -20 & 52 \\ 52 & -20 & 52 & -20 \\ -20 & 52 & -20 & 52 \end{bmatrix}.$$

It will be a challenge to compute the characteristic polynomial, so let us see if we can guess some eigenvectors. First of all, we see that columns 1 and 3 are identical and so are columns 2 and 4. This gives that

$$\begin{bmatrix} 1 \\ 0 \\ -1 \\ 0 \end{bmatrix}, \begin{bmatrix} 0 \\ 1 \\ 0 \\ -1 \end{bmatrix}$$

are eigenvectors at the eigenvalue 0. We also see that each row has the same numbers in it, and thus the sum of the entries in each row is the same. But this means that

$$\begin{bmatrix} 1 \\ 1 \\ 1 \\ 1 \end{bmatrix}$$

must be an eigenvector. Indeed,

$$\begin{bmatrix} 52 & -20 & 52 & -20 \\ -20 & 52 & -20 & 52 \\ 52 & -20 & 52 & -20 \\ -20 & 52 & -20 & 52 \end{bmatrix} \begin{bmatrix} 1 \\ 1 \\ 1 \\ 1 \end{bmatrix} = \begin{bmatrix} 64 \\ 64 \\ 64 \\ 64 \end{bmatrix} = 64 \begin{bmatrix} 1 \\ 1 \\ 1 \\ 1 \end{bmatrix}.$$

Thus we have found three of the eigenvalues (64, 0, and 0), and since the trace of a matrix equals the sum of the eigenvalues we can deduce that

$$\operatorname{tr} AA^* - (64 + 0 + 0) = 208 - 64 = 144$$

is the remaining eigenvalue of AA^*. The corresponding eigenvector must be orthogonal to the other eigenvectors we have found (since AA^* is Hermitian), and using this we find the eigenvector

$$\begin{bmatrix} 1 \\ -1 \\ 1 \\ -1 \end{bmatrix}$$

at eigenvalue 144. For singular values of A, we now get $\sigma_1(A) = \sqrt{144} = 12$ and $\sigma_2(A) = \sqrt{64} = 8$, and thus $\hat{\Sigma} = \begin{bmatrix} 12 & 0 \\ 0 & 8 \end{bmatrix}$. For the vectors $\mathbf{v}_1$ and $\mathbf{v}_2$ in $\hat{V} = \begin{bmatrix} \mathbf{v}_1 & \mathbf{v}_2 \end{bmatrix}$, we divide the eigenvectors of AA^* corresponding to 144 and 64 by their length, giving

$$\mathbf{v}_1 = \begin{bmatrix} \frac{1}{2} \\ -\frac{1}{2} \\ \frac{1}{2} \\ -\frac{1}{2} \end{bmatrix}, \mathbf{v}_2 = \begin{bmatrix} \frac{1}{2} \\ \frac{1}{2} \\ \frac{1}{2} \\ \frac{1}{2} \end{bmatrix}.$$

Finally, to determine $\hat{W} = \begin{bmatrix} \mathbf{w}_1 & \mathbf{w}_2 \end{bmatrix}$, we use $\hat{V}^*A = \hat{\Sigma}\hat{W}^*$, which gives

$$\begin{bmatrix} -6 & 6 & -6 & 6i \\ 4 & 4 & 4 & 4i \end{bmatrix} = \begin{bmatrix} 12\mathbf{w}_1^* \\ 8\mathbf{w}_2^* \end{bmatrix}$$

and, consequently

$$\hat{W} = \begin{bmatrix} -\frac{1}{2} & \frac{1}{2} \\ \frac{1}{2} & \frac{1}{2} \\ -\frac{1}{2} & \frac{1}{2} \\ -\frac{i}{2} & -\frac{i}{2} \end{bmatrix}.$$

In conclusion, we have

$$A = \begin{bmatrix} -1 & 5 & -1 & 5i \\ 5 & -1 & 5 & -i \\ -1 & 5 & -1 & 5i \\ 5 & -1 & 5 & -i \end{bmatrix} = \begin{bmatrix} \frac{1}{2} & \frac{1}{2} \\ -\frac{1}{2} & \frac{1}{2} \\ \frac{1}{2} & \frac{1}{2} \\ -\frac{1}{2} & \frac{1}{2} \end{bmatrix} \begin{bmatrix} 12 & 0 \\ 0 & 8 \end{bmatrix} \begin{bmatrix} -\frac{1}{2} & \frac{1}{2} & -\frac{1}{2} & \frac{i}{2} \\ \frac{1}{2} & \frac{1}{2} & \frac{1}{2} & -\frac{i}{2} \end{bmatrix}.$$

$\square$

8.8 Exercises

Exercise 8.8.1. For $\mathbf{u} = \begin{bmatrix} 1 \\ -4 \\ 3 \end{bmatrix}$ and $\mathbf{v} = \begin{bmatrix} 2 \\ -2 \\ 2 \end{bmatrix}$, find $\langle \mathbf{u}, \mathbf{v} \rangle$ and $\|\mathbf{u}\|$.

Exercise 8.8.2. Given are $\mathbf{u} = \begin{bmatrix} 2 \\ 3 \\ -1 \end{bmatrix}, \mathbf{v} = \begin{bmatrix} 4 \\ \frac{1}{2} \\ 3 \end{bmatrix}$. Compute the following:

(a) $\langle \mathbf{u}, \mathbf{v} \rangle$.

(b) $\|\mathbf{v}\|$.

(c) $\frac{1}{\langle \mathbf{u}, \mathbf{u} \rangle} \mathbf{u}$.

(d) $\mathbf{v}^* \mathbf{u}$.

(e) $\mathbf{u} \mathbf{v}^*$.

Exercise 8.8.3. Prove the Cauchy-Schwarz inequality on $\mathbb{R}^n$ by observing that the inequality $\langle \mathbf{x} + t\mathbf{y}, \mathbf{x} + t\mathbf{y} \rangle \geq 0$, leads to a quadratic polynomial that is nonnegative for all $t \in \mathbb{R}$. Thus its discriminant has to be ≤ 0.

Exercise 8.8.4. Let $\mathbf{x} \neq \mathbf{0}$. Show that $\|\mathbf{x}\| + \|\mathbf{y}\| = \|\mathbf{x} + \mathbf{y}\|$ if and only if $\mathbf{x} = \alpha \mathbf{y}$ for some $\alpha \geq 0$.

Exercise 8.8.5. Prove the following for $\mathbf{u}, \mathbf{v} \in \mathbb{F}^n$.

(a) $\|\mathbf{u} + \mathbf{v}\|^2 + \|\mathbf{u} - \mathbf{v}\|^2 = 2\|\mathbf{u}\|^2 + 2\|\mathbf{v}\|^2$ (parallellogram law).

(b) $\|\mathbf{u} + \mathbf{v}\|^2 - \|\mathbf{u} - \mathbf{v}\|^2 = 4\mathrm{Re}\langle \mathbf{u}, \mathbf{v} \rangle$.

Exercise 8.8.6. The **cosine rule** states that for a triangle with sides of lengths a, b and c, we have that

$$c^2 = a^2 + b^2 - 2ab \cos \theta,$$

where θ is the angle between the sides of lengths a and b.

(a) Use the cosine rule to prove that the angle θ between two nonzero vectors $\mathbf{u}$ and $\mathbf{v}$ in $\mathbb{R}^n$ is given via

$$\cos \theta = \frac{\langle \mathbf{u}, \mathbf{v} \rangle}{\|\mathbf{u}\| \, \|\mathbf{v}\|}. \tag{8.18}$$

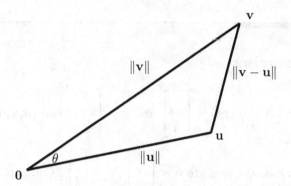

Figure 8.11: The cosine rule gives $\|\mathbf{v} - \mathbf{u}\|^2 = \|\mathbf{u}\|^2 + \|\mathbf{v}\|^2 - 2\|\mathbf{u}\|\|\mathbf{v}\|\cos\theta$.

(b) Use part (a) to justify that orthogonality between two vectors $\mathbf{u}$ and $\mathbf{v}$ is defined as $\langle \mathbf{u}, \mathbf{v} \rangle = 0$.

(c) Provide an alternative proof of (8.18) by projecting the vector $\mathbf{u}$ onto $\mathrm{Span}\{\mathbf{v}\}$.

Exercise 8.8.7. Find the distance between the vectors $\begin{bmatrix} 0 \\ 1 \\ 2 \end{bmatrix}$ and $\begin{bmatrix} 2 \\ 0 \\ 5 \end{bmatrix}$.

Exercise 8.8.8. Find the distance between $\begin{bmatrix} -2 \\ 0 \\ 4 \end{bmatrix}$ and $\begin{bmatrix} -1 \\ 3 \\ -5 \end{bmatrix}$.

Exercise 8.8.9. Find the distance between $\begin{bmatrix} -1 + 2i \\ 4 - 3i \end{bmatrix}$ and $\begin{bmatrix} 1 - i \\ i \end{bmatrix}$.

Exercise 8.8.10. Let $A \in \mathbb{C}^{n \times n}$. Show that $\det A^* = \overline{\det A}$.

Exercise 8.8.11. Let $A \in \mathbb{F}^{n \times n}$. We call A **skew-Hermitian** if $A^* = -A$.

(a) Show that $\begin{bmatrix} 2i & 3 + 4i \\ -3 + 4i & 5i \end{bmatrix}$ is skew-Hermitian.

(b) Show that A is skew-Hermitian if and only if iA is Hermitian.

(c) Use (b) to show that eigenvalues of a skew-Hermitian matrix are purely imaginary.

(d) Let $B \in \mathbb{F}^{n \times n}$. Show that $\frac{1}{2}(B + B^*)$ is Hermitian and $\frac{1}{2}(B - B^*)$ is skew-Hermitian.

Since $B = \frac{1}{2}(B + B^*) + \frac{1}{2}(B - B^*)$, it follows that any square matrix can be written as the sum of a Hermitian matrix and a skew-Hermitian matrix.

Exercise 8.8.12. Let $A \in \mathbb{F}^{m \times n}$. Show the following.

(a) $\mathrm{Nul} A^* A = \mathrm{Nul} A$.
    (Hint: Observe that if $A^* A\mathbf{x} = \mathbf{0}$, then $\|A\mathbf{x}\|^2 = \mathbf{x}^* A^* A\mathbf{x} = 0$.)

(b) $\mathrm{Col}(A^* A) = \mathrm{Col} A^*$.
    (Hint: Use (a) and a dimension argument.)

Exercise 8.8.13. Determine if the following set is orthogonal, orthonormal, or neither.
$$\left\{ \begin{bmatrix} 0 \\ 1 \\ 0 \end{bmatrix}, \begin{bmatrix} \frac{1}{\sqrt{2}} \\ 0 \\ -\frac{1}{\sqrt{2}} \end{bmatrix}, \begin{bmatrix} \frac{1}{\sqrt{3}} \\ \frac{1}{\sqrt{3}} \\ \frac{1}{\sqrt{3}} \end{bmatrix} \right\}.$$

Exercise 8.8.14. Let $W = \mathrm{Span} \left\{ \begin{bmatrix} 0 \\ -3 \\ 3 \\ 1 \end{bmatrix}, \begin{bmatrix} 1 \\ 2 \\ 4 \\ 0 \end{bmatrix} \right\}$. Find a basis for $W^\perp$.

Exercise 8.8.15. Let $W = \mathrm{Span} \left\{ \begin{bmatrix} 3 \\ 3 \\ 1 \end{bmatrix}, \begin{bmatrix} 6 \\ 6 \\ 2 \end{bmatrix} \right\}$. Find a basis for $W^\perp$.

Exercise 8.8.16. Let $W = \mathrm{Span} \left\{ \begin{bmatrix} 1 \\ -i \\ 1-i \end{bmatrix}, \begin{bmatrix} 2 \\ -3i \\ 1 \end{bmatrix} \right\} \subset \mathbb{C}^3$. Find a basis for $W^\perp$.

Exercise 8.8.17.

(a) Let $\mathbf{u} = \begin{bmatrix} 1 \\ -4 \\ -5 \end{bmatrix}$ and $\mathbf{v} = \begin{bmatrix} 2 \\ -2 \\ 2 \end{bmatrix}$. Are $\mathbf{u}$ and $\mathbf{v}$ orthogonal?

(b) Let $\mathbf{y} = \begin{bmatrix} 1 \\ -5 \\ 0 \end{bmatrix}$. Compute $\hat{\mathbf{y}}$, the orthogonal projection of $\mathbf{y}$ onto $W = \mathrm{Span}\{\mathbf{u}, \mathbf{v}\}$.

(c) Show that $\mathbf{y} - \hat{\mathbf{y}} \in W^\perp$.

Exercise 8.8.18. Find the distance of $\mathbf{u} = \begin{bmatrix} 2 \\ -4 \end{bmatrix}$ to the line $3x_1 + 2x_2 = 0$.

Exercise 8.8.19. Find the distance of $\mathbf{u}$ to the subspace $W = \mathrm{Span}\{\mathbf{w}_1, \mathbf{w}_2\}$, where
$$\mathbf{u} = \begin{bmatrix} 1 \\ -2 \\ 0 \\ 1 \end{bmatrix}, \mathbf{w}_1 = \begin{bmatrix} 1 \\ 0 \\ 1 \\ 0 \end{bmatrix}, \mathbf{w}_2 = \begin{bmatrix} -1 \\ 2 \\ -1 \\ 2 \end{bmatrix}.$$

Exercise 8.8.20. Find the distance of **u** to the subspace $W = \text{Span}\{\mathbf{w}_1\}$, where

$$\mathbf{u} = \begin{bmatrix} i \\ -2 \end{bmatrix}, \mathbf{w}_1 = \begin{bmatrix} 3+i \\ 1-i \end{bmatrix}.$$

Exercise 8.8.21. Find the orthogonal projection of **u** onto $\text{Span}\{\mathbf{v}\}$ if $\mathbf{u} = \begin{bmatrix} 2 \\ 3 \end{bmatrix}$ and $\mathbf{v} = \begin{bmatrix} -1 \\ 5 \end{bmatrix}$.

Exercise 8.8.22. Compute the orthogonal projection of $\begin{bmatrix} 1 \\ 3 \end{bmatrix}$ onto the line through $\begin{bmatrix} 2 \\ -1 \end{bmatrix}$ and the origin.

Exercise 8.8.23. Let $W = \text{span} \left\{ \begin{bmatrix} 1 \\ 2 \\ 2 \\ 4 \end{bmatrix}, \begin{bmatrix} -2 \\ 1 \\ -4 \\ 2 \end{bmatrix} \right\}$.

(a) Find the orthogonal projection of $\begin{bmatrix} -3 \\ 4 \\ 4 \\ 3 \end{bmatrix}$ onto W.

(b) Find the distance from $\begin{bmatrix} -2 \\ 6 \\ 6 \\ 7 \end{bmatrix}$ to W.

Exercise 8.8.24. Let $\mathbf{v}_1, \ldots, \mathbf{v}_n$ be nonzero orthogonal vectors in $\mathbb{R}^n$. Show that $\{\mathbf{v}_1, \ldots, \mathbf{v}_n\}$ is linearly independent.

Exercise 8.8.25. Let $W = \left\{ \begin{bmatrix} 1 \\ i \\ 1+i \\ 2 \end{bmatrix}, \begin{bmatrix} 0 \\ -i \\ 1+2i \\ 0 \end{bmatrix} \right\} \subseteq \mathbb{C}^4$. Find a basis for $W^\perp$.

Exercise 8.8.26. Find the orthogonal projection of **u** onto $\text{Span}\{\mathbf{v}\}$ if $\mathbf{u} = \begin{bmatrix} 4 \\ 6 \end{bmatrix}$ and $\mathbf{v} = \begin{bmatrix} -1 \\ 5 \end{bmatrix}$.

Exercise 8.8.27. Let

$$W = \text{span} \left\{ \begin{bmatrix} 1 \\ 2 \\ 0 \\ 1 \end{bmatrix}, \begin{bmatrix} 1 \\ 0 \\ 2 \\ 1 \end{bmatrix} \right\} \subset \mathbb{R}^4.$$

(a) Find an orthonormal basis for W.
(Hint: Use the Gram–Schmidt process).

(b) Find a basis for $W^\perp$.

Exercise 8.8.28. Find an orthonormal basis for the subspace in $\mathbb{R}^4$ spanned by

$$\begin{bmatrix} 1 \\ 1 \\ 1 \\ 1 \end{bmatrix}, \begin{bmatrix} 1 \\ 2 \\ 1 \\ 2 \end{bmatrix}, \begin{bmatrix} 3 \\ 1 \\ 3 \\ 1 \end{bmatrix}.$$

Exercise 8.8.29. Let $A = \begin{bmatrix} \frac{1}{2} & \frac{3}{2} & a \\ \frac{3}{2} & \frac{1}{2} & b \\ c & d & e \end{bmatrix}$. Can a, b, c, d, e be chosen to make A unitary?

Exercise 8.8.30. Show that the product of two unitary matrices is unitary. How about the sum?

Exercise 8.8.31. Show that the following matrices are unitary.

(a) $\frac{1}{\sqrt{2}} \begin{bmatrix} 1 & 1 \\ 1 & -1 \end{bmatrix}$.

(b) $\frac{1}{\sqrt{3}} \begin{bmatrix} 1 & 1 & 1 \\ 1 & e^{\frac{2i\pi}{3}} & e^{\frac{4i\pi}{3}} \\ 1 & e^{\frac{4i\pi}{3}} & e^{\frac{8i\pi}{3}} \end{bmatrix}$.

(c) $\frac{1}{2} \begin{bmatrix} 1 & 1 & 1 & 1 \\ 1 & i & -1 & -i \\ 1 & -1 & 1 & -1 \\ 1 & -i & -1 & i \end{bmatrix}$.

(d) The above are Fourier matrices of sizes $2, 3$ and 4. Can you guess the form of the $n \times n$ Fourier matrix?

Exercise 8.8.32. Find the QR factorization of the following matrices.

(a) $\begin{bmatrix} 2 & 1 \\ 1 & 1 \\ 2 & 1 \end{bmatrix}$.

(b) $\begin{bmatrix} 1 & 2i & 0 \\ 2 & 1 & 1 \\ 0 & 1 & 1+i \end{bmatrix}$.

(c) $\begin{bmatrix} 1 & 1 & 3 \\ i & -i & -i \\ 2 & 1 & 4 \end{bmatrix}$.

(d) $\begin{bmatrix} 5 & 9 \\ 1 & 7 \\ -3 & -5 \\ 1 & 5 \end{bmatrix}$.

(e) $\begin{bmatrix} 1 & 2 & 3 \\ 1 & 0 & 5 \end{bmatrix}$.

Exercise 8.8.33. Let $W \subseteq \mathbb{F}^n$ be a subspace. Show that $(W^\perp)^\perp = W$. (Hint: Show first that $W \subseteq (W^\perp)^\perp$ and then show that both sides have the same dimension.)

Exercise 8.8.34. We defined the dot product on $\mathbb{R}^n$, which satisfies the properties (i)-(iii) in Proposition 8.1.2. On other vector spaces one can define similar operations that also satisfy these properties (which is then called an inner product). As an example, consider the vector space $C[-1, 1]$ of continuous functions with domain $[-1, 1]$ and co-domain $\mathbb{R}$; that is,

$$C[-1, 1] = \{\mathbf{f} : [-1, 1] \to \mathbb{R} : f \text{ is continuous}\}.$$

Addition and scalar multiplication is defined as usual for functions:

$$(\mathbf{f} + \mathbf{g})(x) = \mathbf{f}(x) + \mathbf{g}(x), (c\mathbf{f})(x) = c\mathbf{f}(x).$$

We can now introduce

$$\langle \mathbf{f}, \mathbf{g} \rangle = \int_{-1}^{1} \mathbf{f}(x)\mathbf{g}(x)dx.$$

Show the following for all $\mathbf{f}, \mathbf{g}, \mathbf{h} \in C[-1, 1]$ and $a, b \in \mathbb{R}$:

(a) $\langle a\,\mathbf{f} + b\,\mathbf{g}, \mathbf{h} \rangle = a\langle \mathbf{f}, \mathbf{h} \rangle + b\langle \mathbf{g}, \mathbf{h} \rangle$.

(b) $\langle \mathbf{f}, \mathbf{g} \rangle = \langle \mathbf{g}, \mathbf{f} \rangle$.

(c) $\langle \mathbf{f}, \mathbf{f} \rangle \geq 0$ and $[\langle \mathbf{f}, \mathbf{f} \rangle = 0 \iff \mathbf{f} = \mathbf{0}]$.

This shows that $\langle \mathbf{f}, \mathbf{g} \rangle = \int_{-1}^{1} \mathbf{f}(x)\mathbf{g}(x)dx$ is an inner product on $C[-1, 1]$.

Exercise 8.8.35. For the following, find the least squares solution to the equation $A\mathbf{x} = \mathbf{b}$.

(a) $A = \begin{bmatrix} 1 & 2 \\ 3 & -4 \\ -1 & 2 \end{bmatrix}, \mathbf{b} = \begin{bmatrix} 1 \\ 1 \\ 1 \end{bmatrix}$.

(b) $A = \begin{bmatrix} 2 & 1 \\ 3 & 1 \\ -1 & 1 \\ -2 & 1 \end{bmatrix}, b = \begin{bmatrix} 3 \\ -1 \\ -3 \\ 2 \end{bmatrix}.$

(c) $A = \begin{bmatrix} -2 & 3 \\ -4 & 6 \\ 1 & 2 \end{bmatrix}, b = \begin{bmatrix} 1 \\ 0 \\ 0 \end{bmatrix}.$

(d) $A = \begin{bmatrix} 1 & 1 \\ 2 & 1 \\ 3 & 1 \end{bmatrix}, b = \begin{bmatrix} 3 \\ 5 \\ 4 \end{bmatrix}.$

Exercise 8.8.36. Find the equation of the least squares line that best fits the data points $(-1, -2), (0, 1), (1, 1), (2, 1)$, and $(3, 4)$.

Exercise 8.8.37. Find the equation of the least squares line that best fits the data points $(1, 3), (2, 5)$, and $(3, 4)$. Plot the three points and the line $y = cx + d$.

Exercise 8.8.38. Given points $(-1, -1), (0, -4), (1, 6), (2, 0)$. Find the least squares fit of these points by a parabola $y = ax^2 + bx + c$.

Exercise 8.8.39. Show that the sum of two Hermitian matrices is Hermitian. How about the product?

Exercise 8.8.40. Let

$$B = \begin{bmatrix} 0 & 0 & \sqrt{2} & 0 \\ 0 & 0 & 1 & \sqrt{2} \\ \sqrt{2} & 1 & 0 & 0 \\ 0 & \sqrt{2} & 0 & 0 \end{bmatrix}.$$

(a) Show that the eigenvalues and of B are $2, 1, -2, -1$ with eigenvectors

$$\begin{bmatrix} 1 \\ \sqrt{2} \\ \sqrt{2} \\ 1 \end{bmatrix}, \begin{bmatrix} -\sqrt{2} \\ 1 \\ -1 \\ \sqrt{2} \end{bmatrix}, \begin{bmatrix} 1 \\ \sqrt{2} \\ -\sqrt{2} \\ -1 \end{bmatrix}, \begin{bmatrix} -\sqrt{2} \\ 1 \\ 1 \\ -\sqrt{2} \end{bmatrix},$$

respectively

(b) Write B as $B = UDU^*$ with U unitary and D diagonal.

Exercise 8.8.41. For the following matrices, find the spectral decomposition; that is, for each matrix A find a unitary U and a diagonal D so that $A = UDU^*$.

(a) $\begin{bmatrix} 6 & 2 \\ 2 & 9 \end{bmatrix}.$

(b) $\begin{bmatrix} 2 & i \\ -i & 2 \end{bmatrix}$.

(c) $\begin{bmatrix} 2 & \sqrt{3} \\ \sqrt{3} & 4 \end{bmatrix}$.

(d) $\begin{bmatrix} 3 & 1 & 1 \\ 1 & 3 & 1 \\ 1 & 1 & 3 \end{bmatrix}$.

(e) $\begin{bmatrix} 0 & 1 & 0 \\ 0 & 0 & 1 \\ 1 & 0 & 0 \end{bmatrix}$.

(f) $\begin{bmatrix} 2 & 1 & 0 \\ 1 & 0 & 1 \\ 0 & 1 & 2 \end{bmatrix}$.

(g) $\begin{bmatrix} -1 & 1 & -1 \\ 1 & -1 & -1 \\ -1 & -1 & -1 \end{bmatrix}$. (Hint: -2 is an eigenvalue)

(h) $\begin{bmatrix} 4 & -6 & 4 \\ -6 & 5 & -2 \\ 4 & -2 & 0 \end{bmatrix}$.

(i) $\begin{bmatrix} 4 & 2 & 2 \\ 2 & 4 & 2 \\ 2 & 2 & 4 \end{bmatrix}$.

Exercise 8.8.42. Let $A = \begin{bmatrix} 3 & 2i \\ -2i & 3 \end{bmatrix}$.

(a) Show that A is positive semidefinite.

(b) Find the positive **square root** of A; that is, find a positive semidefinite B so that $B^2 = A$.

(Hint: Use the spectral decomposition of A)

Exercise 8.8.43. A **quadratic form** in two variables is of the form

$$f(x, y) = ax^2 + bxy + cy^2.$$

Consider the associated level curve $\{(x, y) \in \mathbb{R}^2 : f(x, y) = 1\}$.

(a) Show that we can write $f(x, y) = \begin{bmatrix} x & y \end{bmatrix} M \begin{bmatrix} x \\ y \end{bmatrix}$, where

$$M = \begin{bmatrix} a & \frac{b}{2} \\ \frac{b}{2} & c \end{bmatrix}.$$

(b) Show that if $a = 2$, $b = 0$, and $c = 3$, the level curve is an ellipse.

(c) Show that if $a = 2$, $b = 0$, and $c = -3$, the level curve is a hyperbola.

(d) Show that if M has positive eigenvalues, then the curve is an ellipse. (Hint: Write $M = UDU^T$ with U unitary and D diagonal, and perform a change of variables $\begin{bmatrix} x' \\ y' \end{bmatrix} = U^T \begin{bmatrix} x \\ y \end{bmatrix}$.)

(e) Show that if M has a positive and a negative eigenvalue, then the curve is a hyperbola.

(f) What happens in the remaining cases?

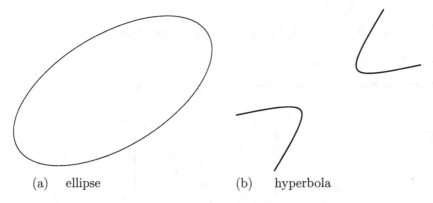

(a)   ellipse   (b)   hyperbola

Figure 8.12: An ellipse and a hyperbola.

Exercise 8.8.44. Let $w \in \mathbb{C}^n$ with $w^*w = 1$. Introduce the so-called **Householder matrix** H_w defined by $H_w = I - 2ww^*$.

(a) Show that w is an eigenvector for H_w. What is the corresponding eigenvalue?

(b) Show that if $v \neq 0$ is orthogonal to w then v is an eigenvector of H_w corresponding to eigenvalue 1.

(c) For $w = \begin{bmatrix} \frac{1}{\sqrt{2}} \\ \frac{-i}{\sqrt{2}} \end{bmatrix}$ compute H_w.

Exercise 8.8.45. True or False? Justify each answer.

(a) If for real vectors u and v we have $\|u + v\|^2 = \|u\|^2 + \|v\|^2$, then u and v are orthogonal.

(b) Every subspace of $\mathbb{R}^n$ has an orthonormal basis.

(c) If $W \subseteq \mathbb{R}^7$ has dimension 3, then $\dim W^\perp = 4$.

Exercise 8.8.46. Define the **cross product** $\mathbf{x} \times \mathbf{y}$ for two vectors $\mathbf{x}, \mathbf{y} \in \mathbb{R}^3$ by

$$\begin{bmatrix} x_1 \\ x_2 \\ x_3 \end{bmatrix} \times \begin{bmatrix} y_1 \\ y_2 \\ y_3 \end{bmatrix} = \begin{bmatrix} x_2 y_3 - x_3 y_2 \\ x_3 y_1 - x_1 y_3 \\ x_1 y_2 - x_2 y_1 \end{bmatrix}. \tag{8.19}$$

Show the following.

(a) $\langle \mathbf{x}, \mathbf{x} \times \mathbf{y} \rangle = \langle \mathbf{y}, \mathbf{x} \times \mathbf{y} \rangle = 0$.

(b) $\mathbf{x} \times \mathbf{y} = -\mathbf{y} \times \mathbf{x}$.

(c) $\mathbf{x} \times \mathbf{y} = \mathbf{0}$ if and only if $\{\mathbf{x}, \mathbf{y}\}$ is linearly dependent.

(d) $\|\mathbf{x} \times \mathbf{y}\|^2 = \|\mathbf{x}\|^2 \|\mathbf{y}\|^2 - (\langle \mathbf{x}, \mathbf{y} \rangle)^2$.

Exercise 8.8.47. Consider the $n \times n$ tri-diagonal matrix

$$A_n = \begin{bmatrix} 2 & -1 & 0 & \cdots & 0 \\ -1 & 2 & -1 & \cdots & 0 \\ \vdots & \ddots & \ddots & \ddots & \vdots \\ 0 & \cdots & -1 & 2 & -1 \\ 0 & \cdots & 0 & -1 & 2 \end{bmatrix}.$$

This is an example of a **Toeplitz matrix** which by definition is a matrix A where the entries satisfy $a_{ij} = a_{i+1,j+1}$ for all possible i and j. Show that $\lambda_j = 2 - 2\cos(j\theta)$, $j = 1, \ldots, n$, where $\theta = \frac{\pi}{n+1}$, are the eigenvalues of A. (Hint: Show that

$$\mathbf{v}_j = \begin{bmatrix} \sin(j\theta) \\ \sin(2j\theta) \\ \vdots \\ \sin(nj\theta) \end{bmatrix}$$

is an eigenvector associated with λ_j. You will need to use rules like $\sin(a+b) = \sin a \cos b + \cos a \sin b$ to rewrite $\sin(kj\theta - j\theta) + \sin(kj\theta + j\theta)$, for instance.)

The matrix $\begin{bmatrix} \mathbf{v}_1 & \cdots & \mathbf{v}_n \end{bmatrix}$ is called the (first) **discrete sine transform**. Note that its columns are orthogonal. The discrete cosine transform (see Exercise 5.5.22) also has orthogonal columns, and the columns of the discrete cosine transform are the eigenvectors of the real symmetric matrix obtained from A_n by making the $(1,1)$ and (n,n) entries equal to 1.

Exercise 8.8.48. Determine the singular value decomposition of the following matrices.

(a) $A = \begin{bmatrix} 1 & 1 & 2\sqrt{2}i \\ -1 & -1 & 2\sqrt{2}i \\ \sqrt{2}i & -\sqrt{2}i & 0 \end{bmatrix}$.

(b) $A = \begin{bmatrix} -2 & 4 & 5 \\ 6 & 0 & -3 \\ 6 & 0 & -3 \\ -2 & 4 & 5 \end{bmatrix}$.

(c) Find also the compact singular value decompositions of the above matrices.

Exercise 8.8.49. Let U be a unitary 3×3 matrix, and let

$$A = U \begin{bmatrix} 1 & 0 & 0 \\ 0 & 2 & 0 \\ 0 & 0 & 3 \end{bmatrix} U^*, \quad B = U \begin{bmatrix} -5 & 0 & 0 \\ 0 & -2 & 0 \\ 0 & 0 & 3 \end{bmatrix} U^*.$$

How can one find the singular value decompositions of A and B without doing much computation?

Exercise 8.8.50. Let U be a unitary $n \times n$ matrix. Show that $|\det U| = 1$.

Exercise 8.8.51. Let A be a $n \times n$ matrix with singular value decomposition $A = V\Sigma W^*$. Notice that since A is square, we have that V, Σ and W are also square matrices. Show that $|\det A| = \det \Sigma = \sigma_1(A) \cdots \sigma_n(A)$, where we allow for $\sigma_j(A) = 0$ when $j > \text{rank} A$.

Exercise 8.8.52. Use (8.15) to prove

$$\sigma_1(AB) \leq \sigma_1(A)\sigma_1(B). \tag{8.20}$$

Exercise 8.8.53. Let $A = \begin{bmatrix} P & Q \end{bmatrix} \in \mathbb{C}^{k \times (m+n)}$, where P is of size $k \times m$ and Q of size $k \times n$. Show that

$$\sigma_1(P) \leq \sigma_1(A).$$

Conclude that $\sigma_1(Q) \leq \sigma_1(A)$ as well.

Exercise 8.8.54. For $\Sigma \in \mathbb{F}^{m \times n}$ as in (8.14) define $\Sigma^\dagger \in \mathbb{F}^{n \times m}$ by

$$\Sigma^\dagger = \begin{bmatrix} \frac{1}{\sigma_1} & 0 & \cdots & 0 & \cdots & 0 \\ 0 & \frac{1}{\sigma_2} & \cdots & 0 & \cdots & 0 \\ \vdots & \vdots & \ddots & \vdots & & \vdots \\ 0 & 0 & \cdots & \frac{1}{\sigma_k} & \cdots & 0 \\ \vdots & \vdots & & \vdots & 0 & \vdots \\ 0 & 0 & \cdots & 0 & \cdots & 0 \end{bmatrix}. \tag{8.21}$$

(a) Show that $\Sigma\Sigma^\dagger\Sigma = \Sigma$ and $\Sigma^\dagger\Sigma\Sigma^\dagger = \Sigma^\dagger$.

For $A \in \mathbb{F}^{m \times n}$, we call $B \in \mathbb{F}^{n \times m}$ a **generalized inverse** of A if $ABA = A$ and $BAB = A$.

(b) Show that if A is invertible (and thus $m = n$), then A^{-1} is the only generalized inverse of A.

(c) For $A \in \mathbb{F}^{m \times n}$ with singular value decomposition $A = V\Sigma W^*$, define $A^\dagger = W\Sigma^\dagger V^*$. Show that $A^\dagger$ is a generalized inverse of A.

(d) For $A = \begin{bmatrix} 1 \\ 0 \end{bmatrix}$, show that for any $b \in \mathbb{F}$ the matrix $B = \begin{bmatrix} 1 & b \end{bmatrix}$ is a generalized inverse. For which value of b do we have $B = A^\dagger$?

The matrix $A^\dagger$ is called the **Moore-Penrose inverse** of A. Note that $AA^\dagger$ and $A^\dagger A$ are Hermitian. This additional property characterizes the Moore-Penrose inverse.

Exercise 8.8.55. Let $A \in \mathbb{F}^{m \times n}$ have rank k, and singular value decomposition $A = V\Sigma W^*$. Show that

(a) $\{\mathbf{v}_1, \ldots, \mathbf{v}_k\}$ is a basis for $\mathrm{Ran}A$.

(b) $\{\mathbf{w}_{k+1}, \ldots, \mathbf{w}_n\}$ is a basis for $\mathrm{Ker}A$.

(c) $\{\mathbf{w}_1, \ldots, \mathbf{w}_k\}$ is a basis for $\mathrm{Ran}A^*$.

(d) $\{\mathbf{v}_{k+1}, \ldots, \mathbf{v}_m\}$ is a basis for $\mathrm{Ker}A^*$.

Exercise 8.8.56. Consider a real matrix A of size $m \times n$. To store this matrix mn real numbers need to be stored.

(a) Suppose that using the singular value decomposition A is approximated by $\hat{A}$ of rank k. Factor $\hat{A} = BC$ where B is $m \times k$ and C is $k \times n$. How many real numbers need to be stored to store B and C (and thus $\hat{A}$).

(b) How would one find factors B and C as in (a) using the singular value decomposition of A?

(c) Suppose now that a black and white image (such as in Figure 8.7) is represented as a matrix of size 768×1024. If this matrix is approximated by a rank 100 matrix, how much does one save in storing the approximation instead of the original matrix?

This represents a simple idea to do image compression. More sophisticated ideas inspired by this simple idea lead to useful image compression algorithms. One can easily search for these algorithms, as well as other image processing algorithms (image enhancement, image restoration, etc.), which make good use of Linear Algebra.

Exercise 8.8.57. Principal Component Analysis (PCA) is a way to extract the main features of a set of data. It is for instance used in facial recognition algorithms. In this exercise we will outline some of the Linear Algebra that is used in PCA. To properly understand PCA one needs to cover several statistical notions, but we will not treat those here. We will simply illustrate some of the mechanics in an example.

Let us say we are studying different cells and try to group them in different types. To do so, we use measurements obtained by sequencing the mRNA in each cell. This identifies the level of activity of different genes in each cell. These measurements are as follows.

$$
X = \begin{array}{c} \begin{array}{ccccccccc} c1 & c2 & c3 & c4 & c5 & c6 & c7 & c8 & c9 \end{array} \\ \begin{bmatrix} 0.4 & 1.6 & 0.3 & 0.3 & 1.5 & 1.5 & 0.3 & 1.3 & 0.9 \\ 1.3 & 0.2 & 1.5 & 1.1 & 0.3 & 0.2 & 1.6 & 0.1 & 1.6 \\ 1.1 & 1.8 & 0.9 & 1.2 & 2. & 01.9 & 0.9 & 1.7 & 0.7 \\ 2.3 & 0.8 & 1.7 & 2.1 & 0.4 & 0.7 & 2.1 & 0.6 & 2.1 \end{bmatrix} \end{array} \begin{array}{l} \text{gene1} \\ \text{gene2} \\ \text{gene3} \\ \text{gene4} \end{array}
$$

Thus in cell 1 we have that gene 4 is most active, while gene 1 is the least active. First we compute the average activity of each gene, which corresponds to the average of the values in each row. We can compute this as

$$
\text{average} = \frac{1}{n} X \begin{bmatrix} 1 \\ 1 \\ \vdots \\ 1 \end{bmatrix} = \begin{bmatrix} 0.9000 \\ 0.8778 \\ 1.3556 \\ 1.4222 \end{bmatrix},
$$

where $n = 9$ is the number of columns of X. Next we subtract the appropriate average from each entry so that the numbers in the matrix represent the deviation form the average. We obtain

$$
Y = X - \begin{bmatrix} 0.9000 \\ 0.8778 \\ 1.3556 \\ 1.4222 \end{bmatrix} \begin{bmatrix} 1 & 1 & \cdots & 1 \end{bmatrix} =
$$

$$
\begin{bmatrix} -.500 & .700 & -.600 & -.600 & .600 & .600 & -.600 & .400 & .000 \\ .422 & -.678 & .622 & .222 & -.578 & -.678 & .722 & -.778 & .722 \\ -.256 & .444 & -.456 & -.156 & .644 & .544 & -.456 & .344 & -.656 \\ .878 & -.622 & .278 & .678 & -1.022 & -.722 & .678 & -.822 & .678 \end{bmatrix}.
$$

In this case, as the set of data is relatively small you may notice that if gene 1 activity is high in a cell then gene 3 activity also tends to be high. This means that the gene 1 activity and the gene 3 activity are positively correlated. On the other hand when the gene 1 activity is high, then gene 2 activity tends to be low. Thus gene 1 and gene 2 are negatively correlated. When two rows are orthogonal to one another, they represent uncorrelated activities. In PCA we are going to replace the rows by linear combinations of them so that the new rows become uncorrelated. In this process we do not want to change the overall variation of the data. To do this, the singular value decomposition is exactly the right tool. In this case, we write $Y = V\Sigma W^*$ and create a new data matrix $C = V^*Y = \Sigma W^*$. We now obtain

$$C = \begin{bmatrix} 1.08 & -1.22 & 0.94 & 0.86 & -1.45 & -1.28 & 1.24 & -1.23 & 1.05 \\ 0.18 & -0.10 & 0.00 & 0.37 & -0.03 & 0.01 & 0.01 & 0.10 & -0.54 \\ -0.24 & -0.19 & 0.39 & -0.13 & 0.17 & -0.08 & 0.12 & 0.08 & -0.12 \\ -0.05 & 0.02 & 0.03 & 0.02 & -0.17 & -0.04 & -0.05 & 0.20 & 0.04 \end{bmatrix}.$$

Then

$$CC^* = \Sigma W^* W \Sigma^* = \begin{bmatrix} 12.1419 & 0 & 0 & 0 \\ 0 & 0.4808 & 0 & 0 \\ 0 & 0 & 0.3323 & 0 \\ 0 & 0 & 0 & 0.0784 \end{bmatrix}.$$

The off-diagonal entries in CC^* being zero corresponds exactly to the orthogonality of the rows of C. In addition, the diagonal entries of CC^* correspond to the variations of the rows of C. These are also the squares of the lengths of each row, and the squares of the singular values of Y. The total variation is the sum of the diagonal entries of CC^*, and thus

$$\text{total variation} = \text{tr } CC^* = 13.0333.$$

By the manner in which the singular values are ordered, the first one is the largest, and the corresponding row in C represents the first principal component of the data. The second is the next largest, and the second row in C represents the second principal component, and so on. In this case, the first principal component has variation 12.1419 out of a total of 13.0333, so the first principal component represents $12.1419/13.0333 \times 100\% = 93\%$ of the total variation. Thus looking at the first row of C, which is a linear combination of the activities of genes 1–4, we already capture a large part of the data. In the first row of C, we see that cells 1,3,4,7,9, all have values in the interval $[0.86, 1.24]$ while cells 2,5,6,8 have values in the interval $[-1.45, -1.22]$. Since these intervals are well separated, it will be safe to say that these two groups of cells are of a different type.

Next we look at the first two principal components combined, and visualize this by the scatter plot in Figure 8.13.

As we had already observed, it is clear that the first principal component

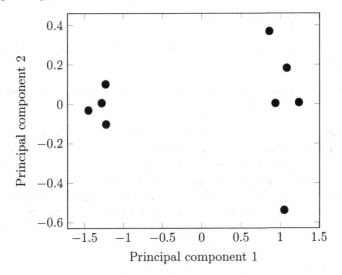

Figure 8.13: Two components capture 96.8% of the total variation.

divides the cells in two different groups. Looking at the second critical component, we see that perhaps cell 9 at coordinates $(1.05, -0.54)$ is separated from cells 1,3,4,7. It is important to realize, though, that the variation of the second component accounts for only $0.4808/13.0333 \times 100\% = 3.6\%$ of the total variation so the apparent separation may not actually be significant.

Let us practice this now on another example. An event organizer has to seat 10 people at 2 different tables, and wants to do this based on common interests. A questionnaire about different activities (how often do you watch sports? how often do you go to the movies? etc.) leads to the following data.

$$X = \begin{array}{c} \begin{array}{cccccccccc} \text{p1} & \text{p2} & \text{p3} & \text{p4} & \text{p5} & \text{p6} & \text{p7} & \text{p8} & \text{p9} & \text{p10} \end{array} \\ \begin{bmatrix} 7 & 11 & 6 & 5 & 10 & 4 & 2 & 8 & 1 & 6 \\ 2 & 4 & 4 & 3 & 4 & 2 & 1 & 2 & 1 & 2 \\ 5 & 8 & 4 & 3 & 7 & 3 & 2 & 6 & 1 & 4 \\ 6 & 9 & 3 & 2 & 9 & 2 & 1 & 8 & 0 & 6 \end{bmatrix} \end{array} \begin{array}{l} \text{activity1} \\ \text{activity2} \\ \text{activity3} \\ \text{activity4} \end{array} .$$

(a) Determine the average of each activity, and determine the matrix Y representing the deviation from the averages.

(b) Use your favorite software to compute the singular value decomposition $Y = V\Sigma W^*$.

(c) Compute the matrix $V^*Y = \Sigma W^*$.

(d) Make a scatter plot based on the first two principal components.

(e) What percentage of the total variation do the first two principal components account for?

(f) Separate the people in two groups based on the scatter plot found under (d).

Exercise 8.8.58. Suppose that we have 10 hours of digital footage of a hallway that is supposed to be empty all the time, and we are looking for those 20 seconds where someone walks through it. How can we use the singular value decomposition to find those 20 seconds? Let us assume that the footage is in black and white, so that each frame corresponds to a matrix of gray scales. For each frame, we will make a vector out of it by just stacking the columns. We call this operation **vec**. For instance,

$$
\text{vec} \begin{bmatrix} 1 & 2 & 3 \\ 4 & 5 & 6 \end{bmatrix} = \begin{bmatrix} 1 \\ 4 \\ 2 \\ 5 \\ 3 \\ 6 \end{bmatrix}.
$$

Now we build a matrix A of the complete footage, where each column corresponds to the vec of a frame; thus

$$
A = \begin{bmatrix} \text{vec(frame 1)} & \text{vec(frame 2)} & \cdots & \text{vec(frame } N) \end{bmatrix},
$$

where $N = 10 \times 3600 \times 59$, if we have 59 frames per second.

(a) If every frame is the same (so nobody walking through the hallway and no light fluctuations), what will be the rank of A?

We can now take the singular value decomposition of A, and look at the first few singular values.

Singular values of A

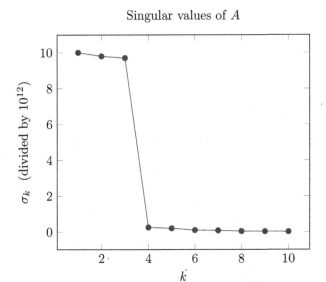

(b) Based on the drop off after the third singular value, what may be the rank of a good low rank approximation of A?

(c) If $\hat{A}$ is obtained from A by applying Proposition 8.7.3 with $l = 3$, explain how $A - \hat{A}$ has most entries be of relatively low absolute value. How could this be useful in finding the frames where the person walks through the hallway?

The above represents an essential idea in a much more complicated algorithm. For more details, search for Robust Principal Component Analysis.

Answers to Selected Exercises

Chapter 1

Exercise 1.6.1. (a) row reduced echelon. (c) neither. (e) row echelon.

Exercise 1.6.2. (a) $\begin{bmatrix} 1 & 0 & 1 \\ 0 & 1 & 2 \\ 0 & 0 & 0 \end{bmatrix}$.

Exercise 1.6.3. (a) $\begin{bmatrix} x_1 \\ x_2 \\ x_3 \end{bmatrix} = \begin{bmatrix} 1 \\ -1 \\ 0 \end{bmatrix} + x_3 \begin{bmatrix} \frac{1}{2} \\ -1 \\ 1 \end{bmatrix}$, with x_3 free.

Exercise 1.6.3. (c) $\begin{bmatrix} x_1 \\ x_2 \\ x_3 \end{bmatrix} = \begin{bmatrix} 3 \\ 0 \\ 4 \end{bmatrix} + x_2 \begin{bmatrix} 1 \\ 1 \\ 0 \end{bmatrix}$, with x_2 free.

Exercise 1.6.3. (e) $\begin{bmatrix} x_1 \\ x_2 \\ x_3 \\ x_4 \end{bmatrix} = \begin{bmatrix} -3 \\ 0 \\ -7 \\ 0 \end{bmatrix} + x_2 \begin{bmatrix} 1 \\ 1 \\ 0 \\ 0 \end{bmatrix} + x_4 \begin{bmatrix} -5 \\ 0 \\ 1 \\ 1 \end{bmatrix}$, with x_2, x_4 free.

Exercise 1.6.5. $\begin{bmatrix} x_1 \\ x_2 \\ x_3 \end{bmatrix} = \begin{bmatrix} \frac{16}{9} \\ \frac{10}{9} \\ \frac{7}{9} \end{bmatrix}$.

Exercise 1.6.8. (a) $\mathbf{b} = \frac{1}{7}\mathbf{a}_1 - \frac{1}{7}\mathbf{a}_2 + \frac{1}{7}\mathbf{a}_3$.

Exercise 1.6.8. (c) $\mathbf{b} = \frac{20}{6}\mathbf{a}_1 + \frac{7}{6}\mathbf{a}_2 - \frac{1}{6}\mathbf{a}_3$.

Exercise 1.6.9. (a) $h = -2$

Exercise 1.6.10. $k = 8$, $h = 2$, $m = 10$.

Exercise 1.6.12. (a) $\begin{bmatrix} 14 \\ 2 \end{bmatrix}$.

Exercise 1.6.12. (c) $\begin{bmatrix} 0 \\ 9 \end{bmatrix}$.

Exercise 1.6.15. No.

Exercise 1.6.17. $h \neq -5$.

Exercise 1.6.18. (a) do not span. (c) do not span.

Exercise 1.6.20. (a) $2KMnO_4 + 16HCl \rightarrow 2KCl + 2MnCl_2 + 8H_2O + 5Cl_2$.

Exercise 1.6.21. (a)

$$\begin{cases} 8I_1 - 2I_2 \quad\quad - 4I_4 &= 4 \\ -2I_1 + 7I_2 - I_3 &= 0 \\ \quad\quad -I_2 + 7I_3 &= -3 \\ -4I_1 \quad\quad\quad + 8I_4 &= 3 \end{cases}.$$

Chapter 2

Exercise 2.6.2. (a) Yes, a subspace; $\left\{ \begin{bmatrix} \frac{3}{2} \\ 1 \\ 0 \end{bmatrix}, \begin{bmatrix} -\frac{5}{2} \\ 0 \\ 1 \end{bmatrix} \right\}$ is a basis for H.

Exercise 2.6.2. (c) Yes, a subspace; $\left\{ \begin{bmatrix} 1 \\ 0 \\ -1 \end{bmatrix}, \begin{bmatrix} -1 \\ 1 \\ 0 \end{bmatrix} \right\}$ is a basis for H.

Exercise 2.6.2. (e) Not a subspace.

Exercise 2.6.4. (a) independent; (c) dependent; (e) dependent.

Exercise 2.6.7. (a) independent; (c) dependent.

Exercise 2.6.9. Yes.

Exercise 2.6.11. (a) Basis for Col A: $\left\{ \begin{bmatrix} 1 \\ 1 \\ 0 \end{bmatrix}, \begin{bmatrix} 1 \\ 0 \\ 2 \end{bmatrix} \right\}$; Basis for Nul A: $\left\{ \begin{bmatrix} -3 \\ 1 \\ 1 \end{bmatrix} \right\}$. Basis for Row A: $\left\{ \begin{bmatrix} 1 & 0 & 3 \end{bmatrix}, \begin{bmatrix} 0 & 1 & -1 \end{bmatrix} \right\}$; rank $= 2$.

Exercise 2.6.11. (c) Basis for Col A: $\left\{ \begin{bmatrix} 1 \\ 2 \\ 0 \\ 0 \end{bmatrix}, \begin{bmatrix} 0 \\ 0 \\ 1 \\ 2 \end{bmatrix}, \begin{bmatrix} 0 \\ 0 \\ 0 \\ 5 \end{bmatrix} \right\}$; rank $= 3$. Basis

for Nul A: $\left\{ \begin{bmatrix} -1 \\ 1 \\ 0 \\ 0 \\ 0 \end{bmatrix}, \begin{bmatrix} -1 \\ 0 \\ -2 \\ 1 \\ 0 \end{bmatrix} \right\}$. Basis for Row A:

$$\left\{ \begin{bmatrix} 1 & 1 & 0 & 1 & 0 \end{bmatrix}, \begin{bmatrix} 0 & 0 & 1 & 2 & 0 \end{bmatrix}, \begin{bmatrix} 0 & 0 & 0 & 0 & 1 \end{bmatrix} \right\}.$$

Exercise 2.6.16. (a) $[v]_B = \begin{bmatrix} 7 \\ -1 \\ 0 \end{bmatrix}$.

Exercise 2.6.17. (a) 3. (c) $\begin{bmatrix} 1 \\ -1 \\ 0 \end{bmatrix}$.

Exercise 2.6.19. (a) 1,1,2,0. (c) 2,2,4,0.

Chapter 3

Exercise 3.7.1. (a) $\begin{bmatrix} -2 & 2 \\ -4 & 5 \\ -2 & 6 \end{bmatrix}$. (c) $\begin{bmatrix} 12 & 16 \\ 0 & -8 \end{bmatrix}$.

Exercise 3.7.2. (a) Ae_1 is the first column of A. $e_1^T A$ is the first row of A. (c) $e_k^T Ae_l$ is the (k, l)th entry of A.

Exercise 3.7.5. (a) k. (c) $mkn + mnl$.

Exercise 3.7.7. (a) undefined as A has 3 columns and B has 2 rows, and $3 \neq 2$. (c) $\begin{bmatrix} -\frac{3}{2} & 1 \\ -\frac{1}{2} & \frac{1}{2} \end{bmatrix}$.

Exercise 3.7.9. (a) $\begin{bmatrix} 1 & -24 & 11 \\ 0 & 7 & -3 \\ 0 & -2 & 1 \end{bmatrix}$. (c) No inverse exists.

Exercise 3.7.12. $X = (B^{-1})^T A^T - C^T$.

Exercise 3.7.23. $X = \begin{bmatrix} 1 & \frac{3}{2} \\ 0 & \frac{11}{2} \\ 0 & 0 \end{bmatrix} + \begin{bmatrix} -4 \\ -2 \\ 1 \end{bmatrix} \begin{bmatrix} x_3 & y_3 \end{bmatrix}$, with $x_3, y_3 \in \mathbb{R}$ free.

Exercise 3.7.25. (a) $L = \begin{bmatrix} 1 & 0 \\ 4 & 1 \end{bmatrix}$, $U = \begin{bmatrix} 1 & -2 \\ 0 & 5 \end{bmatrix}$.

Exercise 3.7.25. (c) $L = \begin{bmatrix} 1 & 0 & 0 \\ 2 & 1 & 0 \\ 0 & -1 & 1 \end{bmatrix}$, $U = \begin{bmatrix} 1 & -1 & 0 \\ 0 & -1 & 1 \\ 0 & 0 & 3 \end{bmatrix}$.

Exercise 3.7.26. (c) $\begin{bmatrix} 1 & 2 & -11 \\ 0 & 1 & -4 \\ 0 & 0 & 1 \end{bmatrix}$.

Chapter 4

Exercise 4.5.1. (a) 14 (c) 114. (e) 12.

Exercise 4.5.7. (a) $x_1 = 0, x_2 = \frac{1}{2}$.

Exercise 4.5.9. -12.

Exercise 4.5.14. $\mathrm{adj}(A) = \begin{bmatrix} -4 & 8 & -4 \\ 1 & -2 & \frac{7}{2} \\ 5 & 0 & -\frac{5}{2} \end{bmatrix}$.

Exercise 4.5.19. (a) $\frac{1}{2}$. (c) $\frac{1}{24}$.

Exercise 4.5.20. 4.

Chapter 5

Exercise 5.5.10. (a) Yes, a subspace. (c) No, not a subspace. (e) No, not a subspace. (g) Yes, a subspace.

Exercise 5.5.11. (a) independent. (c) dependent. (e) independent. (g) independent.

Exercise 5.5.12. (a) Yes.

Exercise 5.5.20. (a) $[\mathbf{v}]_\mathcal{B} = \begin{bmatrix} 2 \\ 7 \\ -19 \\ 17 \end{bmatrix}$. (c) $[\mathbf{v}]_\mathcal{B} = \begin{bmatrix} -1 \\ 5 \\ -7 \\ 1 \end{bmatrix}$. (e) $[\mathbf{v}]_\mathcal{B} = \begin{bmatrix} 8 \\ -9 \\ -3 \\ -1 \end{bmatrix}$.

Chapter 6

Exercise 6.5.1. (a) Not linear. (c) Linear. (e) Not linear.

Exercise 6.5.2. (a) Linear. (c) Linear. (e) Not linear. (g) Linear.

Exercise 6.5.3. (a) $A = \begin{bmatrix} 1 & 0 & -5 \\ 0 & 7 & 0 \\ 3 & -6 & 0 \\ 0 & 0 & 8 \end{bmatrix}$. (c) $h = 8$.

Exercise 6.5.6. (a) $\{(1-t)(3-t)\} = \{3 - 4t + t^2\}$.

Exercise 6.5.16. $\begin{bmatrix} 1 & 1 & 0 & 0 \\ 0 & -1 & 1 & 0 \\ -1 & 0 & 0 & 1 \\ 1 & 0 & 1 & 0 \end{bmatrix}$.

Exercise 6.5.18. $\begin{bmatrix} 2 & 4 & 0 \\ 0 & -1 & 5 \\ 3 & 7 & 10 \end{bmatrix}$.

Chapter 7

Exercise 7.5.1. (a) 2, $\begin{bmatrix} 1 \\ 1 \end{bmatrix}$ and 6, $\begin{bmatrix} -1 \\ 1 \end{bmatrix}$.

Exercise 7.5.1. (c) 2, $\begin{bmatrix} 1 \\ 0 \\ 0 \end{bmatrix}$, 0, $\begin{bmatrix} -\frac{1}{6} \\ \frac{1}{3} \\ 1 \end{bmatrix}$ and -6, $\begin{bmatrix} \frac{1}{24} \\ -\frac{1}{3} \\ 1 \end{bmatrix}$.

Exercise 7.5.3. Yes, Span $\left\{ \begin{bmatrix} -2 \\ 0 \\ 1 \end{bmatrix}, \begin{bmatrix} 0 \\ 1 \\ 0 \end{bmatrix} \right\}$.

Exercise 7.5.7. 1, $\begin{bmatrix} -i \\ 1 \end{bmatrix}$ and 3, $\begin{bmatrix} i \\ 1 \end{bmatrix}$.

Exercise 7.5.10. 12, 7.

Exercise 7.5.16. $\begin{bmatrix} 1 & 0 & 0 \\ 0 & 1 & 0 \\ 0 & 0 & 1 \end{bmatrix}$.

Exercise 7.5.18. We diagonalize A as $A = PDP^{-1}$, with $P = \begin{bmatrix} 1 & -1 \\ 1 & 1 \end{bmatrix}$ and $D = \begin{bmatrix} 1 & 0 \\ 0 & \frac{1}{2} \end{bmatrix}$. Then

$$A^k = P \begin{bmatrix} 1 & 0 \\ 0 & (\frac{1}{2})^k \end{bmatrix} P^{-1} = \begin{bmatrix} \frac{1}{2} + \frac{1}{2^{k+1}} & \frac{1}{2} - \frac{1}{2^{k+1}} \\ \frac{1}{2} - \frac{1}{2^{k+1}} & \frac{1}{2} + \frac{1}{2^{k+1}} \end{bmatrix}.$$

Exercise 7.5.24. (a) $x_1(t) = e^{2t}, x_2(t) = 2e^{2t}$.

Chapter 8

Exercise 8.8.1. 16, $\sqrt{26}$.

Exercise 8.8.7. $\sqrt{14}$.

Exercise 8.8.13. Neither.

Exercise 8.8.16. $\left\{ \begin{bmatrix} -2 - 3i \\ 2 - i \\ 1 \end{bmatrix} \right\}$.

Exercise 8.8.18. $\left\| \begin{bmatrix} -\frac{6}{13} \\ -\frac{4}{13} \end{bmatrix} \right\| = \frac{\sqrt{52}}{13} = \frac{2}{13}\sqrt{13}$.

Exercise 8.8.21. $\begin{bmatrix} -\frac{1}{2} \\ \frac{5}{2} \end{bmatrix}$.

Exercise 8.8.27. (a) $\left\{ \frac{1}{\sqrt{6}} \begin{bmatrix} 1 \\ 2 \\ 0 \\ 1 \end{bmatrix}, \frac{1}{\sqrt{12}} \begin{bmatrix} 1 \\ -1 \\ 3 \\ 1 \end{bmatrix} \right\}$.

Exercise 8.8.32. (a) $Q = \begin{bmatrix} \frac{2}{3} & \frac{1}{\sqrt{18}} \\ \frac{1}{3} & \frac{-4}{\sqrt{18}} \\ \frac{2}{3} & \frac{1}{\sqrt{18}} \end{bmatrix}$, $R = \begin{bmatrix} 3 & \frac{5}{3} \\ 0 & -\frac{\sqrt{2}}{3} \end{bmatrix}$.

Exercise 8.8.41. (a) $D = \begin{bmatrix} 5 & 0 \\ 0 & 10 \end{bmatrix}$, $U = \frac{1}{\sqrt{5}} \begin{bmatrix} -2 & 1 \\ 1 & 2 \end{bmatrix}$.

Exercise 8.8.48. (a) $V = \begin{bmatrix} -\frac{1}{\sqrt{2}} & -\frac{1}{2} & \frac{1}{2} \\ -\frac{1}{\sqrt{2}} & \frac{1}{2} & -\frac{1}{2} \\ 0 & -\frac{i}{\sqrt{2}} & -\frac{i}{\sqrt{2}} \end{bmatrix}$, $\Sigma = \begin{bmatrix} 4 & 0 & 0 \\ 0 & 2 & 0 \\ 0 & 0 & 2 \end{bmatrix}$,

$W = \begin{bmatrix} 0 & -1 & 0 \\ 0 & 0 & 1 \\ i & 0 & 0 \end{bmatrix}$.

Exercise 8.8.57. (f) Table 1 would have persons 1,2,5,8,10 and Table 2 persons 3,4,6,7,9.

Appendix

A.1   Some Thoughts on Writing Proofs

Mathematics is based on axioms and logical deductions, which we refer to as proofs. In this course you may for the first time writing your own proofs. In this section we hope to give some helpful tips, to get you started on this. We will start 'If P then Q' statements. Most statements in this course can be phrased in this way, and we do not want to overload you with too much information. There are of course plenty of resources available where more information can be found.

A.1.1   Non-Mathematical Examples

In this subsection we will try to make some main points without getting into mathematical details.

Let us start with the following non-mathematical statement.
> Statement 1: *If it rains then the streets get wet.*

A proof of this statement may be the following:
· *Proof 1: Suppose it rains. By definition this means that water is falling from the sky. Due to gravity this water will fall on the streets. A surface being wet means having liquid on it. Since the streets have water on them, and water is a liquid, the streets are wet. This proves the statement.*

This is an example of a **direct proof**. In a direct proof of an 'If P then Q' statement, you start with assuming P and then you use logical deductions until you reach the statement Q. In this example we assumed that it rained, and then we argued until we reached the conclusion that the streets are wet. Along the way we used definitions of the terms used, and we used some prior knowledge (gravity exists).

As you will find out, whenever you write a proof there will be a question about what we can assume to be prior knowledge and which parts require explanation. In general, the purpose of a proof is to convince readers that the statement is true. Thus, it depends on who the readers are where you would draw the line between prior knowledge and what needs explanation. For the purpose of this course, it is probably the best to consider the readers of your proof to be your peers. Thus, you do not need to explain that real numbers x and y satisfy $xy = yx$, for example.

An alternative way to prove an 'If P then Q' statement, is to prove the equivalent statement 'If (not Q) then (not P)', which is called the **contrapositive** statement. The contrapositive of Statement 1 is

Statement 1': *If the streets are not getting wet then it is not raining.*

Statements 1 and 1' are equivalent: they are saying the same thing. When we discuss truth tables, we will formalize this more. Thus a proof of Statement 1 could be

Proof 2: Suppose the streets are not getting wet. This means that there is no liquid falling on the streets. Thus no water could be falling down on the streets. Thus it is not raining.

A proof of the contrapositive statement is sometimes referred to as an **indirect proof** of the original statement. So Proof 2 is an indirect proof of Statement 1.

A third way to prove a 'If P then Q' statement is to prove it **by contradiction.** This means that you are assuming that both P and (not Q) hold, and then you argue until you get some obviously false statement. Thus a third proof of Statement 1 may go as follows.

Proof 3: Suppose it rains and the streets are not getting wet. This means that there is water falling from the sky but that the water does not reach the streets. This means that above the streets there is a gravity free zone. But that is false, and we have reached a contradiction.

In summary, Proofs 1, 2 and 3 are all proofs of Statement 1. In general, direct proofs are the easiest to understand so there is some preference to do it that way. However, in some cases direct proofs are not easy to construct and we either have to do an indirect proof or a proof by contradiction. In the next section we will give some mathematical examples of these.

Let us also look at a 'for all' statement. For instance,

All humans live on the same planet.

If you are proving a 'for all' statement it is important to have a reasoning that applies to all. Thus it is not enough to say that 'all my friends live on the same planet, and therefore it is true'. You need some argument that applies to *all* humans. So it could go something like this:

A human needs oxygen in the atmosphere to live. There is only one planet that has oxygen in the atmosphere, namely Earth. Thus for every human there is only one planet where they can live, namely Earth. Thus every human lives on Earth, and consequently every human lives on the same planet.

We will also encounter the situation that a 'for all' statement is false, and we have to prove that it is false. For instance, consider the statement

All students love Linear Algebra.

To disprove such a 'for all' statement, you just have to come up with one example that contradicts the statement. In this instance, you can, for example, take a picture of a fellow student holding a sign that says 'I do not like Linear Algebra'. We call this a **counterexample**. With a counterexample, it is important to be very specific. It is much more convincing than throwing your hands up in the air and say 'how could all students possibly love Linear Algebra?'.

Let us now move on to some mathematical examples.

A.1.2   Mathematical Examples

Proposition A.1.1. *The sum of two odd numbers is even. In other words: If n and m are odd numbers then $n + m$ is even.*

Proof. Suppose that n and m are odd. Then there exist integers k and l so that $n = 2k + 1$ and $m = 2l + 1$. This gives that $n + m = 2k + 1 + 2l + 1 = 2(k + l + 1)$. Thus $n + m$ is divisible by 2, which means that $n + m$ is even. $\square$

This is a direct proof, which makes the most sense in this case. Notice that we rephrased the original statement as an 'If P then Q' statement, as it made it clearer where to start and where to end.

Proposition A.1.2. *Let n be an integer. If n^2 is even, then n is even.*

Proof. Suppose that the integer n is not even. Then n is odd, and thus there exist an integer k so that $n = 2k + 1$. Then $n^2 = 4k^2 + 4k + 1 = 2(2k^2 + 2k) + 1$. This shows that $n^2 - 1 = 2(2k^2 + 2k)$ is even, and thus n^2 is odd. $\square$

In this proof we proved the contrapositive, which was easier. If we wanted to prove it directly we probably would have had to rely on prime factorization.

Proposition A.1.3. $\sqrt{2}$ *is irrational.*
In other words, if x is a positive real number so that $x^2 = 2$ then x is irrational.

Proof. Suppose that x is a positive real number so that $x^2 = 2$ and suppose that x is rational. Since x is positive and rational we can write $x = \frac{p}{q}$ where p and q are positive integers without common factors; otherwise we would just divide out the common factors. Thus, the greatest common divisor of p and q is 1. Then we obtain $\frac{p^2}{q^2} = x^2 = 2$, and thus $p^2 = 2q^2$ is even. Since p^2 is even we obtain by Proposition A.1.2 that p is even. Thus $p = 2k$ for some integer k. Now we obtain that $4k^2 = (2k)^2 = p^2 = 2q^2$, and thus $q^2 = 2k^2$ is even. Again by Proposition A.1.2 we then obtain that q is even. But if p and q are even, then they do have a common factor (namely 2). This contradicts the choice of p and q. $\square$

This was a proof by contradiction, and seems to be the only way to prove this statement.

We will also have 'P if and only if Q' statements. These are really two statement in one, namely both 'If P then Q' and 'If Q then P'. A proof of an 'if and only if' statement therefore typically has two parts. Let us give an example.

Proposition A.1.4. *Let n be an integer. Then n^2 is even if and only if n is even.*

Proof. We first prove the 'only if' part, namely we prove 'n^2 is even only if n is even'. In other words, we need to prove 'If n^2 is even, then n is even'. This follows from Proposition A.1.2.

Next, we prove the 'if' part, which is 'n^2 is even if n is even'. In other words we need to prove 'if n is even, then n^2 is even'. Thus assume that n is even. Then there is an integer k so that $n = 2k$. Then $n^2 = (2k)^2 = 4k^2 = 2(2k^2)$. Thus n^2 is divisible by 2, and thus n^2 is even. $\square$

As you notice it takes a bit of effort to break it down into the 'If P then Q' and 'If Q then P' format, but hopefully you will get used to doing this.

Remark. A common mistake is that instead of writing a proof for 'If P then Q', a proof for 'If Q then P' is written. This is a mistake as 'If Q then P' is a different statement. For instance 'If the streets get wet then it is raining' is not a true statement. There could be flooding going on, for instance.

Let us look at the following statement: '*If $x^2 > 9$ then $x > 3$ or $x < -3$*'.

It could be tempting to start with $x > 3$, and conclude that $x^2 > 9$, and do the same for $x < -3$. This, however, would not be a valid proof of the above statement. A correct proof would be

Proof 1. Suppose $x^2 > 9$. Then $x^2 - 9 > 0$. Thus $(x-3)(x+3) = x^2 - 9 > 0$. The only way the product of two real numbers is positive is when they are both positive or both negative. Thus we get $(x-3 > 0$ and $x+3 > 0)$ or $(x-3 < 0$ and $x+3 < 0)$. In the first case we get that $x > 3$ and $x > -3$, which yields $x > 3$. In the second case we get $x < 3$ and $x < -3$, which yields $x < -3$. In conclusion, we obtain $x > 3$ or $x < -3$.

One may of course also prove the contrapositive, and get to the following alternative proof.

Proof 2. Suppose that $-3 \le x \le 3$. Then $0 \le |x| \le 3$. Thus $x^2 = |x|^2 \le 9$, which finishes this indirect proof.

Finally, let us give an example of a counterexample.

Claim: all even integers are divisible by 4.

This claim is false. Take for instance the integer 2, which is an even integer, but $\frac{2}{4} = \frac{1}{2}$ is not an integer, and thus 2 is not divisble by 4.

A.1.3   Truth Tables

When you have a few statements (P,Q,R, etc.) you can make new statements, such as 'If P then (Q and R)'. In this sections we will give you the building blocks for new statements, along with their truth tables. The truth tables tell you whether the statement is true, based on the truth values of P,Q,R, etc. The four basic operations are 'not', 'and', 'or', 'implies'. Here are their truth tables:

P	not P $\neg P$
True	False
False	True

P	Q	P and Q $P \wedge Q$	P or Q $P \vee Q$	If P then Q P implies Q $P \Rightarrow Q$
True	True	True	True	True
True	False	False	True	False
False	True	False	True	True
False	False	False	False	True

On the left of the double lines you have all the possibilities of the variables (P, Q) and on the right of the double line you have the corresponding value for the new statement. So, for instance, the truth table for 'and' (which has symbol $\wedge$) shows that $P \wedge Q$ is only true when both P and Q are true. As soon as one of P and Q is false, then $P \wedge Q$ is false. One thing that could take some getting used to is that $P \Rightarrow Q$ is a true statement if P is false. For instance, 'If grass is blue then water is dry' is a true statement (since grass is not blue).

We can now make more involved statements and their truth tables. Below are, for instance, the truth tables of the statements '$P \wedge (Q \vee R)$' and '$(P \wedge Q) \Rightarrow (P \wedge R)$'. Note that we use T and F for True and False, respectively.

P	Q	R	$Q \vee R$	$P \wedge (Q \vee R)$	$P \wedge Q$	$P \wedge R$	$(P \wedge Q) \Rightarrow (P \wedge R)$
T	T	T	T	**T**	T	T	**T**
T	T	F	T	**T**	T	F	**F**
T	F	T	T	**T**	F	T	**T**
T	F	F	F	**F**	F	F	**T**
F	T	T	T	**F**	F	F	**T**
F	T	F	T	**F**	F	F	**T**
F	F	T	T	**F**	F	F	**T**
F	F	F	F	**F**	F	F	**T**

Recall that we also have the 'P if and only if Q' statement, which we denote as $P \Leftrightarrow Q$, and which is short for $(P \Rightarrow Q) \wedge (Q \Rightarrow P)$. Its truth table is

P	Q	$P \Rightarrow Q$	$Q \Rightarrow P$	$P \Leftrightarrow Q$ $(P \Rightarrow Q) \wedge (Q \Rightarrow P)$
T	T	T	T	**T**
T	F	F	T	**F**
F	T	T	F	**F**
F	F	T	T	**T**

We say that two statements are **equivalent** if they have the same truth tables. For instance, $P \Rightarrow Q$ and $\neg Q \Rightarrow \neg P$ are equivalent:

P	Q	$P \Rightarrow Q$	$\neg Q$	$\neg P$	$\neg Q \Rightarrow \neg P$
T	T	**T**	F	F	**T**
T	F	**F**	T	F	**F**
F	T	**T**	F	T	**T**
F	F	**T**	T	T	**T**

A.1.4   Quantifiers and Negation of Statements

Some statements have the form

'*for all* $x \in S$ *the statement* $P(x)$ *is true*'   in shorthand: '$\forall x \in S : P(x)$',

while others have the form
'*there exists an* $x \in S$ *so that* $Q(x)$ *is true*'   in shorthand: '$\exists x \in S : Q(x)$'.

We call 'for all' ($\forall$) and 'there exists' ($\exists$) **quantifiers**. Here are some examples

Proposition A.1.5. *For every natural number* $n \in \mathbb{N}$ *we have that* $n^2 + n$ *is even.*

Proof. Let $n \in \mathbb{N}$. Then either n is even or $n + 1$ is even. But then $n^2 + n = n(n+1)$ is even. $\square$

Proposition A.1.6. *For every integer* $n \in \mathbb{Z}$ *there exists an integer* $m \in \mathbb{Z}$ *so that* $n + m = 0$.

Proof. Let $n \in \mathbb{Z}$. We can now put $m = -n$. Then $m \in \mathbb{Z}$ and $n + m = 0$. $\square$

The last statement actually uses both quantifiers. Notice that to prove the existence of m we were very specific how we should choose m. In general, when you prove an existence statement you should try to be as specific as possible.

Some statements P that you will encounter will be false statements. In that case $\neg P$ will be a true statement. We call $\neg P$ the **negation** of P. So when you prove that P is false, you actually are proving that $\neg P$ is true. Thus, it will be helpful to write $\neg P$ in its most convenient form. Here are some rules

- $\neg(\forall x \in S : P(x))$ is equivalent to $\exists x \in S : \neg P(x)$.

- $\neg(\exists x \in S : Q(x))$ is equivalent to $\forall x \in S : \neg Q(x)$.

- $\neg(\neg P)$ is equivalent to P.

- $\neg(P \wedge Q)$ is equivalent to $(\neg P) \vee (\neg Q)$.

- $\neg(P \vee Q)$ is equivalent to $(\neg P) \wedge (\neg Q)$.

- $\neg(P \Rightarrow Q)$ is equivalent to $P \wedge (\neg Q)$.

To see the equivalence of the last item, for instance, you can write down the truth tables for the statements and see that they are the same:

P	Q	$P \Rightarrow Q$	$\neg(P \Rightarrow Q)$	$\neg Q$	$P \wedge (\neg Q)$
T	T	T	F	F	F
T	F	F	T	T	T
F	T	T	F	F	F
F	F	T	F	T	F

If you write down the truth table for $(\neg(P \Rightarrow Q)) \Leftrightarrow (P \wedge (\neg Q))$ you get

P	Q	$(\neg(P \Rightarrow Q)) \Leftrightarrow (P \wedge (\neg Q))$
T	T	**T**
T	F	**T**
F	T	**T**
F	F	**T**

A statement that is always true is called a **tautology**. Thus

$$(\neg(P \Rightarrow Q)) \Leftrightarrow (P \wedge (\neg Q))$$

is a tautology. Other examples of tautologies are

- $P \vee (\neg P)$
- $(\neg(P \wedge Q)) \Leftrightarrow ((\neg P) \vee (\neg Q))$.
- $(\neg(P \vee Q)) \Leftrightarrow ((\neg P) \wedge (\neg Q))$.

Consider the statement

There exists a vector $\mathbf{u} \in \mathbb{R}^2$ that is orthogonal to every $\mathbf{v} \in \mathbb{R}^2$.

This is a true statement.

Proof. Let $\mathbf{u} = \begin{bmatrix} 0 \\ 0 \end{bmatrix} \in \mathbb{R}^2$. Then for every $\mathbf{v} = \begin{bmatrix} v_1 \\ v_2 \end{bmatrix} \in \mathbb{R}^2$ we have $\langle \mathbf{u}, \mathbf{v} \rangle = 0 \cdot v_1 + 0 \cdot v_2 = 0$, and thus $\mathbf{u}$ is orthogonal to $\mathbf{v}$. $\quad\square$

If we change the statement slightly, it is no longer true:

There exists a nonzero vector $\mathbf{u} \in \mathbb{R}^2$ that is orthogonal to every $\mathbf{v} \in \mathbb{R}^2$.

This is a false statement. Thus its negation must be true. The negation of this statement is:

For every nonzero vector $\mathbf{u} \in \mathbb{R}^2$ there exists a $\mathbf{v} \in \mathbb{R}^2$ so that $\mathbf{u}$ is not orthogonal to $\mathbf{v}$.

Proof. Let $\mathbf{u} = \begin{bmatrix} u_1 \\ u_2 \end{bmatrix} \in \mathbb{R}^2$ be a nonzero vector (thus $u_1 \neq 0$ or $u_2 \neq 0$). Choose $\mathbf{v} = \mathbf{u}$. Then $\langle \mathbf{u}, \mathbf{v} \rangle = u_1^2 + u_2^2 > 0$. Thus $\mathbf{u}$ is not orthogonal to $\mathbf{v}$.

A.1.5 Proof by Induction

For statements of the form

For every $n \in \{n_0, n_0 + 1, n_0 + 2, \ldots\}$ we have that $P(n)$ is true

we have a proof technique called **proof by induction** that can be helpful. The proof will have the following form

- **Basis step**: Show that $P(n_0)$ is true.

- **Induction step**: Show that the following statement is true: 'If $P(m)$ is true for some $m \geq n_0$, then $P(m+1)$ is true'.

The idea is that now we have that $P(n_0)$ is true, due to the basis step. And then, if we apply the induction step with $m = n_0$, we obtain that $P(n_0 + 1)$ is true. Now we can apply the induction step with $m = n_0 + 1$, and we obtain that $P(n_0 + 2)$ is true. Now we can apply the induction step with $m = n_0 + 2$, and we obtain that $P(n_0 + 3)$ is true. Etc. Thus repeated use of the induction step yields that $P(n)$ is true for all $n \in \{n_0, n_0 + 1, n_0 + 2, \ldots\}$.

Let us look at an example.

Proposition A.1.7. *For every $n \in \mathbb{N} = \{1, 2, 3, \ldots\}$ we have that $n^3 - n$ is divisible by 6.*

Proof. We prove this by induction.

- **Basis step**: Let $n = 1$. Then $n^3 - n = 0 = 6 \cdot 0$, thus the statement is true for $n = 1$.

- **Induction step**: Suppose that for some $m \in \mathbb{N}$ we have that $m^3 - m$ is divisble by 6. Thus $m^3 - m = 6k$ for some $k \in \mathbb{Z}$. Then $(m+1)^3 - (m+1) = (m^3 + 3m^2 + 3m + 1) - (m+1) = (m^3 - m) + 3(m^2 + m) = 6k + 3(m^2 + m)$. By Proposition A.1.5 we have that $m^2 + m = m(m+1)$ is even; thus $m^2 + m = 2l$, for some $l \in \mathbb{Z}$. Now we find that $(m+1)^3 - (m+1) = 6k + 3(2l) = 6(k+l)$. And thus $(m+1)^3 - (m+1)$ is divisible by 6. This takes care of the induction step.

$\square$

There is also a 'strong form' of proof by induction, which is as follows:

- **Basis step**: Show that $P(n_0)$ is true.

- **Induction step**: Show that the following statement is true: 'If $P(k)$, $k \in \{n_0, n_0 + 1, \ldots, m\}$ are true for some $m \geq n_0$, then $P(m + 1)$ is true'.

It is not very often (probably never in this course) that you need induction in this strong form, but it is good to know that you have the option to use this variation.

A.1.6   Some Final Thoughts

We hope that this appendix helps you starting off writing proofs. Writing proofs is not an easy task and everyone struggles with it from time to time. It takes a lot of practice. We hope that the proofs in the texts are helpful in giving you more ideas. Now sometimes, especially with long proofs, one's mind starts to wander off and one gets lost. What may be helpful, when you find yourself in that situation, is to think about parts of the proof. You can ask 'which parts of this proof are easy' and 'which parts are hard'. There are certainly proofs in this course where you might think 'I would have never thought of that'. What we can tell you is that you are not alone in that. Even the best mathematicians see proofs that they would have never thought of. The point, though, of reading other people's proofs is that it gives you new ideas. When you think 'I would have never thought of this myself', you should not despair. What you should do is to store the idea in your brain somewhere, so that if you have to prove something similar, you have the idea handy. So, when you read or listen to proofs, look for ideas to add to your 'bag of mathematical tricks'.

A.2   Complex Numbers

The complex numbers are defined as

$$\mathbb{C} = \{a + bi \; ; a, b \in \mathbb{R}\},$$

with addition and multiplication defined by

$$(a + bi) + (c + di) := (a + c) + (b + d)i,$$

$$(a + bi)(c + di) := (ac - bd) + (ad + bc)i.$$

Notice that with these rules, we have that $(0 + 1i)(0 + 1i) = -1 + 0i$, or in shorthand $i^2 = -1$. Indeed, this is how to remember the multiplication rule:

$$(a + bi)(c + di) = ac + bdi^2 + (ad + bc)i = ac - bd + (ad + bc)i,$$

where in the last step we used that $i^2 = -1$. It may be obvious, but we should state it clearly anyway: two complex numbers $a + bi$ and $c + di$, with $a, b, c, d \in \mathbb{R}$ are equal if and only if $a = c$ and $b = d$. A typical complex number may be denoted by z or w. When

$$z = a + bi \text{ with } a, b \in \mathbb{R},$$

we say that the **real part** of z equals a and the **imaginary part** of z equals b. The notations for these are,

$$\text{Re } z = a, \quad \text{Im } z = b.$$

A complex number z is called **purely imaginary** if $z = ib$, where $b \in \mathbb{R}$. In other words, z is purely imaginary if and only if $\text{Re } z = 0$ (if and only if $z = -\overline{z}$, as we will see).

It is quite laborious, but in principle elementary, to prove that $\mathbb{C}$ satisfies all the field axioms (see the next section for the axioms). In fact, in doing so one needs to use that $\mathbb{R}$ satisfies the field axioms, as addition and multiplication in $\mathbb{C}$ are defined via addition and multiplication in $\mathbb{R}$. As always, it is important to realize what the neutral elements are:

$$0 = 0 + 0i, \quad 1 = 1 + 0i.$$

Another tricky part of this is the multiplicative inverse, for instance,

$$(1 + i)^{-1}, \ (2 - 3i)^{-1}. \tag{A.1}$$

Here it is useful to look at the multiplication

$$(a + bi)(a - bi) = a^2 + b^2 + 0i = a^2 + b^2. \tag{A.2}$$

This means that as soon as a or b is not zero, we have that $(a + bi)(a - bi) = a^2 + b^2$ is a nonzero (actually, positive) real number. From this we can conclude that

$$\frac{1}{a + bi} = (a + bi)^{-1} = \frac{a - bi}{a^2 + b^2} = \frac{a}{a^2 + b^2} - \frac{b}{a^2 + b^2}i.$$

So, getting back to (A.1),

$$\frac{1}{1 + i} = \frac{1}{2} - \frac{i}{2}, \quad \frac{1}{2 - 3i} = \frac{2}{13} + \frac{3i}{13}.$$

Now you should be fully equipped to check all the field axioms for $\mathbb{C}$.

As you notice, the complex number $a - bi$ is a useful 'counterpart' of $a + bi$, so that we are going to give it a special name. The **complex conjugate** of $z = a + bi$, $a, b \in \mathbb{R}$, is the complex number $\overline{z} := a - bi$. So, for example,

$$\overline{2 + 3i} = 2 - 3i, \quad \overline{\frac{1}{2} + \frac{6i}{5}} = \frac{1}{2} - \frac{6i}{5}.$$

Thus, we have

$$\operatorname{Re} \overline{z} = \operatorname{Re} z, \quad \operatorname{Im} \overline{z} = -\operatorname{Im} z.$$

Finally, we introduce the **absolute value** or **modulus** of z, via

$$|a + bi| := \sqrt{a^2 + b^2}, \ a, b, \in \mathbb{R}.$$

For example,

$$|1 + 3i| = \sqrt{10}, \quad \left|\frac{1}{2} - \frac{i}{2}\right| = \sqrt{\frac{1}{4} + \frac{1}{4}} = \frac{\sqrt{2}}{2}.$$

Note that we have the rule

$$z\overline{z} = |z|^2,$$

as observed in (A.2), and its consequence

$$\frac{1}{z} = \frac{\overline{z}}{|z|^2}$$

when $z \neq 0$.

We also have

$$\overline{z + w} = \overline{z} + \overline{w}, \qquad \overline{zw} = \overline{z}\,\overline{w}.$$

A complex number is often depicted as a point in $\mathbb{R}^2$, which we refer to as the **complex plane**. The x-axis is the 'real axis' and the y-axis is the 'imaginary axis.' Indeed, if $z = x + iy$ then we represent z as the point (x, y) as in Figure A.1.

The distance from the point z to the origin corresponds to $|z| = \sqrt{x^2 + y^2}$. The angle t the point z makes with the positive x-axis is referred to as the **argument** of z. It can be found via

$$\cos t = \frac{\operatorname{Re} z}{|z|}, \quad \sin t = \frac{\operatorname{Im} z}{|z|}.$$

Thus we can write

$$z = |z|(\cos t + i \sin t).$$

The following notation, due to Euler, is convenient:

$$e^{it} := \cos t + i \sin t.$$

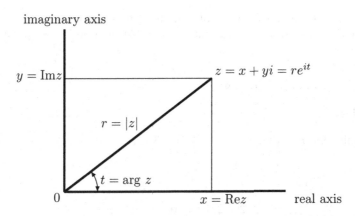

Figure A.1: A complex number in the complex plane.

Using the rules for $\cos(t+s)$ and $\sin(t+s)$, one can easily check that

$$e^{it}e^{is} = e^{i(t+s)}.$$

In addition, note that

$$\overline{e^{it}} = e^{-it}.$$

Thus for $z = |z|e^{it} \neq 0$, we have that $z^{-1} = \frac{1}{|z|}e^{-it}$.

We will be solving quadratic equations. It is often most convenient to complete the square. Let us illustrate this with an example.

Example A.2.1. Solve $z^2 + 4z + 9 = 0$. Completing the square, we get $(z+2)^2 - 4 + 9 = 0$. Thus $(z+2)^2 = -5$. This yields $z + 2 = \pm i\sqrt{5}$, and we find $z = -2 \pm i\sqrt{5}$ as the solutions. □

Next, let us solve $z^n = \rho$.

Example A.2.2. Solve $z^n = \rho$. Let us write $\rho = re^{i\theta}$ with $r \geq 0$ and $\theta \in \mathbb{R}$. Then we have $z^n = re^{i(\theta+2k\pi)}$, where $k \in \mathbb{Z}$. Now we find $z = \sqrt[n]{r}e^{\frac{i(\theta+2k\pi)}{n}}$, $k \in \mathbb{Z}$. This leads to the solutions

$$z = \sqrt[n]{r}e^{\frac{i(\theta+2k\pi)}{n}}, \quad k = 0, \ldots, n-1.$$

□

Exercise A.2.3. In this exercise we are working in the field $\mathbb{C}$. Make sure you write the final answers in the form $a + bi$, with $a, b \in \mathbb{R}$. For instance, $\frac{1+i}{2-i}$ should not be left as a final answer, but be reworked as

$$\frac{1+i}{2-i} = \left(\frac{1+i}{2-i}\right)\left(\frac{2+i}{2+i}\right) = \frac{2+i+2i+i^2}{2^2+1^2} = \frac{1+3i}{5} = \frac{1}{5} + \frac{3i}{5}.$$

Notice that in order to get rid of i in the denominator, we decided to multiply both numerator and denominator with the complex conjugate of the denominator.

(i) $(1+2i)(3-4i)-(7+8i)=$

(ii) $\frac{1+i}{3+4i}=$

(iii) Solve for x in $(3+i)x+6-5i=-3+2i$.

(iv) Find $\det\begin{bmatrix}4+i & 2-2i \\ 1+i & -i\end{bmatrix}$.

(v) Compute $\begin{bmatrix}-1+i & 2+2i \\ -3i & -6+i\end{bmatrix}\begin{bmatrix}0 & 1-i \\ -5+4i & 1-2i\end{bmatrix}$.

(vi) Find $\begin{bmatrix}2+i & 2-i \\ 4 & 4\end{bmatrix}^{-1}$.

A.3   The Field Axioms

A **field** is a set $\mathbb{F}$ on which addition and multiplication

$$+:\mathbb{F}\times\mathbb{F}\to\mathbb{F},\quad\cdot:\mathbb{F}\times\mathbb{F}\to\mathbb{F},$$

are defined satisfying the following rules:

1. **Closure of addition:** for all $x,y\in\mathbb{F}$ we have that $x+y\in\mathbb{F}$.

2. **Associativity of addition:** for all $x,y,z\in\mathbb{F}$ we have that $(x+y)+z=x+(y+z)$.

3. **Commutativity of addition:** for all $x,y\in\mathbb{F}$ we have that $x+y=y+x$.

4. **Existence of a neutral element for addition:** there exists a $0\in\mathbb{F}$ so that $x+0=x=0+x$ for all $x\in\mathbb{F}$.

5. **Existence of an additive inverse:** for every $x\in\mathbb{F}$ there exists a $y\in\mathbb{F}$ so that $x+y=0=y+x$.

6. **Closure of multiplication:** for all $x,y\in\mathbb{F}$ we have that $x\cdot y\in\mathbb{F}$.

7. **Associativity of multiplication:** for all $x,y,z\in\mathbb{F}$ we have that $(x\cdot y)\cdot z=x\cdot(y\cdot z)$.

8. **Commutativity of multiplication**: for all $x, y \in \mathbb{F}$ we have that $x \cdot y = y \cdot x$.

9. **Existence of a neutral element for multiplication**: there exists a $1 \in \mathbb{F} \setminus \{0\}$ so that $x \cdot 1 = x = 1 \cdot x$ for all $x \in \mathbb{F}$.

10. **Existence of a multiplicative inverse for nonzeros**: for every $x \in \mathbb{F} \setminus \{0\}$ there exists a $y \in \mathbb{F}$ so that $x \cdot y = 1 = y \cdot x$.

11. **Distributive law**: for all $x, y, z \in \mathbb{F}$ we have that $x \cdot (y + z) = x \cdot y + x \cdot z$.

We will denote the additive inverse of x by $-x$, and we will denote the multiplicative inverse of x by x^{-1}.

Examples of fields are:

- The real numbers $\mathbb{R}$.

- The complex numbers $\mathbb{C}$.

- The rational numbers $\mathbb{Q}$.

- The finite field $\mathbb{Z}_p$, with p prime. Here $\mathbb{Z}_p = \{0, 1, \ldots, p-1\}$ and addition and multiplication are defined via modular arithmetic.

  As an example, let us take $p = 3$. Then $\mathbb{Z}_3 = \{0, 1, 2\}$ and the addition and multiplication table are:

+	0	1	2
0	0	1	2
1	1	2	0
2	2	0	1

.	0	1	2
0	0	0	0
1	0	1	2
2	0	2	1

  So, in other words, $1 + 1 = 2$, $2 + 1 = 0$, $2 \cdot 2 = 1$, $0 \cdot 1 = 0$, etc. In fact, to take the sum of two elements we take the usual sum, and then take the remainder after division by $p = 3$. For example, to compute $2 + 2$ we take the remainder of 4 after division by $p = 3$, which is 1. Similarly for multiplication. Regarding additive and multiplicative inverses, note for instance that $-1 = 2$, $2^{-1} = 2$.

  The tables for $\mathbb{Z}_2$ are:

+	0	1
0	0	1
1	1	0

.	0	1
0	0	0
1	0	1

The tables for $\mathbb{Z}_5$ are:

+	0	1	2	3	4
0	0	1	2	3	4
1	1	2	3	4	0
2	2	3	4	0	1
3	3	4	0	1	2
4	4	0	1	2	3

·	0	1	2	3	4
0	0	0	0	0	0
1	0	1	2	3	4
2	0	2	4	1	3
3	0	3	1	4	2
4	0	4	3	2	1

In $\mathbb{Z}_5$, we have for instance $-1 = 4$, $-2 = 3$, $2^{-1} = 3$, $4^{-1} = 4$.

Index